Arms Spending, Development and Security

Arms Spending, Development and Security

Edited by :

Manas Chatterji

Professor of Management, Binghamton University, State University of New York, United States

Jacques Fontanel

Professor of Economics, Universite Pierre Mendes-France

Akira Hattori

Professor of Economics
Fukuoka University, Japan

APH PUBLISHING CORPORATION
5 ANSARI ROAD, DARYAGANJ
NEW DELHI-110 002

Published by :

S.B. Nangia
for **APH Publishing Corporation**
5, Ansari Road,
Darya Ganj
New Delhi - 110002
Tele. : 3274050

ISBN 81-7024-691-1

First Published : 1996

Laser typeset by :
Computer Codes,
M-1, Virat Bhawan, Comml. Complex,
Mukerjee Nagar, Delhi-110009
Ph. : 7241200

Printed at :
Efficient Offset Printers,
215, Shahzada Bagh Indl. Complex.
Phase-II, Delhi - 110 035
Tel. : 533736, 533762
New Delhi

To the Health and Happiness of the World's Children

CONTRIBUTIONS FROM PEACE ECONOMICS AND PEACE SCIENCE

General Editor : Manas Chatterji
Binghamton University, Binghamton, N.Y. U.S.A

Arms Spending, Development, and Security edited by Manas Chatterji, Jacques Fontanel and Akira Hattori

Preface

After the cold war ended, significant changes took place in the worlds of politics and economics. There is only one superpower now, but it is becoming difficult to contain regional conflicts. The role of the United Nations is becoming increasingly crucial, but due to financial and other reasons it, is becoming ineffective. Since many developing countries in the UN receive protection from superpowers, they are trying to build up their military strength. This encourages some countries to export arms aggressively, fueling the arms race in developing countries.

One good thing that has resulted from the fall of the former Soviet Union is the conversion of a military economy to a civilian one. Different experiments are now being conducted in many countries with respect to this economic conversion. Nuclear proliferation has become an important issue. If the regional conflicts cannot be handled properly, nuclear war is still possible particularly if nuclear material falls into the hands of terrorist and irresponsible nations. What is urgently needed is an institutional structure within the United Nations which can effectively work for Peace Keeping and Peace Making. This structure should have financial stability and military power. For this purpose, it should also have the power to raise money through taxes from rich nations.

In recent years, there have been significant contributions in the area of peace, security and development by political scientists, economists and other social scientists. Some of them are historical and descriptive, while others and some are theoretical and quantitative. This book contains a mixture of such articles contributed by distinguished scholars. They are not only analytical but also policy oriented. I hope they can contribute to peace and security in the world. I would like to thank James Hurban for editorial assistance and Maureen Whitney for secretarial support.

Manas Chatterji
Binghamton, New York
September 1995

ALSO BY MANAS CHATTERJI

URBAN AND REGIONAL MANAGEMENT IN COUNTRIES
IN TRANSITION (co-editor) forthcoming

ENVIRONMENT AND HEALTH IN DEVELOPING
COUNTRIES (co-editor) forthcoming

REGIONAL SCIENCE PERSPECTIVES FOR THE FUTURE
(editor) forthcoming

REGIONAL SCIENCE FOR DEVELOPING COUNTRIES (co-
editor) forthcoming

THE ECONOMICS OF INTERNATIONAL SECURITY (co-
editor)

ANALYTICAL TECHNIQUES IN CONFLICT
MANAGEMENT

DISARMAMENT, ECONOMIC CONVERSION AND
MANAGEMENT OF PEACE (co-editor)

DYNAMICS AND CONFLICT IN REGIONAL STRUCTURAL
CHANGE (co-editor)

ECONOMIC ISSUES OF DISARMAMENT (co-editor)

ENERGY AND ENVIRONMENT IN THE DEVELOPING
COUNTRIES (co-editor)

ENERGY, REGIONAL SCIENCE AND PUBLIC POLICY (co-
editor)

ENVIRONMENT, REGIONAL SCIENCE AND
INTERNATIONAL MODELING (co-editor)

HAZARDOUS MATERIAL DISPOSAL:Siting and Management
(editor)

MANAGEMENT AND REGIONAL SCIENCE FOR
ECONOMIC DEVELOPMENT

NEW FRONTIERS IN REGIONAL SCIENCE (co-editor)

SPACE LOCATION AND REGIONAL DEVELOPMENT
(editor)

SPATIAL, ENVIRONMENTAL AND RESOURCE POLICY IN
THE DEVELOPING COUNTRIES (co-editor)

TECHNOLOGY TRANSFER IN THE DEVELOPING
COUNTRIES (editor)

HEALTH CARE COST CONTAINMENT POLICY : AN
ECONOMETRIC STUDY

Contents

PART II—ARMS PRODUCTION, EMPLOYMENT AND ECONOMIC CONVERSION

PART III—SECURITY AND DEVELOPMENT

(xiii)

List of Tables and Figures

Notes on Contributors

Allision Astorino-Courtois is Assistant Professor of Political Science, Texas A & M University.

Vladimir Bondarev is a Research Associate with the Institute of Economic Transition, Moscow.

Jurgen Brauer is Assistant Professor of Economics at Augusta College, Augusta, Georgia.

Manas Chatterji is Professor of Management at Binghamton University (SUNY) in Binghamton, New York; Honorary Distinguished Professor of Management, Indian Institute of Management, Calcutta, India; and Guest Professor, Peking University, Beijing, China.

Roland de Penanros is Professor of Economics, Centre d'Economie Sociale, Universite de Bretagne Occidentale, Brest, France.

Lloyd Dumas is Professor of Political Economy and Economics at the University of Texas (Dallas), Arlington, Texas.

Robert Eisner is William R. Kenan Professor of Economics, (Emiritus) Northwestern University, Evanston, Illinois. He is a past President of the American Economics Association.

Dietrich Fischer is Professor of Computer Science at Pace University, White Plains, NY.

Toshitaka Fukiharu is a Professor of Economics at Kobe University, Japan.

Nehemia Geva is Visiting Assistant Professor of Political Science at Texas A & M University.

David Gold is with the UN Center on Transnational Corporations, New York.

Keith Hartley is Director of the Centre for Defence Economics at the University of York, England.

Nagahuru Hayabusha is a Senior Staff Writer for the Asahi Shimbun, Tokyo, Japan.

(xx)

Albrecht Horn is a Principal Officer at MSPA/DESIPA at the United Nations, New York.

Michael Intriligator is Professor of Economics and Political Science at University of California, Los Angeles.

Walter Isard is Professor of Economics (Emiritus) at Cornell University. He is the founder of the social science disciplines of Regional Science and Peace Science.

Lawrence R. Klein is Benjamin Franklin Professor of Economics (Emiritus) at University of Pennsylvania. He is a Nobel Laureate in Economics.

Hiroyuki Kosaka is a Professor of Administrative Science at Keio University, Japan.

Peter Lock is at the Free University of Berlin and EART (European Association for Research on Transformation), Hamburg, Germany.

Serge V. Malakhov is a Research Associate with the Institute of Economic Transition, Moscow.

Alex Mintz is the Cullen-McFadden Professor of Political Science, Director in Foreign Policy Decision Making Program, Texas A & M University; and Professor of Political Science, University of Haifa, Israel.

Isamu Miyazaki is the Chairman of Daiwa Institute of Research, Tokyo, Japan.

Marie Noelle Le Nouail is Professor of Economics, Centre d'Economie Sociale, Universite de Bretagne Occidentale, Brest, France.

Petra Opitz is Professor of Economics, University of Oldenburg, Germany.

Thierry Sauvin is Professor of Economics, Centre d'Economic Sociale, Universite de Bretagne Occidentale, Brest, France.

Robert Schwartz is a trustee and the Treasurer of Economists for Aruns Reduction (ECAAR), NY, NY.

Corlos Seiglie is Associate Professor of Economics at Graduate School of Management, Rutgers University; and Associate Professor of Economics at Binghamton University.

Ron Smith is Professor of Economics at Birkbeck College, London University, England.

Yoshio Suzuki is Chairman of the Board of Counselors at Nomura Research Institute, Ltd, Japan. He is a former Executive Director of the Bank of Japan.

James Tobin is Professor of Economics (Emiritus) at Yale University. New Haven, Connecticut. He is a Nobel Laureate in Economics.

Seiji Tsutsumi is President of the Saison Foundation, Tokyo, Japan.

Hubert van Tuyll is Assistant Professor of History at Augusta College, Augusta, Georgia.

William Weida is Professor of Economics at Colorado College, Colorado Springs, Colorado.

Susan Willet is Research Associate at the Centre for Defence Economics at the University of York, England.

Conflict, Arms and Peace

Manas Chatterji

Introduction

The Cold War is over. For the U.S. and its Western allies an era of peace is emerging. After many years, it appears that not a single nation or a group of nations is opposing them. However, victory of democracy over totalitarianism may lead to complacency and there is no guarantee that there will be no conflict in the future just because they face no danger now. This is an interdependent world. Instant communication and technological development have reduced greatly the geographical distance. Past enemies are now allies and the future will depend on the world power configuration and the definition of strategic interest. Although democracies never went to war amongst themselves, the concept of democracy and free enterprise was always dependent on perception of economic interest and security. The Cold War led to a specific geographic, political, and economic structure in the world. Now that it is over, after some initial disequilibrium and conflicts (often violent) a new type of structure will emerge based on some centripetal and centrifugal forces.

Sources of Conflict

In recent years, there have been a number of wars. Some of them are:

1. War in Argentina

2. Nicaragua

3. El Salvador

4. Conflict in Ethiopia, Somalia, Angola, etc.

5. Afghanistan war (one million dead, five million refugees- still unsettled)

6. Cambodia (millions dead)

7. Iraq-Iran war ($400 billion spent, millions dead): Stalemate in 1988

8. US-Iraq war

9. Civil war in Sri Lanka, Yugoslavia, former Soviet republics, and many other countries, etc.

Some say that this is just the beginning of regional conflicts and wars.

There are a number of factors associated with conflict. The most important one often cited is religion. It is an irony that while the prophets of all religions spread the gospel of peace, it is often seen as a source of violent conflict throughout human history. It is being repeated in Lebanon, Middle East, Hindu-Muslim riots in India, Northern Ireland etc. However, there are ample evidences to the contrary. The United States, although a secular nation is a religious country. In Indian Subcontinent Hinduism and Islam remain a uniting social and political force. This is also true in the Arab world. In Latin America, the rigidity of Catholicism is continually challenged by forces of modernization and more liberal protestant theology.

In modern times, the issues of peace and war have come to entail more than ethical or religious questions; they are interwoven in the social, economic, and political fabric of the global community. Therefore, it is vital that we develop a theoretical basis for conflict management and peace analysis and indicate how the techniques used in the social sciences can be applied to solve practical, real-life problems in such areas as personal-family conflicts; societal problem such as race relations; ethical problems; and planning problems such as housing, transportation, and so forth; and of course, international relations.

World governments spend vast amounts of financial and other resources on their military sectors. In 1988, for example, world-wide gross national product was estimated at some $20.64 trillion, with world military expenditures at $1.03 trillion or about 5 per cent of world resources. In 1988, also, there were only three trillion-dollar economies: those of the United States ($4.8 trillion), Japan ($2.8 trillion) and the former West Germany ($1.2 trillion). Cutting world military expenditures by a mere 10 percent would release an annual $100 billion fund, larger than the gross national product (in 1988) of all but a handful of countries (Canada, Spain, the UK, Italy, France, the USA, Germany and Japan, as well as China and Brazil). Brauer and Chatterji (1993).

If we look at the data of arms spending for the developing countries in 1980 and 1990, we observe that the stocks of weapons have greatly increased in the Middle East and North Africa followed by E. Asia and China.

Although Sub-Saharan Africa has been low in terms of absolute level of weapons most rapid growth of weapons has taken place there. For each geographical region in the world, two or three states have led the rapid increase in arms. A relatively stable level of weapons has been reached in most developing world although the level has declined a little in the 1980's due to recession. The 'between region' difference is also significant indicating various conflict or military situations existing in different parts of the developing world. Table 1 gives some information about the military expenditures.

TABLE 1

Selected Military Indicators, by Region, 1989

	Military Expenditure ('000$U.S.)	Milex/GNP (percent)	Milex/capita ($U.S.)	Armed forces (thousands)	Population (millions)	Soldiers/ thousand
Africa	15,300	4.5	26	1,659	584.1	2.8
East Asia	83,200	2.0	47	8,432	1,752.4	4.8
Latin America	14,500	1.5	33	1,601	434.6	3.7
Middle East	53,200	12.0	296	3,203	179.8	17.8
South Asia	11,500	3.4	10	2,017	1,106.6	1.8
North America	314,900	5.5	1,145	2,329	275.0	8.5
Oceania	7,100	2.2	289	89	24.5	3.6
NATO Europe	147,400	3.2	429	3,284	343.8	9.6
Warsaw Pact	365,700	10.9	911	4,952	401.3	12.3
Other Europe	22,400	2.1	229	726	97.7	7.4
Developed	867,400	5.0	749	10,040	1,158.1	8.7
Developing	167,700	4.3	41	18,250	4,041.7	4.6
WORLD	1,035,100	4.9	199	28,290	5,199.8	5.4

Source: U.S. ACDA, *World Military Expenditures and Arms Transfers, 1990* Washington:.ACDA, 1991), 47-50

Although it is hard to find any person in the world who is against peace, some people are afraid of peace because they think that it will lead to scaling down employment and security. They do not buy the argument that military spending is an unproductive use of resources and that the international arms race has in fact decreased both national and world security. Thus it is necessary to show why their apprehension is unfounded and how the adverse effects of disarmament can be averted by proper advanced planning for appropriate economic conversion.

It is difficult to obtain any precise estimate of the percentage of people working in the military establishment. The average figure varies between 3 and 4 per cent in the developed countries where 2 to 3 per cent of the gross domestic product (GDP) are involved in the military purchase of goods and services. A conservative estimate holds that 15 million workers are involved directly or indirectly with military spending in the Organization for Economic Cooperation and Development (OECD) countries. If this spending is reduced by one-third, about 4 million people or about 1 per cent of the labour force will be unemployed. Although this figure is not substantial, it may have grave regional and sectoral implication. In New York State, for example, about 8 to 10 per cent of manufacturing jobs are dependent on military spending.Long Island's and Binghamton's (Broome Country in New York State) dependence is very high. The same may be true in other countries such as England, France, and Sweden. All sectors of the economy are not equally dependent on military expenditure. The regional dependence in employment can be as high as 60 per cent (say in Brest, France) when the direct when indirect impact are considered. It is believed that in the United States, on dollar of public expenditure will generate about an equal number of military and civilian jobs, whereas the personal consumption expenditure produced twice as many civilian jobs than military jobs.

The majority of workers in defense industries are engineers and technicians. The workers and the manager, are the ones who may have difficulty in finding alternative employment due to cut in military expenditures. Clerical and other nonskilled workers will have no difficulty in finding jobs, although there will be some problems for older persons and minority workers. Although production workers may need training to switch, such training is also required for civilian workers when new products are introduced. Defense industries are driven by difficult considerations. Prices are guaranteed, there is no competition, and'the specifications are more stringent compared to that civilian work. But, taking into consideration worker shortages and the worker characteristics, defense workers as a group will have less trouble getting new employment compared to other workers as a result of economic conversion.

One problem area of conversion is in the area of research and development (R&D). It is a small percentage of total military spending, but diversion of these released resources to such areas as electronics may have significant impact. Military spending for R&D is about 70 per cent of the U.S. federally funded R&D and about 30 per cent of the total R&D spending. There are some products such as submarines and guided missiles that do not have civilian counterparts. But there are many products where there is a huge overlap. In the military, engineers and technicians are interested in building large, costly, high-performance system that are contracted in small numbers. These products require a labor-intensive production process. If we can use such a method in the civilian arena (civilian space project, nuclear energy, transportation, etc.) then the adjustment will be smooth.

The impact of defense cut can be in terms of (1) sectoral change, (2) regional impact, (3) sociological and psychological problems, and (4) conversion problems. A considerable amount of research material is available on the subject of estimating regional impact, particularly using some analytical techniques such as input-output analysis.

Disarmament, peace, environment, and development are all interlinked. A large amount of environment pollution is directly and indirectly related to military good, particularly nuclear weapons. The potential for heath hazard is evident in the former Soviet Union and disaster in space from nuclear weapons exist. The production and possible use of chemical weapons creates serious environmental problems. War resulting in the destruction of oil fields (such as in the Middle East) will affect the ozone layer. Conversion from the production of military to civilian goods will lead to serious environmental problems involving the destruction and storage of hazardous materials. The diversion of resources to military from civilian needs will leave no money left for environmental protection in the development countries. It is very important for us to discuss these issues and link them with mainstream analytical tools of economic analysis.

Brauer and Chatterji (1993) suggest a more economical, resource-saving method for groups of families to pool their resources into a common insurance fund. Each family pays a small, affordable portion of its current income into the pool that covers the expenses of those affected by unfortunate events. Insurance has greatly stimulated economic activity precisely because it simultaneously set free vast amounts of resources for productive *and* for protective use. In the context of international security and world peace, the idea of insurance is most relevant. So far, most nations acted as if they were individual family units insuring themselves against misfortune by setting aside about 5 per cent of their annual resources for the military sector.

One unresolved question in the area of military spending leads to economic development. From the point of economic theory, when resources are used to build a weapon, say a tank, it does not generate any income. On the other hand, if the same funds are used for an economic good, say in manufacturing or education, it generates further income. However, it depends whether the expenditure is made on soldiers, road building, infrastructure or on weapons. Obviously, in many countries military is contributing to the task of national building. There is the use of dual technology production namely an industrial unit producing both the civilian and military planes and computers. Military is often used to help police to control internal disturbances.

There is no conclusive evidence on the relationship between economic development and military spending. Researchers used empirical evidence for a single country or a group of countries and used different econometric techniques to estimate this relationship. But they have obtained conflicting evidence. That does not mean that there is no such relationship — positive or negative. It requires further studies involving more sophisticated techniques and comprehensive specifications.

Another question is the link between arms and security. What is the security? Is this the absence of war? The former Soviet Union was highly armed and still is, but what happened to its security? Security is a broader concept not only involving military but also economic, social and political factors. Some such questions have been dealt with in Chatterji, et al. (1994). The articles included in this book extend that material.

There is a long history of Peace Research. In recent years the discipline has moved into some distinct directions. First, there is the new area of Peace Science, where the researchers try to develop theoretical structures of conflict without any references to actual conflict situation. Some such theoreticians and quantitatively oriented political scientists do occassionally use empirical analysis. Recently, there also has been significant effort in integrating macroeconomics with issues of arms spending and macroeconomic stability and business conditions.

This book starts with Tsutumi discussing the question of arms spending from the point of view of a successful business man, Tobin and Eisner link budget deficits with arms spending whereas Klein addresses the topic of arms shipment from the developed to the developing countries. Yoshio Suzuki extends his discussion. Walter Isard, David Gold and Michael Intriligator look at the international aspect of arms production. The implication of reducing arms production and channelling it to civilian use, has been discussed by Dumas, Hartley, Smith and others in reference to the economies of some specific countries. In articles related to security and development, the authors emphasized that it is the economic security rather than military that is relevant. Scholars like Seiglie, Fukiharu develop some abstract theoretical models whereas Mintz, Hayabusha present some practical aspects of security and economics. Thus, the articles in this volume links conflict, arms, development and security both from the theoretical and empirical viewpoints.

References

Brauer, J. and Manas Chatterji. (1993). (Edited) *Economic Issue of Disarmament*. Macmillan, London.

Chatterji, M., Henk Jager and Annemarie Rima. (1994). (Edited) *The Economics of International Security*. Macmillan, London.

Part I

Disarmament and Restructuring The World Economy

The first part of this book contains some papers presented at the first meeting of the Japanese affiliate of ECAAR (Economist Allied for Arms Reduction) held in Tokyo, Japan on June 21, 1993.

(Editors' Note)

Chapter 1

A Message

Seiji Tsutsumi

I recall how troubled I was when Professor Lawrence Klein, during his visit here sometime ago to prepare for ECAAR Japan's activity, asked me what kind of debates are going on here in Japan concerning peace dividends by Japanese businessman and economists. There is very little awareness in this country to seriously discuss how the new world system should be built and what kind of role Japan should play, even though the Japanese were pleased about the end of the Cold War. It has been the consistent attitude of Japan since the end of World War II that the United States or the United Nations will come up with the decisions about world affairs. And this attitude basically has not changed until today even though Japan has become an economic superpower.

I thought a little and I had to answer that since Japan has prepaid the peace dividend, there is a prevalent feeling that the need to discuss anew the question of peace dividend is slight. I am saying all this in this welcome address for the following reasons. First, I believe that Japan must take part in the building of the new world system with a voice of its own. Toward this goal, I ardently hope that as many people as possible from various sectors will seriously participate in debates on what kind of role Japan should and can play in the reconstruction of the world economy and disarmament in the post-cold War era. I hope very much that this symposium will initiate the process for this objective. Needless to say, peace and disarmament are not something to be claimed from an ideological stance. This is a theme that is endorsed intellectually, encouraged by specific recommendations, and becomes effective through action.

Immediately before this symposium, Japan plunged into the general elections campaign. It signifies the end of a political regime established in 1955 in which there was just one dominant conservative party against a group of opposition parties that had a little more than one-third seats, barely enough to prevent Constitutional revisions. It signifies the beginning of a new political season. However, no body knows what kind of political force will emerge thereafter. One thing must be clarified. That, it is the dove of disarmament, the immortal bird of peace, which will fly out of the ashes of the 1955 regime, and should not be a night hawk or an owl. Of course, Minerva's owl is an exception. Therefore, it is extremely significant for ECAAR to hold its first symposium in Japan at such a time.

Again, I would like to reiterate my sincere appreciation to your interest and participation in this symposium, on behalf of the sponsors. Thank you very much.

Chapter 2

Defence Spending, the Budget Deficit, and the US Economy

James Tobin

The Decline of Defence Spending

Defence expenditures are claiming an ever smaller share of US Potential Gross Domestic Product. ('Potential' means adjusted for cyclical departures from full employment). This year the defence share of the nation's real capacity output will fall below 5 percent for the first time since the beginning of the Korean War. Within President Clinton's term, defence will be claiming a smaller percentage of the nation's capacity output than at any time since 1940. Before the end of the decade the defence share of the economy will be below 3 percent.

During the Second World War, the military effort took over 40 percent. Drastic demobilization from 1945 to 1950 brought the share down to 5 percent. Then came Korea, the Soviet nuclear threat, and the cold war. Defence reached 10 percent of a growing GDP. In the 1960s a modest fall in the percentage was reversed by the Vietnam war, after which the downward trend resumed, to 5 percent in 1980. President Carter began a new buildup in response to the Soviet incursion in Afghanistan.

President Reagan sharply accelerated the buildup. His plan would have added 2 or 2-1/2 points to the percentage. But he had cut taxes sharply in 1981 and his Administration was not able to find enough politically acceptable cuts in civilian expenditures to pay for both the defence build up and the tax cuts. The result was explosive deficit spending. A chronic impasse occurred between the Reagan Administration and the Democratic Congress regarding what to do about it. One of the outcomes was that Reagan was able to raise defence spending only half as much as he wanted.

A decline in the defence/GDP ratio began in 1985, when the Reagan buildup reached its peak, at about 6 percent of potential GDP. Thereafter defence spending began growing more slowly than GDP, both actual GDP and potential GDP. President Bush's last budget contemplated cuts that would bring the Pentagon's share of the nation's output down almost to 3 percent by 1997. President Clinton proposes additional cuts that will bring the figure below 3 percent. Defence has declined in real terms, that is in physical volume, 29 percent since 1985 and the Clinton budget will cut it another 17 percent by 1997. It will even decrease in nominal dollar outlays.

The size of the armed forces, which reached 3.5 million persons in 1968 during the Vietnam war, is now down to 1.8 million and will fall to 1.4 million by 1997 in the Clinton program. Similar cuts are being made in civilian employment in the Defence Department. Whole army divisions will be disbanded ; the Navy will have fewer aircraft carriers and submarines ; the Strategic Defence Initiative—Reagan's 'star wars'—will be drastically cut and confined to ground-based defensive weapons ; many bases at home and abroad will be shut down ; weapons programs,—research, development, and procurement,—will be discontinued, curtailed, or slowed.

If these trends continue, if these budget proposals are implemented, then by the turn of the century the U.S. defence effort, relative to the size of the economy, will not be so far out of line with the defence efforts and burdens of other G-7 countries.

The Clinton Administration, specifically Defence Secretary Les Aspin, is in the process of revising its overall strategic design for the new international environment. The details of future budgets will depend on the outcome of this rethinking. Certainly the requirement that the U.S. be prepared to fight two major wars at one time has been abandoned. The focus on nuclear deterrence of a rival superpower is of course obsolete, and the crises that arise throughout the world are no longer viewed through the lens of cold war. The emphasis now is on control of the spread of nuclear weapons, and on participation in U.N. or other multilateral efforts to keep or restore peace and order. The

Pentagon increasingly emphasises small mobile forces that can be rapidly deployed anywhere in the world. `

Economic Consequences

The economic consequences of the shrinkage in defence programs are extremely important. They are painful to many individuals : servicemen who had counted on military careers but now will be involuntarily returned to civilian life ; civilian employees who will also be looking for other jobs ; employees of the vast network of contractors and sub-contractors who depended on Pentagon business ; other residents of areas where bases are being shut or businesses are losing defence-related orders. The direct impacts on employment are estimated to amount to about 1.2 million jobs, roughly 1 percent of the labor force.

The demobilization of soldiers, sailors, and air force personnel has some unintended unfortunate side-effects. The military services have been in the forefront of equal opportunities for minorities. Those opportunities have been especially important for young black and Hispanic men, many of who were growing up in urban neighbourhoods of crime, drugs, and poverty, often without fathers or other adults who could serve as male role models. Schools often failed them, or they failed to take advantage of the schools. The army and the other services gave them an environment with discipline and order, taught them useful skills and habits of work, and showed them that their own efforts could pay off in promotion and self-respect. They could be role models for younger brothers or neighbours. The loss of these opportunities will be felt, unless governments can find a use for their talents in non-military settings.

For the nation as a whole, however, the economic effects should be manageable. The national economy absorbed in stride relatively larger cutbacks in the late 1940s after World War II, in the early 1950s after Korea, and in the 1970s after the Vietnam war.

The trouble is that the macroeconomic environment was more favorable in those days. The present demobilization is impinging on a macroeconomically weak U.S. economy, indeed on a weak world economy. The American economy has been soft for five years now, beginning in the second quarter of 1989.

It is true that the official business cycle arbiters of the National Bureau of Economic Research recognize only the three quarters 1990.3 through 1991.1 as recession. But for eleven quarters before and at least five quarters after, the economy was growing more slowly than the sustainable rate at which potential GDP—the capacity of the economy at an inflation—safe rate of unemployment of about 5-1/2 percent—can grow with normal increases in labour force and advances in productivity. This rate is between 2 and 2-1/2 percent. By that standard, we have been in a growth recession—I guess that's the only kind you have in Japan—for 19 quarters. As a result, unemployment rose from 5.5 percent in 1988 to 7.5 percent in 1992, has declined slowly in 1993.

A discouraging statistic is the scarcity of vacant jobs. There are fewer of them than would normally be expected at 7 percent unemployment, as few as there were with unemployment rates two or three points higher in the recession troughs of 1975 and 1982. The contractions of defence-related industries are part of the story, but by no means all of it. Many other U.S. enterprises, including some of the largest and most successful firms in the past, are downsizing, reducing permanently their work free. An extraordinary high proportion of the unemployed are job losers with no prospects of returning to their former jobs or employers.

Consumer confidence and business confidence are, after an upward spurt after the presidential election and inauguration, low. The central bank, having lowered the money-market interest rates it controls to 3 percent in December 1992, seems in no mood to stimulate a stronger and faster recovery. The President and Congress preoccupied by deficit reduction, are in no mood to give the economy a fiscal stimulus. I shall discuss this situation shortly.

The Clinton Administration budget includes $20 billion for defence reinvestment and transition assistance, including dual-use technology programs and high-technology development grants. But for the most part adjustment are left to the enterprises, workers, and communities affected. Inevitably most of the adjustment will involve different firms, products, and locations. The opportunities for conversion of existing facilities tailor-made for defence contracting are limited.

One unfortunate result is that many firms seek salvation in arms exports, and for several reasons—saving jobs, improving the trade balance, keeping defence production capacity intact—the Defence Department and the Administration will help them. This, of course, is happening even more in Russia. The arms trade is a problem requiring international attention and action. It should be a major preoccupation of ECAAR.

The Vanishing Peace Dividends

Five years ago, and especially when the end of the cold war seemed to be approaching, Americans enjoyed the anticipation of "peace dividends", the fruits of victory in the four-decade duel. Savings from the defence budget would liberate funds for balancing the budget, activating long underfunded social and public investment programs, cutting taxes, increase aid to the Third World, protecting the environment, supporting the arts, and doing many other good things. Alas, the peace dividends are here, but somehow they are not available for any of those wonderful purposes.

How come ? Part of the answer is that Ronald Reagan gave up income tax revenues of 5 percent of GDP in 1981. Despite several tax increases during the Reagan-Bush years, that lost revenue was never recouped. At the same time, the income tax was made much less progressive, and the deductions, exemptions, and brackets were all indexed for inflation. The result was to eliminate the previous automatic tendency of revenues to rise as a proportion of GDP as either real growth or inflation increased GDP. Social security payroll taxes were increased in 1983 enough to bring the social security account into surplus, but the government simply counts that surplus as a reduction in its deficit.

However, Reagan-Bush anti-tax politics is not the whole story. For the federal budget the culprit is entitlements, for example transfers to people who satisfy the rules of eligibility for prescribed benefits. Outlays for entitlements are open-ended, depending not on annual budgetary decisions and congressional appropriations but simply on the numbers of legitimate beneficiaries and the size of the transfers for which they are eligible.

Entitlements rise inexorably year after year. This is not because Congress has become more generous—that hasn't happened since the Nixon years. It is for two other reasons : The number of eligible beneficiaries, notably elderly persons and poor persons, rises faster than the population as a whole. The cost of the benefits for which they are eligible rises faster than the general price level ; this is particularly and dramatically true of health care.

The two federal health care programmes, Medicare for the elderly and Medicaid for the poor, took 2.5 percent of potential GDP in 1970, 3.5 percent in 1980, and 4.9 percent in 1990. If nothing is done, they will take an additional 2.5 percent by 2002. Health care costs have been rising relative to overall prices by 6 percent a year. After account is also taken of the increase in the number of eligible beneficiaries, almost all the rises in the federal health care share of GDP are explained. The average client is getting very little more service—although the service is doubtless of higher quality.

If defence was the chief burden on the economy and the federal budget in the 1950s and 1960s, health care has taken over that role for the 1990s and the first decade of next century. That is why the health care reform which Mrs. Clinton, as task force chairman, is now delivering to him, will be the President's most important fiscal program, even more decisive than the fiscal and economic package now before the Congress. Unless a way is found to keep the relative cost of health care from perpetually rising, the prospects for controlling and reducing the deficit are nil.

Deficit Reduction and Gridlock

President Clinton inherited from his two predecessors a host of formidable economic maladies. On the macroeconomic scene, he faced two critical problems, one a long-run condition and the other a short-run difficulty.

The symptoms of the first are that labour productivity and real wages have stagnated for two decades, through cyclical recessions and prosperities alike. This trend endangers the standards of living of future Americans. Clearly the society has been consuming too much and investing too little in the future. The share of GDP put into business plant and equipment, new housing, public infrastructure, science and technology and education has been too small. A major contributing factor has been federal budget deficits incurred to reduce taxes for affluent consumers and to build up armaments.

That is the rationale, really the only rationale, for giving high priority to both deficit reduction and to public investments in infrastructure and education. Clinton proposed to do both.

Unfortunately his two predecessors also bequeathed to him an economy in prolonged growth recession, as discussed above, with stubbornly higher unemployment and desperately scarce jobs. Clinton owed his election to the public's disaffection from George Bush, whom they perceived to be doing nothing to pep up the weak and slack economy.

Clinton's dilemma was that fiscal policies to deal with either one of these maladies might well make the other one worse. He could spend money, even reduce taxes, to speed recovery and create jobs, but in the process he would add to budget deficits, at least temporarily. If his stimulus package was oriented to private and public productivity-enhancing investments, he could argue that he was doing good things for future Americans even if he borrowed to pay for them. But that was too subtle an argument for politicians, reporters, and voters, especially with Ross Perot on the sidelines calling simplistically for deficit reduction for its own sake.

Alternatively, Clinton could give first priority to deficit reduction, with the risk that raising taxes and cutting spending would weaken the fragile recovery, perhaps even trigger another recession, and leave him facing the electorate in 1996 with no better economy than Bush had in 1992.

Confronting this dilemma, the new President's economic team was divided. With his eyes open, Clinton embraced the austerity side of the dilemma. His state of the union message February 17, 1993, delivered in person to the Congress and on TV to a vast national audience, was a stunning success. As a token recognition of the immediate macroeconomic needs, his program did include a small fiscal stimulus for 1993-94, tax incentives for business equipment investments, and grants to cities for needed public investment projects that would create jobs quickly.

The stimulus package failed in Congress. The Administration was unable to explain away its apparent inconsistency with the primary deficit reduction objective of the larger proposal. The large Republican Minority in the Senate had decided to oppose Clinton's budget proposals at every turn, and they showed their power by filibustering the stimulus package to death. The Investment Tax Credit proposal just withered away ; its supposed business beneficiaries took no interest in it.

Clinton's deficit reduction plan aimed to cut the accumulation of federal debt over the five years 1993-97 by $500 billion (about one-eighth of the outstanding debt), relative to what it would be otherwise. The annual deficit in 1997 would be reduced to $200 billion, 3 percent of GDP, an amount small enough to stop the debt from growing faster than the economy. The $500 billion cumulative deficit reduction was to achieved roughly by $250 billion in new taxes and $250 billion in cuts of expenditures (mostly defence) net of modest new investment and social initiatives Clinton had long advocated.

In the end, the budget Congress enacted reached the $500 billion target for deficit reduction, equally divided between expenditures cuts and new revenues. But Clinton's cherished new investments and social initaitives were severely squeezed. Congress would not devote either governmental economies or new taxes to these purposes.

Standard macroeconomics suggests that the deficit reduction package, however useful for the long run, will damage the prospects of recovery unless the Federal Reserve takes active measures to reduce interest rates for all maturities, short as well as long. Chairman Alan Greenspan and his colleagues were not prepared to do that. In fact, there is danger that the "Fed" will tighten credit and raise money market rates.

The logic of politics and public opinion has led Clinton and his economic team to claim that deficit reduction, without any initial temporary stimulus at all, is really the proper medicine for both of the economy's ills, short-run recovery and long-run growth. Heartened by the declines in long term bond interest rates that have occurred since November, they argue that these are sufficient stimulus and will more than offset the direct negative impacts on demand of tax increases and spending cuts. This is a very dubious proposition, and it could turn out to be embarrassing if the recovery falters or bond interest rates rise.

During the twelve years 1981-92 the federal government was stgmied by gridlock because Congress and the White House were of opposite parties. In 1992-93 gridlock continued even though they were of the same party.

By May Clinton's honeymoon period had expired, and the euphoria with which his tough deficit reduction budget was received in February had evaporated. Although the Democrats have majorities in both, neither the President nor the congressional party leadership can enforce party discipline. Individual members threatened to vote against the budget unless it is amended to the liking of their particular constituents. They had bargaining power because the Republican minorities stuck 100% together against anything the President proposed, having decided their long run political interest lies with "No New Taxes" !

On the crucial vote in the House of Representatives, the deficit reduction package won by a single vote. Gridlock was narrowly averted. Clinton tirelessly waged a skillful campaign of lining up support in Congress, member by member. He did the same to obtain passage of NAFTA, against the odds. On that issue Democratic party discipline collapsed completely; most of the leaders of the party in the House opposed NAFTA. It was saved because most Republicans supported it; after all, it was initially a Bush project.

Clinton's leadership skills looked better at the end of his first year than they did at the beginning, and so did the chances of avoiding gridlock. The big test in 1994 will be health care reform, where several Democratic and Republican plans will complete with the Clintons' own proposal. Meanwhile, the gradual winding down of military spending will continue as planned. The budget cannot afford any slowdown in this process, white increasing alarms and threats scattered throughout the world, and the shortages of jobs for workers displaced by defence cuts, will prevent further and faster cuts.

POSTSCRIPT January 1995

Preparing papers for conferences on events in progress is a hazardous pastime, especially when the papers are destined for later publication. Between the conference in June 1993 and my revision of this paper for publication in December 1993. I changed some paragraphs of my paper that had become obsolete. In the year 1994, between submission of that revision and the appearance of page proofs, more and more important developments have occurred. Doubtless this paper and this postscript will be obsolete by the time the book is published.

The U.S. economy recovered from cyclical recession in 1994. The growth of both GDP and employment exceeded all expectations, and unemployment fell to the lowest rates since 1972. What was even more surprising, given the remarkable real growth, was that price and wage inflation rates actually declined. There was no fiscal stimulus package; indeed fiscal policy was moderately contractionary, devoted wholly to deficit reduction. Moreover, the Federal Reserve raised short-term interest rates by 250 basis points during the year. The fears I expressed in the paper, that the economy would suffer from deficient aggregate demand in these circumstances, turned out to be quite unjustified. Of course, what will happen in 1995 is uncertain.

Another unusual outcome was that President Clinton and his party reaped no political gain from the excellent macroeconomic performance of 1993-1994, or from the significant reduction in the budget deficit they engineered. Instead, they suffered a stunning defeat when the midterm elections gave the Republicans control of both houses of congress. One reason was that both Congress and the public turned against the Clintons' proposal for health care reform, and no compromise or alternative passed the Congress either. The President had staked too much of his reputation on this initiative, and its protracted demise in the last half year of the 103d Congress overshadowed many successes, including NAFTA and GATT as well as deficit reduction.

The Republican agenda, Speaker Gingrich's "Contract with America", promises wholesale retrenchment in federal spending to finance both tax cuts and balancing the budget. The one place where the Republicans advocate greater expenditures is the Pentagon, perhaps as much as $25 billion a year, much of it to revive Reagan's "star wars" dream of active anti-missile defence.

Chapter 3

Arms Trade : The Next Step in Arms Reduction

Lawrence R. Klein

Dominance of the Arms Race between the Superpowers

The United States and the Soviet Union dominated the military situation of the world for many years. Their confrontations reached a peak in the late 1980s, just before the breakup of the USSR. In 1989, the global military situation was as shown in Table 3.1.

Table 3.1 : *Military Expenditures and Armed Forces 1989*

	Expenditures	Personnel
US	$304.1 bn.	2,241 thous.
USSR	$311.0 bn.	3,700 thous.
World	$1035.1 bn.	28,290 thous.

Source : ACDA

Thus 60 percent of the world military expenditures were estimated to have been incurred by the two superpowers. NATO Europe accounted for another $147.4 bn., while the Warsaw Pact countries allied to the USSR accounted for $54.7 bn. If these two figures are added to the US and USSR totals, we see that 80 percent was tied up in the Cold War stand-off.

In terms of people in the armed forces, the percentages are less daunting. The superpowers' share of world people under arms was little more than 20 percent. All NATO and WTO forces amounted to about 37 percent of the world total.

Many countries had their armies, but they were not equipped with such formidable and costly instruments of war. That is where much of the spending of the US and USSR went in their race for superiority.

Every nation needs some minimal defence posture ; so in this post-cold war environment we might look for only some modest arms reductions, at the outset, at least, among other countries. Most of the reduction will have to come from the superpowers.

The US peaked in 1986 at $311.7 bn. (measured in 1989 prices) ; there has thus been a real decline in the defence establishment. The Soviet peak is estimated by ACDA to have occurred in 1988, and the decline in just one year—the year of the break-up in Eastern Europe—was almost $20 bn., much larger than the American decline. After 1989, the decline has continued in both the US and the USSR, dramatically so in the latter. Were it not for the Persian Gulf War of 1990-91, the early decline would have been even larger in the US.

European NATO countries have not made similar reductions, but the WTO countries in Eastern Europe have reduced their military outlays, although they were not very large in the first place.

For the rest of the world, the trends are mixed. Developing countries, as a whole, are estimated to have made reductions of about $25 bn. (1989 prices) since 1983. We know, however that in this period, some developing countries armed heavily.

It will be possible for the world to realize some Peace Dividend, whether it goes into federal budgetary funds for such things as deficit reduction, or into specific public spending (for the betterment of ordinary citizens) it is still to be counted as a Dividend. If it is used for deficit reduction, it helps to bring down interest rates and stimulate capital formation, of a peaceful sort. If it goes into health support, education, public infrastructure, or into cultural foundations it benefits people directly, as well as indirectly.

US military expenditures are falling year-by-year, in real terms. At the end of the restructuring process, the reduction may amount to 20-30 percent from the peak point, cited above. In addition, there is progress in the closing of military bases, the reorientation of military R & D (away from Star Wars) and large scale downsizing among military contractors. All-in-all, the USA is sliding towards a higher fraction of civilian (vs. military) economic activity.

The change is even more abrupt in the USSR. Troops are being recalled from WTO Countries ; the military establishment is relatively inactive because of a dearth of funding. WTO countries are striving to produce consumer and other civilian goods. They will cut back the military sector, just in order to grow, in a healthy, civilian way.

It' is more challenging to convince the developing countries to scale back military spending. Some equate power and prestige with a strong military posture. They fear attacks from neighbours, when animosity has prevailed for many years, or centuries.

There certainly has been no arms reduction in the former Yugoslavia, nor in Somalia or other warring nations.

Within the two superpowers, there is one path for adjustment to the new strategic situation that is very attractive from an economic point of view, namely conversion ; that is the process of shifting production from military to civilian use. Budget deficit reduction and easier credit conditions help to ease the burdens of conversion and speed up the process. It is an alternative to increased social spending or simply to allowing the market mechanism to work. This is especially so for the US and NATO allies.

Favourable and unfavourable examples exist for successful conversion. Both the United States and the USSR made good conversions after the Second World War. Actually, it was very much a case of reconversion, especially for the United States. In the US case, most civilian industries were on the trend line of their expansion paths, that were established prior to the War, by 1950. The advent of the Korean War did not upset conversion because it was not an all-out industrial effort.

Two aspects of the economic environment favoured conversion in the late 1940s. In the first place it meant a return to familiar civilian activities that were interrupted by the War. This is the reconversion aspect. Secondly, there was exceedingly strong demand to absorb the increased supply of civilian goods. This was effective demand for the US, i.e. demand that was supported by purchasing power, in the form of accumulated liquid savings.

The present conversion effort is taking place in the midst of an adverse cyclical phase-weak recovery in the US and drastic recession in the former Soviet Union (FSU). Effective demand is, therefore, lacking.

During the last 40-50 years, the military production sector had built a close working relationship with the US Department of Defence. Heavy dependence on that relationship left private military contractors ill-prepared to meet the competition of the market place for producing and distributing civilian goods. In some notable cases, military contractors have succeeded in shifting to civilian activity, while others, especially among giant corporations, have had dismal failures.

There are two possible strategies. Large corporations may diversify between military and civilian activities ; or they may produce military goods as long as they are in strong demand and then convert to civilian production when the political situation changes. In the former case, the state of the economy will have a large bearing on whether diversification works. In a poor cyclical phase, such as the one that prevails now, it will be difficult to reap the benefits of diversification. Also, weak business conditions can interfere with the operation of a newly converted industrial facility.

The outcome is a matter of timing. People were prepared to be patient and wait during the five years following the end of the Second World War. The public is always impatient, but that is a matter of degree. The country and the world survived 1945-50, and soon after, every conceivable goods or service was available in the market place. The United States enjoyed the full benefits of conversion sooner than others. Outside the US and outside some few countries who were unscathed or not involved there was gradual reconstruction of the destroyed facilities, dropping of rationing, and the start of enjoyment of the output of the new postwar era.

There is no good reason to believe or assert that conversion will be less satisfactory now, if given a five-year period to set the markets straight. There may well be frustrations now, but there should not be such a high degree of impatience, and by 1995, civilian output should have been fully converted to the extent needed.

Arms Trade and Transfers

When we turn from the large scaling-back that is taking place in the FSU and USA, we find that the next major problem is the arms business—trading and transferring arms. First let us take a look at the main countries involved.

Table 3.2 : *Arms Transfers, 1989*

	Imports	Exports
World Total	$45,320 mill.	$45,320 mill.
Developed countries	10,680	41,010
Developing	34,630	4,415

Source : ACDA

The advanced industrial countries are major exporters, and the developing countries are importers. The sums are quite large, and the developing countries appear to be well supplied by the developed countries.

NATO and the WTO were major exporters, while Middle East and other Asian countries were major importers.

Table 3.3 : *Regional Arms Transfers, 1989*

	Imports	Exports
Africa	$3,990 mill.	$70 mill.
East Asia	5,335	2,640
Latin America	2,555	285
Middle East	12,060	1,020
South Asia	7,890	20
NATO	8,145	18,820
WTO	3,150	21,480

Source : ACDA

Some of the leading countries involved were as follows : The USA, UK, France, and Germany were major NATO suppliers. The USSR was the major WTO exporter. Indeed, arms constituted a major export of the Soviet Union—almost a 20 percent factor in their total merchandise exports.

Outside NATO and WTO, China was a major exporter.

Table 3.4 : *Major Exporters of Arms, 1989*

	World Total	To LDCs[*] (1990 prices)
USSR	$19,600 mill.	10,869
US	11,200	3,454
UK	3,000	1,968
France	2,700	2,065
China	2,000	865
Germany	1,200	173
Czechoslovakia	875	221
World Total	45,430	

Source : ACDA; SIPRI

[*]   In comparison with the tabulated figures for 1989 in current prices, these estimates for 1989 in 1990 prices look low for the USSR, China, Germany because the world totals in the 1992 SIPRI yearbook are lower than the corresponding totals in the table.

It is evident that a few major suppliers account for most of the arms traffic. Also, substantial portions of the exports are destined for developing countries. In total, nearly three-fourths of total exports went to LDCs in 1989.

Table 3.5 : *Major Importers of Arms, 1989*

Saudi Arabia	$4,200 mill.
Afghanistan	3,800
India	3,500
Greece	2,000
Iraq	1,900
US	1,600
Japan	1,400
Vietnam	1,300
Iran	1,300
Cuba	1,200
Turkey	1,100
Syria	1,000

Source : ACDA

There are some striking differences between the list of large importers and large exporters. The export list is highly concentrated, and the whole distribution across many countries is skewed. The seven countries are dominant in the world total, and the size of exports drops off rapidly after the seventh on the list (Czechoslovakia).

Apart from China, there are no developing countries in the export list, while the import list is dominated by developing countries.

Not all importers are LDCs, however, the US and Japan are, perhaps, the richest countries in the world, and they import arms on a significant scale. In fact, Japan, where low defence spending has been so well studied and considered to be responsible for the good civilian growth record, is moving up smartly, as well as steadily, in total defence spending and in arms imports.

The Middle East is well represented in the list of the top 12 importers for 1989, with countries like Israel, Egypt, Libya, UAE, and Kuwait in the second tier of importers, with amounts that are close to $1000 million in many cases.

These figures fluctuate a good deal from year-to-year. For example, Iraq is estimated to have spent $9,100 mill. on arms imports in 1984. Aside from the fact that many countries hide arms traffic statistics, many prepare for military action over several years and, naturally, the figures are large for cases like Afghanistan when a shooting war was in progress.

Many of the countries from the WTO are drastically reducing arms expenditures as a result of changed conditions since 1989. They are devoting much attention to civilian restructuring of their economies. In general, military spending is declining throughout the world, but, unfortunately, not everywhere.

Yugoslavia was not a member of WTO, but has some similarities to the WTO neighbours. Their imports reached $625 mill. in 1987, but were generally less than $300 mill. per year. Also, their exports were fairly large. How are all the factions in the former Yugoslavia able to field such awesome and ruthless forces? Arms control specialists claim that they are using Yugoslavia's inventory and are also producing some fresh arms. This is hard to believe as the complete answer. The various factions must be tapping into the world traffic.

Are they getting surplus arms from the FSU? Are they getting supplies from other WTO members? Czechoslovakia admirably renounced arms production and supply to the outside world, but soon succumbed to the lure of financial gain. First, they agreed to honour existing contracts, then they appeased the poorer provinces of Slovakia by permitting resumption of production and export. Now, with the country divided, it is generally supposed that a good portion of the industry, which was significantly present in Slovakia, is back in the arms business and trafficking.

Policies for Containment of the Arms Trade

The discussion thus far has dealt with military spending and arms transfers, mainly from a descriptive point of view. We are experiencing difficulty in realizing the peace dividend and in achieving a more peaceful world. As for the peace dividend, it is unfolding in the sense that military spending is being reduced, gradually in the United States and sharply in the FSU. Spending is not being reduced everywhere, but in enough places to be able to detect a slowing down of the military sector of the entire world economy. The processes of downsizing of defence industries, size of the armed forces, total military spending, and conversion take time. If by 1995 the process is nearly complete, we should be able to say that the effort has been successful, and that, in itself, is a major part of the peace dividend. The rest should take the form of a steadily increasing flow of civilian goods and services. There is no basis for claiming that this goal was not achieved, or that the sequel to the ending of the Cold War was any more delayed than the sequel to World War II.

The bothersome issues are the outbreak of limited conflicts such as the Gulf War, the Yugoslav ethnic battles, or the Somalian tribal conflicts. These have surely been fueled by arms transfers, mainly from the superpowers, and other industrial countries to developing or poorly developed countries. The issue before us now is to block the supply routes and otherwise halt the transfers of arms for the purpose of carrying on limited engagements.

Some policies for containment are :

1. Implementation of nuclear non proliferation treaties, extended to cover weapons of mass destruction, in general.

2. The use of conditionality in the granting of Official Development Assistance (ODA).

3. The acquisition and disposal of surplus weapons, i.e. those that become available as the Cold War unwinds.

4. The use of UN inspection and peace keeping operations.

1. Not all nuclear powers have agreed to non-proliferation agreements, but they can be held in restraint if the major powers practice abstention. As long as the United States, for example, refrains from nuclear testing, other nuclear powers also do so. It is important for the US, the FSU, and others who possess significant stockpiles of weapons with nuclear war heads to continue the test ban, for if they resume testing, it can be assumed that lesser powers will also resume and that aspiring nuclear powers will speed up their nuclear research and development programs.

Nuclear testing and nuclear facilities can more readily be detected than can some other weaponry of mass destruction, such as biological and chemical weapons; therefore the ban on nuclear testing is important but does not provide ironclad guarantees for peace-loving. Restraint on the part of the United States from testing is of the utmost importance now.

2. The Japanese government, the largest ODA provider in the world, has issued a white paper on the description and criteria used for their decisions. It is interesting that *conditions* have been laid down in terms of restraints on military spending, military use of grant funds, the military composition of imports, and the support of research into methods of mass destruction in order for developing countries to qualify for ODA grants. This is an exemplary statement and deserves to be adopted on a much wider scale by as many grantors as possible. In particular, the World Bank and similar multilateral institutions should pursue the same kind of policy line.

3. Weapons of mass destruction are *one* thing and conventional arms are *another*. Of course, great effort must be expanded on containing the former; that is why the continuation of the nuclear testing ban is so important, but also the limited regional conflicts are proving to be devastating, and they must be stopped by blocking the supply to aggressors of conventional arms through trade or other transfer.

It would be wise for the advanced powers who are financially able to do so to acquire conventional weapons, say from Soviet withdrawals in Europe or from stockpiles of the FSU in their home boundaries, and destroy the arms. It is admirable for Western companies and individuals to purchase the human efforts of displaced military supply technicians in order to channel their activities towards peaceful pursuits.

All these are expensive activities, but what price can we put on limitation of open conflict?

4. The United Nations provide the only hope that the world can be policed with some degree of commitment and force. In recent years, the UN have gained status as a peacekeeping force, but their power is limited. It must be extended to the point at which they could move into places like Yugoslavia or Somalia, be respected, and enforce the peace. Also they should have the power of inspection. World scientists can monitor radiation and the use of nuclear devices. On site inspection is also needed, but it is absolutely essential for monitoring possible ventures into chemical and biological weaponry. This inspection power should be vested in a UN force.

Chapter 4

The Economics of Disarmament as an Investment Process

Michael Intriligator

1. Disarmament as an Investment Process

To analyse the economic effects of disarmament, it is useful to treat disarmament as a type of investment process. An investment process involves an expenditure at the present time which is followed eventually by a return on this expenditure. The investment could, for example, involve purchasing a financial instrument like a bond, which yields some designated return until its maturity. Alternatively, it could involve the construction of physical capital in the form of plant or equipment which eventually generates some return from the sale of the goods or services produced by this capital along with complementary inputs.

In terms of its economic aspects, disarmament can be considered as such an investment process. It involves initial conversion costs, including the direct adjustment costs of retraining workers and soldiers, retooling or building new capital, and developing the capability to produce non-defence goods and services. In addition, there are the direct and opportunity costs of unemployment of labour, capital, and other inputs into the production of military goods and services. Both types of conversion costs are incurred over a transition period, which may last years. They are then followed by benefits as inputs are reallocated to the production of civilian goods and services, which provides the ultimate peace dividend. Both the initial conversion costs and the ultimate peace dividend could be measured in financial terms or, even more appropriately, in real terms, as the real cost of the initial resource unemployment and reallocation and the real benefit of the ultimate additional civilian output. A return from disarmament can then be obtained as the implied social rate of return, taking explicit account of both the real costs and the real benefits of disarmament.

This way of analysing the economic effects of disarmament is useful in dispelling two myths, which, unfortunately, are held as strong beliefs by their proponents. The first myth is that there would be an immediate peace dividend that can be paid out to the citizens of the disarming country or used in some other way, such as paying off the national debt, building or rebuilding infrastructure, funding social services, reducing taxes, and so, forth the presumption being that reallocation of resources of labour, capital, and other inputs (including energy and other natural resources, services, and material inputs of intermediate goods) can be made instantaneously and costlessly. This naive view treats military expenditures as a category of social spending and simply shifts it to another such category, like shifting money from one pocket to another. It ignores the fact that the process of conversion entails a fundamental reallocation of resources in the economy, with real adjustments to be made in employment patterns, capital utilisation and the industrial structure. While there are potentially major gains from disarmament, particularly over the long term, it would be a mistake to ignore the short-term adjustment costs of disarmament, including particularly the potential for unemployment of labour, capital, and other resources stemming from reductions in military expenditure.

The second myth is that disarmament would lead to an economic downturn that would be irreversible, the presumption being that the economy is completely dependent on military spending. This cynical view ignores past successful conversions following war periods, showing that a national economy will eventually recover from disarmament and reap the rewards of greater civilian output. It also ignores the fact that, at least for most Western industrialised market economies, military spending represents a relatively small percent of output, amounting to less than five percent of GDP.

These two myths are mirror images of one another and both contain some element of truth. The first myth, however, ignores the initial costs of disarmament, while the second myth ignores the ultimate benefits of disarmament. By contrast, interpreting disarmament as an investment process allows explicitly for both the initial conversion costs and the ultimate benefits of the peace dividend.

2. Future Prospects for the Economic Effects of Disarmament

The future prospects for the economic effects of disarmament depend crucially on the expected pattern of costs and benefits stemming from the process of disarming. Depending on this pattern, the social rate of return from disarmament, treated as an investment process, could either be very high or very low, or possibly even negative. If the future pattern involves relatively low conversion costs, a short transition time to benefits, and relatively high benefits from disarming, then the social rate of return from disarmament would be very high. By contrast, if the future pattern involves relatively high conversion costs, a long transition time to benefits, and relatively low benefits from disarming then the social rate of return from disarmament would be very low or possibly even negative. These alternative patterns for the economic effects of disarmament themselves depend on the particular situation facing the disarming nation. It is useful to distinguish three different world areas: the industrialised market economies of North America and Western Europe, the former socialist economies of the former Soviet Union and Eastern Europe, and the developing economies. Within each, the return from disarmament depends on both general economic conditions and the policies chosen by the disarming nation.

3. Future Prospects : The Industrialised Market Economies

For the industrialised market economies of North America and Western Europe, the potential exists for a high return from disarmament, particularly in the case of long-term sustained and gradual reductions in military expenditures under favourable market conditions and government policies. Favourable market conditions would be those of economic expansion, with tight labor markets and large sums available for investment in new plant and equipment. Favourable government policies would include manpower retraining, economic assistance to military personnel and defence workers in finding new jobs, and economic assistance to defence-oriented industries and plants in identifying new market opportunities. In such circumstances there would be low conversion costs, a short transition time, and high conversion benefits, implying a high return from disarmament. A historical example was the successful conversion of the US economy from military to civilian production in the post World War II period and, to a lesser extent, following the Korean and

Vietnam wars. There is, however, also the potential for a low or even negative return from disarmament if there are short-term episodic reductions in military expenditures under conditions of economic recession, with high rates of unemployment, with relatively small sums available for investment in new plant and equipment, and with either a lack of governmental action to address these problems or wasteful government bail-out type subsidies to unemployed workers and impacted industries or regions. Under these conditions there would be high conversion costs, a long transition time, and low conversion benefits, leading to a low or negative return from disarmament. Unfortunately, it would appear that this is where the US and other NATO countries are now heading, with essentially no peace dividend unless market conditions and government policies change considerably. Changed government policies would aid conversion indirectly by fostering overall economic expansion, tight labour markets, and capital formation and directly by providing for retraining, retooling, and information on new market opportunities.

4. Future Prospects : The Former Socialist Economies

For the former socialist economies of the former Soviet Union and Eastern Europe, there is the potential for a high return from disarmament, as in the case of the successful conversion of the USSR economy from military to civilian production in the post World War II period under supportive central planning policies directing resources to civilian uses. There is also, however, the potential for a low or even negative return from disarmament if there are abrupt and precipitous declines in military expenditures under conditions of economic crisis of falling output and large-scale unemployment with little available for investment and with a lack of either central planning or markets and continued wasteful support of defence industries and personnel, which, unfortunately, appears to be where the former USSR and other former socialist countries are heading. It will require an unusual combination of focused national economic policies, together with technical assistance from other nations and international economic organizations, to foster the creation of market institutions so as to avoid potentially disastrous consequences. In the absence of central planning, markets are needed for the reallocation of the resources released due to reduced military spending.

5. Future Prospects : The Developing Economies

For the developing economies, there is the potential for a high return from disarmament, particularly if there are long-term sustained gradual reductions in military expenditures in the Middle East, South Asia, Africa, and in other regions, when undertaken in conditions of economic expansion and high rates of investment stemming from high prices of exports, such as oil, and foreign assistance, and when combined with supportive national macroeconomic and trade policies. The results of these economic conditions and government policies would be a high return from disarmament. By contrast, there is a potential for a low or even negative return from disarmament if there are abrupt and precipitous declines in military expenditures under conditions of economic decline stemming from low prices of exports, reduced foreign assistance, and poor macroeconomic and trade policies. The results of these rather different economic conditions and government policies would be a low return from disarmament, as unfortunately, would appear to be the situation today in several developing countries . This negative outcome will prevail unless countered by international technical and financial assistance and informed national macroeconomic, trade, and defence economics policies.

6. The Role of Arms Control and of Arms Exports

In all three major world regions there are substantial potential economic benefits that could be achieved as a result of disarmament, but only if appropriate policies are pursued, particularly at the national level. Such policies could be greatly facilitated by a new round of arms control agreements, negotiated bilaterally or multilaterally among both former foes and former allies. Former arms control agreements, particularly those involving the US and the USSR were negotiated and concluded with the goal of achieving strategic stability in avoiding wars. They had only relatively minor effects on defence expenditures, which continued to rise. In the current period defence expenditure are falling in the US, the former Soviet Union, and the nations of NATO and the former Warsaw Pact, which taken together, account for a substantial portion of world-wide military expenditure. A new set of arms control agreements could help reduce the costs of disarmament, shorten the transition period between costs and benefits, and increase the benefits of disarmament, thus greatly increasing the economic return from disarmament. Such agreements might be negotiated via the North Atlantic Cooperation Council, the affiliate of NATO that includes both members of NATO and members of the former Warsaw Pact. Such agreements would call for gradual, steady, and predictable reductions in defence expenditures, reductions in military establishment personnel, reductions in military procurement and stocks, reductions in military R&D, etc. They would also call for mutual and cooperative programs to assist in retraining and relocating displaced military and civilian personnel and in reallocating other resources affected by disarmament.

An important aspect of this new round of arms control agreements would be limitations on arms exports to regions of potential conflict, particularly the Middle East. In the absence of enforceable multinational agreements there will be enormous temptations on the part of any country reducing defence expenditures to increase arms exports so as to keep production facilities going and workers employed, to avoid economic disruptions, and to earn foreign exchange.

Yet another important aspect of this new round of arms control agreements would be the tightening and formalisation of previous formal and informal agreements not to export weapons of mass destruction, including nuclear weapons, chemical and biological weapons, missiles, and the technologies required for such weapons, so as to limit their proliferation.

7. Conclusion : The Economic Return from Disarmament

The economic effects of disarmament depend on general economic conditions and national and international policies. With informed national macroeconomic, trade, and defence policies, supplemented by a new round of arms control agreements, particularly those covering arms exports and weapons of mass destruction, there could be substantial economic returns from disarmament, involving minimum conversion costs and maximum eventual benefits of disarmament.

Chapter 5

The Political Economy of Nuclear Weapons and the Economic Development after the End of the Cold War

William J. Weida

Introduction

Nuclear weapons have spawned both a mind set and a dedicated production sector that will make their passage from the arms inventory particularly difficult and prolonged. Nuclear arms functioned as a deterrent because they were terrible, indiscriminate weapons that would cause any adversary to carefully weight the cost of military adventures. Their destructive power fitted nicely with the inaccuracy of early delivery systems. However, as increasingly rapid, accurate and efficient methods were developed to deliver nuclear weapons. Response by the US and the former Soviet Union was becoming almost automatic. Both countries found this condition to be untenable.

The actual contributions to defence by nuclear weapons were never measured, except in rough terms of trade offs between nuclear and conventional forces. The real worth of nuclear weapons was assumed to exist in the deterrence they created, but the necessary components of nuclear deterrence could not be clearly defined and, obviously, could not be tested. Instead, domestic political and economic criteria often played decisive roles in the development and construction of both nuclear weapons and their delivery systems, and the worth of these weapons could often only be measured in terms of their ability to create technological development and regional income and jobs.

However, as the number of nuclear weapons increased, and the response time required to keep a nuclear force viable decreased, the likelihood of war also increased. As a result, the deterrent value of nuclear weapons decreased. Ideas about 'acceptable losses' of territory of portions of populations during a nuclear defence of a nation also meant the benefits of that defence were unevenly shared, depending on the territory, population group, and so on. Further, pollution produced as a byproduct of nuclear weapons was unevenly shared by regions in each nuclear nation. Thus, neither the benefits nor the costs of deterrence were evenly distributed across the population in any nuclear power.

As the number of nuclear warheads decreases, one would assume increasing amounts of the resources dedicated to nuclear weapons would by available for reallocation. However, reallocating nuclear warhead production and operations assets is a different problem then converting nuclear delivery systems production and operations assets. Economic development linked to nuclear weapon reductions will be different for each type of asset and, as an adjunct, for some regions compared to others and to the nation as a whole.

Strategic and Tactical Nuclear Weapon Assets Available for Reallocation

Two scenarios are probable as a result of international agreements to limit the number of nuclear weapons. Conversion of most nuclear warhead production facilities would be realized only if Scenario 2 becomes a reality.

Scenario 1: Stopping or Limiting Nuclear Warhead Production

Agreement to either curtail the production of new nuclear warheads or to limit the total number of nuclear warheads to a fixed amount without additional agreements concerning the disposition of all existing warheads are the least practical way of reducing nuclear arms. However, they bear some similarity to parts of the Strategic Arms Reduction Talks (START). As Henry Kissinger noted, in START, unlike INF, the production lines for strategic weapons would remain open.

In scenarios of this type, the need to supply tritium determines the future shape of the nuclear warhead industry. Tritium has a short half life, and its production must continue as long as warheads are retained. Under Scenario 1, few important changes would take place in the nuclear warhead industry. New contracts would be let, some old facilities would be shut down or combined, and new facilities would be built to increase production efficiency as the number of warheads decreased. Employment reductions at most nuclear production sites could be handled through attrition or transfer. Delivery systems would still be built and refined to accommodate the reduced supply of warheads, but a decrease in the production would occur. Some aerospace industries would not survive.

Scenario 2 : Stopping Nuclear Warhead Production and Disposing Of Existing Warheads

In this scenario, all nuclear warhead production stops and all warheads are disposed of. Production of nuclear materials would not be necessary and all defence nuclear production plants except for those involved in dismantlement could be shut down. Delivery systems for nuclear warheads may or may not be shifted to conventional roles.[2] In any case, a decrease in the production of strategic systems would occur and, in most countries, the market system would govern how resources were reallocated. A number of aerospace industries would probably not 'survive.

This scenario would deal a potentially devastating economic blow to those regions with nuclear production facilities. Further, stopping nuclear weapon production would threaten the governmental agencies tasked with building warheads and systems in each country. These agencies would actively search for new or redefined roles.

Reallocating System and Nuclear Warhead Resources

There is little evidence to support the notion that producers of strategic delivery systems could easily engage in other kinds of production. Military producers stress performance at any cost and are, at best, only tangentially associated with the market system. Success in the commercial market will require so many changes in attitudes about cost, reliability and marketing that few industries will be able to make the transition, a fact born out by past unsuccessful attempts of defence contractors to diversify into commercial markets.[3] Thus, the outcome of any conversion effort will be decided by market forces.

There is no experience with converting facilities for producing nuclear warheads to other uses, and there is good reason to suspect that such efforts would be extremely difficult. Resources used in manufacturing nuclear warheads are so specialized—and the facilities have been so highly contaminated—that even converting the land on which production plants reside is unlikely. To further complicate matters, while the primary purpose of defence-related projects should never be to provide employment, political reality dictates that nuclear production sites must be treated in a manner sensitive to regional economic issues. These plants were purposely located in isolated areas possessing few other economic alternatives, and this has created a significant degree of government obligation to help maintain the economic base.

Since nuclear production sites are too polluted to attract acceptable alternative employment, and since the facilities and the land on which they reside are too dangerous to be abandoned without long term monitoring and upkeep, nuclear waste management and storage probably present the only reasonable medium range economic alternatives at these sites. Thus, while the reallocation of resources from delivery systems to civilian uses will be primarily governed by market forces, the reallocation of resources from nuclear warhead production will be managed in some other way.

Specific Nuclear Resources Available for Reallocation to Non-defence Uses

Assets assigned to the construction and operation of nuclear weapons are accounted for with varying degrees of openness across the nuclear powers. The US and its NATO Allies, each of which is a different kind of nuclear power, can often be used as rough proxies for Russia, the Ukraine and China, where data are poor. Further, the assets present in 1988 appear to be a representative example of the total assets dedicated to nuclear forced before reductions began to take effect.

National Spending Patterns

The US built and deployed strategic nuclear weapons and this meant the United Kingdom and France either did not have to build them or that they could build and operate them in relatively small numbers. It further meant Great Britain and France could have smaller conventional forces. This reduced the amounts the allies needed to spend on both conventional and nuclear weapons and it allowed

European nuclear powers to devote funds to social instead of military programs.

Thus, defence expenditures in most European countries were, in some measure, replaced by a reliance on US nuclear forces. As David M. Abshire, the US Ambassador to NATO, noted,

> the new attention to conventional capabilities is directly related to fear of nuclear war......Some of it is rooted—rightly—in anxieties that our deterrent depends more on nuclear weapons than it should.[4]

For this reason, in the West, the bulk of nuclear assets available for reallocation belong to the United States, and France, Great Britain and the other non-nuclear NATO allies may face higher conventional defence expenses as nuclear reductions continue. These conventional expenses may consume any assets freed up by nuclear reductions.

Former Soviet Union : Spending and Opportunity Costs

Real opportunity costs of Soviet spending were not reflected in defence budgets of the former Soviet Union and this made Soviet spending appear smaller than Western spending for three reasons: first, opportunity costs of building Western systems were more fully expressed through the bidding action of the free market. Second, US opportunity costs were lower than opportunity costs in the Soviet economy, because the US GNP was larger. And third, Western opportunity costs were lower because Western countries operated closer to their production possibility frontiers than the Soviets did to theirs. Soviet opportunity costs increased even further when weapons were technologically advanced (as nuclear weapons were) and were in direct competition with projects needed to modernize the Soviet economy.

The economic strain of defence expenditures was so great in the former Soviet Union that nuclear disarmament could potentially provide a greater economic benefit for Russia than for other nuclear powers. How Russia will balance its perceived strategic weapon needs with the recognition that diversion of resources to defence may further damage its economy is not clear. A Defence Policy Panel report notes that

> economic concerns, namely the need to shift resources from defence to civilian use, appear to be driving....the [Russian]

military towards a smaller, more defensively oriented force structure.[5]

The Nuclear Weapon Infrastructure

A large infrastructure, including research and development laboratories, warhead production facilities, and facilities for the production of delivery systems is required to support the operation and production of nuclear weapons. Only delivery systems were produced at privately owned facilities in the United States, France, and Great Britain. In the former Soviet Union and China, everything was produced in government-owned facilities.

The Infrastructure in the United States

In the 1980's, United States nuclear weapon operations directly employed between 115,000 and 120,000 people, most of whom were military personnel at about 50 US bases and 160 bases in Europe and the Pacific Basin.[6] In Fiscal Year 1988, the US budget for strategic nuclear operations and maintenance was about $40 to $45 billion. Nuclear-specific US costs in NATO, based on manpower allocations, were about $12 billion.[7] At the same time, the US Department of Energy spent about $6.5 billion and employed about 65,000 people to design and build nuclear weapons : 20,000 at three national laboratories and 45,000 at as many as seventeen nuclear warhead production and nuclear materials processing facilities.[8]

The Institute for Policy Studies claims that expenses for nuclear forces accounted for 30 percent of the US R&D budget, 40 percent of intelligence and communications budget, 5 percent of the general purpose forces budget and 10 percent of the support forces budget—all areas not generally associated with nuclear spending.[9] This agrees closely with an estimate by the Center for Defence information that claims the cost was $29.8 billion in 1980, $51.8 billion in 1983, and almost $90 billion by 1987.[10] Total resources devoted to strategic weapons were about 22 percent of all funds spent on defence (including Department of Energy funds, FEMA, and so on or about 6 percent of US federal budget.

In addition, between one-third and one-half of recent US aircraft production, and almost all missile production was for nuclear-capable delivery system.[11] Employment in the aerospace industry included about 550,000 production workers, and a total work force of about 1,151,000, including management, research and development and support staff.[12]

The Infrastructure in the Former Soviet Union

In the former Soviet Union, nuclear weapon operations employed about 420,000 : 300,000 in the Strategic Rocket Forces, 20,000 in the Navy nuclear forces, and 100,000 in Strategic Aviation. The fraction of the Soviet military budget accounted for by the strategic nuclear forces was about comparable to the United States.[13]

Soviet nuclear weapons were produced by the Medium Machine Building Ministry in facilities whose size and level of technology were roughly equivalent to the US production complex including private sector contractors.[14] The Soviet military effort consumed between a quarter and a half of the Gross National Product of the former Soviet Union, and the burden was particularly heavy in the machine building branch of industry, where the defence industrial ministries absorbed about 60 percent of the output. Nine defence industrial ministries conducted most of the applied research and development for new military capabilities. About 50 design bureaus developed major military systems and were, in turn, supported by almost 250 sub-system and component design bureaus.[15]

The Infrastructure in Great Britain

In Great Britain, nuclear weapon operations employed about 6,000 persons.[16] The British nuclear warhead production complex consists of three primary research and production facilities employing 8,000 people. The British have one laboratory to design nuclear warheads and nuclear materials are produced mainly at Calder Hall, Chapelcross, and Windscale. Final warheads are assembled in facilities at Cardiff and Burghfield.[17] The British have developed and deployed a force of ballistic missile submarines. The missiles on these ships were produced in the US, but the warheads were produced in Britain. The British military aircraft industry also produced nuclear capable fighter-bombers. Defence spending in Great Britain was about $35 billion in 1988, accounting for about three percent of GNP. Production of nuclear warhead delivery systems accounted for about $9 to $12 billion of this amount.[18]

The Infrastructure in France

French nuclear weapon operations employed about 18,000 people.[19] French laboratories included the Saclay and Grenoble Centers for Nuclear Studies and certain laboratories under the CEA's military applications branch. Ten French locations were identified as nuclear warhead production sites, but little in publicly known about specific facilities for warhead fabrication. Nuclear materials are produced at Marcoule, Miramas and Pierrelatte.[20] The French produce land-based ballistic missiles, ballistic missile submarines, and nuclear-capable aircraft. Defence spending in France, was about $25 billion in 1988, accounting for about six percent of GNP. Production of nuclear warhead delivery system accounted for about $6 to $8 billion of these expenditures.[21]

The Infrastructure of China

In China, nuclear weapon operations are conducted by the Strategic Rocket Forces, by naval units assigned to operate and maintain a single ballistic missile submarine, and by air force units. The numbers of people assigned to strategic nuclear forces are not publicly known.[22] About forty Chinese locations have been identified as uranium mining and enrichment facilities. Chinese nuclear materials are produced at Lanzhou, Yumen, Baotou, Hong Yuan, Jiuquan, and Urumaqui. Nuclear weapon production and assembly plants have been identified at Lanzhou, Baotou, and Haiyen.[23] Not much is publicly known about the Chinese defence industrial base, however, China produces short and medium range ballistic missiles, ballistic missile submarines, and nuclear capable aircraft.

Competition for General Resources

In light of both the spending patterns of the nuclear powers and the resources available for reallocation, previous claims on these resources give one clue about how they would be reallocated if nuclear forces no longer existed.

General Competition between Nuclear and Social Programs

National resources allocations are usually made to please constituencies with conflicting goals. When adequate priorities do not exist, real national concerns may be altered as regional economic interests attempt to influence budgetary decisions. In the absence of clear requirements, national leaders often found an appeal to regional economic interests was necessary to sell nuclear weapons, and when programs are sold by appealing to regional economic interests, opportunity costs to the nation cannot be adequately evaluated.

The emphasis on programs to address social concerns has been a long-standing priority debate in most countries. It is difficult to say what will happen if the nuclear powers now elect to make different choices because, as Boulding put it.

We cannot simply assume that if something goes down when the military goes up that this was a cause-and-effect relationship...If we were to bring the war industry down to the proportionate level of the thirties, assuming that we maintained full employment...It is not altogether easy to say what would expand.[24]

There is a big difference in the strength of the constituencies for defence and social spending in most countries. It is much harder to form a constituency for transfer payments because of the diverse impact of these expenditures. As Robert DeGrasse point out :

> [These] benefits are paid to anyone who qualifies, no matter where they live. On the other hand, contracting for a tank or building a hydroelectric dam directly affects a specific area. The difference is politically important.[25]

Competition for Resources in R&D Programs

Military R&D is a large part of research in all major powers, and it is a critical element in maintaining nuclear forces. In the US, military R&D is 70 percent of all federally funded R&D and 30 percent of total R&D. In the former Soviet Union it was estimated to be more than 50 percent of all R&D.[25]

Military R&D creates both civilian and military technologies. In fact, the development of civilian technology 'spin-offs' is a major justification for the continued production of arms in each of the nuclear powers. One line of thought is that research is research, and whether or not the competition of research money is won by defence projects, the outcome will be about the same. However, military R&D has three unique characteristics that set it apart from other kinds of R&D.

1. It diverts scientists and engineers from civilian pursuits.
2. It distorts new technology by encouraging expensive applications with little marketability in the private sector.
3. It may lead to governmental control of scientific and technological information.[27]

Military research may not satisfy the requirements of commercial operations or other governmental sectors whose projects were not funded. Most military R&D is applied research for weapons, and little goes toward basic research. Only 12 percent of the budget for US military R&D is spent for technology base programs (excluding the SDI programs).[28] And while total military research and development in the US rose from $ 17.7 billion in 1981 to $30.7 billion in 1989 (in constant 1982 dollars), the amount spent on basic research actually fell from $2.79 billion to $2.59 billion over the same period.[29]

If the benefits of government funded R&D outweighed the costs, one would expect to find a growing technological superiority in the United States and the former Soviet Union. However, this has not been the case. American firms experienced their greatest losses to foreign industries in those areas like aircraft, electronics and machine tools where military R&D predominates. US productivity has also declined—something one would not expect if the technology developed through defence contracts really enhanced products and production methods.[30] In the former Soviet Union, the results have been far worse.

Competition for Resources with other Defence Applications

In the 1970's, all nuclear powers encountered sharply rising costs as the technological level of arms increased. These costs made it more difficult to maintain armed forces, and they affected various countries in different ways with respect to the use of nuclear weapons. For many, rising defence costs made nuclear weapons more desirable because of the perceived cheapness of these weapons.

In a 1984 interview, George Keyworth, the Science Advisor to the President, discussed the historical rationale for the use of tactical nuclear weapons and the opportunities that may arise for their replacement :

> Tactical nuclear weapons have on primary advantage that made them attractive in the postwar years. And that is that they are...cheap means of achieving a significant military capability...[Now we] can develop a super smart munition that can have an accuracy over great distances...This is a better military weapon because it can be used without threat of escalation.[31]

While the concept of "cheap" does not apply when all costs of nuclear weapons are considered, it does apply in government budgeting where each branch of a government works within a very narrow budget arena, and this led to a competition for resources between nuclear weapons and conventional forces.

Tactical nuclear weapons can replace a large number of conventional forces in a major was as long as escalation is not an important consideration. On the other hand, neither tactical nor strategic nuclear weapons can perform the military tasks usually required of conventional US forces (fighting in small, limited engagements in poorly defined conflicts where escalation must be avoided). In addition, strategic forces are designed to be used only as a last resort, and often represent pure additions to the military budget, replacing no older weapons, freeing up no other parts of the military budget, and rendering most other military weapons inconsequential if the strategic forces are used. Carter notes that

> with the advent of nuclear weapons, conventional military forces...generally [became] the stepchild of United States defence concerns...[although] about 70 percent of the US defence budget goes for the purchase, maintenance, and operation of conventional forces[32]

Conventional forces would be the sole heirs to the remaining military budget if nuclear disarmament was achieved. Since there is no indication that the uses of conventional arms are decreasing among the nuclear powers, these forces will present stiff competition with any other potential use of the resources devoted to nuclear weapons.

CONCLUSION

Possibilities for Reallocating Nuclear Weapon Resources

Almost all resources discussed in this paper share one common attribute—they represent weapons, installations, infrastructure, raw materials, and human endeavors that have virtually no other uses. Over the last forty years, much of what was spent for general defence in each of the nuclear powers did not contribute to the strength of the economy as a whole. This situation was even worse in the case of nuclear arms where little conversion or reallocation is possible. From this point forward, resources devoted to nuclear weapons are most accurately considered in a 'future negative' sense; countries who retain nuclear weapons the longest will suffer the largest 'future negative' resource loss.

Thus, nuclear assets available for reallocation are limited, and most reallocation will simply involve redirecting future spending and employment—often to conventional defences. Most resources purchased with the nuclear budget represent a non-recoverable, sunk cost. Further, the degree to which deficit financing was involved in nuclear expenditures will be directly related to how many resources can now be diverted to other uses and how many will be allocated to decrease the outstanding debt. With this in mind, countries such as Great Britain and France, who devoted fewer resources initially and who accumulated less nuclear-related debt than the United States (recorded in dollars) and the former Soviet Union (recorded in a multitude of non-monetary ways) will do best at reallocation.

In summary, viewing the reallocation of nuclear expenditures in the context of economic development implies that resources can actually be redirected. For nuclear weapons, a better approach is to view the resources as a sunk cost and to recognize the obligations incurred by building and operating the weapons. These are obligations both to clean up the mess created by weapon production and to dismantle existing weapons. Since the infrastructure developed to build and support nuclear weapons has little in the way of residual assets to add to these tasks, they will require an obligation of funds that could surpass the amount previously spent on nuclear arms. The total US nuclear budget was between $50 and 100 billion in 1988 depending on how nuclear spending is defined. Cleanup will cost at least $200 billion—not counting the cost of dismantlement—and may go as high as $400 billion. The Russian situation will be much more expensive.[33]

In addition, the US plans to maintain a force of roughly 3500 nuclear warheads for the foreseeable future. It will also spend large amounts to dismantle old warheads. Most larger nuclear facilities will remain open, and all of this is likely to consume at least three quarters of the present nuclear defence budget. If the US force was now at the 3500 warhead level, and if meaningful cleanup started this year, it would still take at least eight years and perhaps as many as forty years before any nuclear funds could be reallocated to economic development in the US. The situation will not be markedly different in France and Great Britain, and it will be far worse in the former Soviet Union. Thus, the final legacy of nuclear arms is likely to be a period of prolonged costs for cleanup and dismantlement that roughly matches in length the period over which nuclear weapons were deployed. And, implied in this legacy is the realisation that reductions in nuclear arms will have almost no positive impact on economic development.

Notes

1.   Kissinger, Henry, quoted in the *Washington Post*, April 24, 1988.

2.   For example, on September 5, 1988, a Titan II vehicle, originally designed as a strategic missile, was used to place an Air Force satellite in orbit. This missile was the first of 12 converted missiles ordered for this task.

3. See Weidenbaum, Murray L., "Industrial Adjustments to Military Shifts and Cutbacks", in *The Economic Consequences of Reduced Military Spending*, Bernard Udis, ed., Lexington Books, Lexington, MA, 1973, pp. 253-287. and President's Economic Adjustment Committee and the Office of Economic Adjustment, OASD (MI&L), *Economic Adjustment/Conversion* Department of Defence, Washington, DC., 1985, Chapter 7, and Lynch, John E., ed., *Economic Adjustment and conversion of Defence Industries*, Westview Press, 1987.

4. Abshire, David M., "NATO : Facing the Facts", *Washington Post*, December 31, 1984, p. 13.

5. Adams, Peter, "Study : Soviet Buildup Persists Despite Economic Concerns", *Defence News*, September 19, 1988, p. 42,43.

6. Carter, A., Steinbruner, J., and Zraket, C., *Managing Nuclear Operations*, The Brookings Institution, Washington, DC., 1987.

7. Gertcher, F., and Kroncke, G.T., *US Aerospace Industry Space Launch Vehicle Production*, a Preliminary Report for the National Defence University by R&D Associates, RDA-TR-301200, Colorado Springs, Colorado, December 1985 ; also see Carter, et all, op. cit.

8. Carter, et all, op. cit.

9. Klare, Michael, "Defence budget hides as much as it reveals about true nuclear spending", *Oakland Tribune*, April 29, 1984, p. 87.

10. For a complete discussion of the assumptions and methodology employed by the CDI, see *The Defence Monitor*, Vol. XII, No. 7, Center for Defence Information, Washington, DC., 1983, pp 1-16 and "Hard Military Choices", *New York Times*, May 28, 1985, p. 19.

11. Gertcher and Kroncke, op. cit.

12. Gertcher and Kroncke, op. cit. ; also see US Department of Commerce, *US Industrial Outlook*, US Government Printing Office, Washington, DC., January 1987.

13. Air Force Association, "The Military Balance 1985-86", *Air Force Magazine*, Volume 69, Number 2, Arlington, Virginia, February 1986 ; also see Carlucci, F., *Soviet Military Power*, US Government Printing Office, Washington, DC., 1988.

14. Arkin, W. and Fieldhouse, R., *Nuclear Battlefields ; Global Links in the Arms Race*, Ballinger Publishing Company, Cambridge, Massachusetts, 1985.

15. Arkin and Fieldhouse, op. cit.

16. Air Force Association, op. cit. : also see Arkin and Fieldhouse, op. cit.

17. Ibid.

18. US Arms Control and Disarmament Agency, *World Military Expenditures and Arms Transfers 1985*, ACDA Publication 123, Washington, DC., August 1985.

19. Air Force Association, op. cit. ; also see Arkin and Fieldhouse, op. cit., and Campbell, C., *Nuclear Facts ; A Guide to Nuclear Weapon systems and Strategy*, The Hamlyn Publishing Group, Limited, New York, 1984.

20. Air Force Association, op. cit. ; also see Arkin and Fieldhouse, op. cit.

21. US Arms Control and Disarmament Agency, op. cit.

22. Arkin and Fieldhouse, op. cit.

23. Arkin and Fieldhouse, op. cit.

24. Boulding, Kenneth E., "The Deadly Industry ; War and the International System," in *Peace and the War Industry*, Kenneth E. Boulding, ed., Aldine Publishing Company, 1970, pp. 6-7.

25. DeGrasse, Robert W., *Military Expansion, Economic Decline*, M.F. Sharp, Inc., Armonk, New York, p. 6.

26. Hollaway, David, *The Soviet Union and the Arms Race*, 2nd. ed., Yale University Press, New Haven, Conn. 1984. p. 134.

27. DeGrasse, op. cit., p. 12.

28. Reppy, Judith, "Conversion From Military R&D : Economic Aspects", 38th Pugwash Conference, Dagomys, USSR, 29 August-3 September, 1988, p. 2.

29. Thompson, Mark, "Research, development for defence takes beating", *Gazette Telegraph*, Colorado Springs, Colorado, April 3, 1988.

30. DeGrasse, op. cit., p. 13-14.

31. Keyworth, George A. II, in an interview with Roger Fontaine in the *Washington Times*, October 19, 1984.

32. Carter, Barry E., "Strengthening Our Conventional Forces", in *Rethinking Defence and Conventional Forces*, Center for National Policy, Washington, DC., 1983, p. 19.

33. Alvarez, Robert, Senate Government Affairs Committee, Speech to IEER Technical Training Workshop, June 5, 1993, Washington, DC.

Chapter 6

The Economic Perspective of Disarmament

Isamu Miyazaki

When World War II ended, people pinned high hopes in the arrival of genuine peace. In fact, just before the world came to an end, the allied forces had drawn up the Yalta concept on 'the postwar order.' In the economic sphere, debates also had taken place on the Bretton Woods concept to stabilise the international monetary system and the General Agreement on Tariffs and Trade (GATT) to protect the free trade system so as to make the best use of the lost cause that led to the two world wars.

With the end of the war in 1945, these concepts were realised to the creation of the United Nations, the International Monetary Fund (IMF) and the GATT. But the arrival of peace was temporary and the honeymoon period for the United States and the Soviet Union was short lived. The Iron Curtain was drawn and the Cold War began. In 1950, the hot Korean War erupted, mounting tensions and confrontations between the East and the West, which plunged the world into a futile and spectacular arms race. Such an arms race created a psychological equilibrium of terror or threats under Democles' sword. Politically, it created mutual suspicion and economically, the following costs has to be paid.

First, the massive military spending became the cause of deteriorating fiscal situation in most countries, thereby an inflationary factor. Needless to say, inflation erodes the real level of peoples' standards of living. From the policy perspective, it constantly strained excessively monetary policies. Moreover, it undermined, for example, the effectiveness of monetary policies. In this connection, the world military expenditure five years ago is considered to be approximately $1 trillion. The United States appropriated about 6 percent of its GNP, and the Soviet Union, between 15 and 25 percent of its GNP to military spending, together accounting for about 60 percent of total world military expenditure. The military expenditure has been decreasing in the past four years. In particular, in 1992, it is said that this figure declined about 15 percent from the previous year, but continues to be an extremely atrocious amount.

Second, as a result of slanted allocation of financial resources and labor force to military technology, the private sector technology progress was delayed, weakening the competitiveness of even the United States, the greatest economic power. This first and second factors led the U.S. twin deficits to expand tremendously.

Third, in arms trade, the ordinary pricing principle, or market principle, does not function. Thus, the rigidity was created in the overall free economy or the market economy. Socially, as President Eisenhower said in his special speech, it created the industrial-military complex. Power concentrated in one place, restraining in real terms the democratic freedom in various aspects.

Against such a background, the end of the Cold War and the dissolution of the East-West confrontation were something to be welcomed unconditionally. But that does not immediately lead to progress in disarmament, or peace dividend. In proceeding with disarmament or disarmament process, we must pay the following cost.

First is the various costs to realise disarmament, among which is the cost to examine and inspect whether disarmament is actually taking place. The cost to inspect the nuclear non-proliferation is inevitably included. Another is the cost to discard or decompose the existing weapons and ammunitions. Even dismantling a tank entails enormous amount of cost.

Dismantling or treating chemical or nuclear weapons, which entails not only danger but large costs, also requires enormous amount of money when safe-keeping of related facilities or preservation of the environment accompanies these efforts. For some time, it is projected that the cost of dismantling nuclear weapons will be larger than that of developing them.

Second, regrettably disarmament does not mean total abolishment of military forces. Even if proliferation or an increase of nuclear weapons can be prevented, the total nuclear weapons that exist on earth are more than enough to overkill mankind. And even the maintenance of a minimum military force entails a large amount of cost. In most countries, taking the change of the disarmament, there is a strengthening move to promote modernization, particularly of military equipment. Thus, the cost of modern weapons and related fields will reach a massive amount. In the case of Japan, while its defence expenditure itself is restrained in recent years, the cost to modernise its equipment along with personnel costs continue to increase. In the case of China, while it is the basic attitude of the authorities to welcome disarmament, and the number of soldiers is being reduced, the military expenditure itself in increasing, which is believed to be the outcome of modernizing equipment.

Third, in a number of small countries, there is a move to strengthen their military power, resulting from the intensifying regional conflicts and ethnic confrontations. And this trend is seen particularly in Central Europe, the Middle East, and the border between India and Pakistan. It is said that in 1992, there were about 32 arms conflicts in various parts of the world.

When transition from military demand to private demand faces difficulty, particularly when it coincides with sluggish economic activity, extreme difficulty is brought about. And thus the effects of disarmament on reducing the burden on the economy will be further delayed.

Therefore, the end of the Cold War does not immediately bring about benefits from disarmament or peace dividend. From the humanitarian or social standpoint, however, this of course is welcomed. And ultimately, it is likely to produce positive effects economically. Of course, in discussing the disarmament issue, the economic problem is merely one measurement of a value and thus there is no such rationale as disarmament should not be realized, because it costs money. But, I would like to line up the economic effects of disarmament.

First is the fiscal soundness resulting from decreased military expenditures. If the reduced portion cannot be transferred to other uses, they should take the form of tax cuts, or diminish government borrowing. In any case, it should take some burden off the private sector. But, actually, the saved resources are allocated to other uses, hardly leading to directly lighten the burden of the private sector. For example, in the United States, it is appropriated for reviewing of the social security system or improving the social infrastructure, which will most likely be ineffective in reducing the overall fiscal deficit. However, although in reallocating resources, permeation effects of different expenditure items are all different. The economic permeation effects of private capital spending or public investment could be sufficiently larger than these of military spending.

Second is the liberation of technology. Unlike the pre-World War II years, I believe that the permeation effects of technology on the economy is larger in private technology than military technology. Therefore, disarmament will likely accelerate the overall technological progress. The military industry in Japan itself is very small, but the technological progress in the civilian sector was very rapid. And, in some sense, this was adopted in various countries, particularly in the United States by the arms industry. For example, as seen during the Gulf War, it enabled the qualitative improvement of armed forces. Japanese technology developed mainly in the civilian sector.

Third is the increase in economic development assistance in place of military spending, and it is expected to promote the disarmament of the third world. Today, the total world military spending is approximately $1 trillion, and the Official Development Assistance (ODA) extended from the industrialized countries to developing countries is about the twentieth of that total, or about $50 billion. Therefore, if the cut in military spending can be used as assistance to the third world, it can have an extremely large effect.

Finally, the fourth point is that the decrease in military spending will promote economic rationalism and the revival of the price mechanism, through which the industrial military complex will be more quickly dissolved. In this sense, the economic effects of disarmament can sufficiently be expected, but there are some specific problems. I would like to cite two or three. The first is the problems related to the conversion of military demand into private demand.

The military industry that supported arms expansion is regionally wide, with very wide-ranging fields related to parts procurement, and the people affiliated with the industry, including the military staff, engineers, the white-collared workers, general workers and other personnel of related services are extremely wide-ranging. Therefore, transition from military to private demand affect a very wide spectrum of regions and industries as well as people, requiring elaborate planning. Particularly, as I mentioned earlier, a large number of countries are in economic difficulties today, and thus transition will be laborious. And since the people involved with the military sector are normally neither knowledgeable nor well trained with the market economy, transition to private demand is very difficult.

The next problem is related to recessions. In many countries, there are moves to change directions toward arms exports rather than decreasing their military industries. I can only say that it is extremely regrettable that the major industrialised countries, in particular the United States, France and the former Soviet Union are pursuing this option. Of course, some developing countries are sharply increasing arms exports, but in their case, one can sympathise that they have to obtain foreign currencies.

Another point is the relation between disarmament and development assistance. There are some developing countries which continue to expand military capability on the one hand but on the other receive development assistance from the industrialised countries. This is basically a very contradictory story. Therefore, there is a need to clarify the relation between economic assistance and disarmament, and to increase assistance to those countries which promote disarmament.

Finally, given such disarmament problems, how should Japan cope with this issue in the future ? Japan's Constitution stipulates that it will not solve any international conflict by use of force. Thus, up to now, Japan has spared around one percent of GNP to military expenditure. This basic policy should be maintained in the future. We have also banned the export of weapons, which should also be continued. Moreover, we have the three non-nuclear principles and this also must be adhered to.

In addition to such bases, disarmament is proceeding world-wide with the end of the Cold War, so I would like to mention three related points.

First is to make international cooperation on the conversion problem. What I mean by conversion is from military to private demand. For example, with the Japanese technological level or the competence of our engineers, Japan has the capability of inspecting whether disarmament is being conducted, and whether the weapons are being treated or discarded according be plans. And so I believe cooperation to supervise and survey disarmament should be strengthened.

Second is the fact that today Japan is disbursing the largest amount of economic cooperation expenditure among the DAC member countries. From the standpoint of curtailing international arms trade as I mentioned earlier, we need to effectively use this economic assistance. Japan is an extremely large importer of weapons and thus how convincing Japan can be remains to be seen.

Third is the so-called intellectual assistance. After World War II, Japan resolutely implemented private demand conversion. This experience of conversion, some cases were successful and some were failures, but we can teach those countries which are now in the conversion process through furnishing of information.

Finally, I would like to tell you a recent experience of mine. I visited China in May, during which I discussed with a number of people on the future of China. At present, China is in the midst of an extremely dynamic economic development. There is a projection that by the year 2002, China's GNP will double that of Japan, and triple that of Germany. And there was a person who expressed concern that if China becomes an economic power, it will eventually become a military power. But one Chinese leader responded that it is questionable for China to become such an economic power in less than 10 years. Of course there is the question of how you consider the exchange rate, but at the same time, the leader also said that there exists the problem of a widening discrepancy within China even though its economic growth rate is a high, and there is also inflationary concern and thus China cannot become an economic power. And even if China became an economic power, it has no room to become a military power.

I would like to believe his words, but some of the other people, particularly the Asians who were present when this discussion was taking place expressed concern that Japan is more likely to turn from an economic power into a military power. I have always believed this to be impossible, but the fact that there are neighboring countries or those countries which have close relations with Japan feel that there is such a threat, is a grave problem. And I felt strongly that Japan must take this point into full consideration when deliberating on what should be our future direction, and particularly how our economy ought to be.

Thank you very much.

MODERATOR : Thank you Mr. Miyazaki. Since Mr. Miyazaki is the only Japanese keynote speaker, I would like to invite Professor Tobin to comment on this speech.

TOBIN : Mr. Miyazaki certainly covered a lot of territory. He made many very interesting points and suggestions in his talk.

I think he was right to emphasize that administering disarmament on a global basis is not a free good, that some outlays will be required, and some efforts on an international basis will be required. One suggestion that appeals to me and some other people in the United States is to use our resources, technology, and money to help the Russians and other military powers in the former Soviet Union disarm and destroy their excess nuclear and conventional weapons. That would be a source of hard currency form them. It would contribute more to our defence and to the security of the world than any other things that might be done. All these costs could easily consume at least three quarters of the current Department of Energy nuclear production budget.

If the United States, by some form of magic, was able to drop to 3,500 warheads tomorrow, and if meaningful clean up started tomorrow, it would take between eight and 40 years before any nuclear funds would be available to be reallocated to economic development either in the United States or any developing country around the world. This situation will not be appreciably different for France and the United Kingdom, it will be far worse in the former Soviet Union. Thus, the final legacy of nuclear arms is likely to be a period of prolonged, costly clean up and dismantlement, and this period is likely to roughly match the length of time over which nuclear weapons have been developed and deployed. Implied in this legacy is the realization that reductions of nuclear arms will have almost no positive impact on economic development aside from the regional effects of federal spending that are focused on the clean up.

Chapter 7

Arms Reduction and Containment of Arms Exports : Some Comments

Yoshio Suzuki

Arms Exports attracting attention along with the nuclear issue

All three keynote papers are indicative of renewed interest in arms reduction since the end of the Cold War.

In the past, discussions on arms reduction centered on the reduction and ultimate abolition of strategic weapons, primarily nuclear weapons, and the impact of such moves on the world economy. Even today, when the Cold War is a thing of the past, the importance of nuclear weapons had not decreased. Rather, it has become an even more realistic and practical issue. In the general discussion, Professor Weida impassionedly argued for the control of nuclear weapons, prevention of nuclear proliferation, and the ultimate abolition of nuclear weapons, and talked about the huge economic waste stemming from the nuclear arms race.

Now that the Cold War is over, the question of arms reduction is not limited to nuclear arms and other strategic weapons. The question has expanded to include the containment of the arms trade and arms transfers, including conventional weapons, the prevention of ethnically or religiously motivated regional strife, and the likely impact of comprehensive, global arms reduction, including conventional weapons, on the structure and the development of the world economy as well as on the global environment. I was extremely impressed that all of the keynote papers touched on these issues.

The question of nuclear arms reduction, a question that is at once old and new, and the question of the containment of the arms trade and arms transfers, including conventional arms, and deeply interrelated in today's post-Cold War World. This is because in order to ease the deflationary impact of reductions in nuclear and other strategic arms on their own economies, advanced industrialized countries, the former Soviet Union, and China—all of which possess nuclear weapons—are exporting huge volumes of conventional weapons to developing countries in strife-ridden regions of the world. Tables 7.1 and 7.2 show figures quoted by Professor Klein. These figures show the magnitude of weapons exports from nuclear nations to the developing countries in war-torn areas.

Table 7.1 : Arms Exports to Developing Countries by Major Exporters

(1989, $ million)

	Value of arms exports	(To developing countries)
Former Soviet Union	19,600	(10,869)
United States	11,200	(3,454)
United Kingdom	3,000	(1,968)
France	2,700	(2,065)
China	2,000	(865)
World total	445,430	

Table 7.2 : Value of Arms Imports by Major Importers

(1989, $ million)

Saudi Arabia	4,200
Afghanistan	3,800
India	3,500
Greece	2,000
Iraq	1,900
United States	1,600
Japan	1,400
Vietnam	1,300
Iran	1,300
Cuba	1,200
Turkey	1,100
Syria	1,000

On this subject, Professor Tobin argued that now the problem of arms exports must arose international concern and action, and that the members of Economists Allied for Arms Reduction should make this their major task. His arguments gained the sympathy of many people who attended the discussion.

Arms reduction in the context of world integration and disintegration

Our concern and interest are thus expanding to include the problem of the trade in and transfers of conventional weapons in addition to the problem of nuclear weapons. This can be attributed to a dramatic change in the world situation since the end of the Cold War. The end of the confrontation between the United States and the Soviet Union, the two military superpowers, and the collapse of the Iron Curtain have resulted in the rapid unification of the world. At the same time, however, countries that were previously under the umbrellas provided by the two military superpowers emerged from under these umbrellas and have begun to assert their ethnic, cultural or religious identities. This is to say that in today's world, two conflicting elements, namely centripetal force and centrifugal force, are simultaneously at work.

These two trends are reflected in today's arms reduction question. The reason that the problem of the trade in and transfers of conventional arms has come into the limelight is that despite the on-going unification of the world in the area of trade, serious political and military diversification has emerged. Professor Intriligator pointed out ten characteristics of the recent development of the arms reduction issue, but all of them can be accounted for by these two major trends.

The following three perspectives are the ones that I consider most important.

First, in the Cold War era, the problem of nuclear arms reduction was a subject that was dealt with in the context of the confrontation between the United States and the Soviet Union, or the confrontation between East and West. Therefore, this question was a matter of negotiation between the two countries or between two forces and was directed to specific nuclear countries. However, today's arms reduction problem has global implications and covers all the countries in the world, regardless of whether it relates to nuclear nonproliferation or the trade in and transfers of conventional arms.

This leads us to my second point. In the past, negotiations towards arms reduction were a bilateral issue between the two superpowers, or between two power blocs. Today, however, the problem of arms reduction cannot be solved without a multilateral approach including debate at the United Nations. In this sense, the scope for a contribution by Japan, which has a war-renouncing constitution, has expanded substantially.

Third, progress toward arms reduction can no longer be made through the influence of specific countries. It must be based on the achievement of balance among many countries. Japan is the only advanced industrialized country in Asia. Despite its status as the world's second largest economic superpower, it is a non-nuclear nation. With its four faces, namely, as an advanced industrialised country providing economic assistance to less developed countries, as the world's largest creditor nation, as a non-nuclear nation, and as a member of Asia, Japan, if it wishes, can play a major role as a mediator and push forward arms reduction efforts.

Japan can make a greater contribution to arms reduction

I believe that these characteristics of Japan should enable it to make a major contribution to nuclear non-proliferation and the containment of the arms trade.

Let me cite an example. Professor Klein discussed whether the philosophy of nuclear nonproliferation might not be also applied to the containment of the arms trade. I argued against his view.

As is well known, the nuclear nonproliferation treaty is an unequal treaty that seeks to prevent the development of nuclear weapons by inspecting non-nuclear nations, despite the existence of military superpowers that possess nuclear weapons. The reason that this treaty has gained international support despite this shortcoming is that the countries that support the treaty, including Japan, hope to reduce the nuclear arsenals of nuclear countries in stages, with the ultimate aim of achieving the complete abolition of such arms.

However, to carry out international inspections of the conventional weapons arsenals of importing nations in order to keep these nations from accumulating arms, as Professor Klein suggests, is unfair to these nations compared with the countries that are already military superpowers and are exporting their surplus weapons. This is because, unlike nuclear weapons or other weapons of mass destruction, the possession of a certain amount of conventional arms is essential for the defence of a nation.

If our aim is to contain the arms trade and arms transfers, then, international inspections of developing countries that import arms are not the answer. Rather, the export volumes of military superpowers that sell arms should be placed under international control, and export quotas should be reduced annually with the ultimate aim of reducing exports to zero. Japan, which is not an arms exporter, has proposed the establishment of a system under which arms exports would be registered with the United Nations. With these records, Japan could take a more aggressive stance and act more positively to contain the arms trade and arms transfers.

If the containment of the arms trade and arms transfers brought global progress toward the broad-based reduction of not only nuclear and other strategic arms but also conventional weapons, the peace dividend for the economy would ultimately be very substantial. However, the path to the achievement of this end will still be long and rocky.

The countries that would clearly benefit from the peace dividend today are the United States and the former Soviet Union.

In the United States, however, the expansion of budget expenditure under the Reagan and Bush administrations ate up the yet-to-be realized peace dividend. As a result, the budget deficit has not decreased despite a cut in defence spending. It is still uncertain whether the Clinton Administration will be able to reduce the budge deficit sufficiently to enable the country to benefit from the peace dividend.

The countries of the former Soviet Union are now in the process of transition to market economies. The conversion of defence industries into civilian industries, which is necessary to enable these countries to enjoy the peace dividend, goes hand in hand with the difficult problem of privatizing state-run enterprises. Therefore, the reduction of defence expenditure alone will not lead directly to the peace dividend. These countries will have to make sustained efforts to convert their industrial structures and the privatization of state-run enterprises. In addition, there will be new burdens stemming from the destruction or peaceful use of the nuclear weapons that they now possess. This will also impose heavy burdens on the advanced industrialized countries of the West, which will need to extend assistance to the former Soviet Union.

Prior to the Tokyo Summit in July last year, US President Bill Clinton proposed the provision of a gratis privatisation fund amounting to 4 billion dollars to Russia. However, European participants in the Summit failed to respond, because these nations were in serious recession and said that helping their own unemployed was more important than helping Russian workers who had lost their jobs in the defence industries. Finally, the G-7 nations managed to agree on a package of 3 billion dollars, consisting of 500 million dollars in grants and 2.5 billion dollars in loans, at the meditation of Japan, which chaired the Summit.

It is not at all certain when the former Soviet Union will be able to reap the peace dividend. This shows the enormity of the economic waste resulting from the arms race under a socialist, planned economy.

The United States has similarly wasted enormous amounts of resources. This is a basic cause of the decline in the relative position of the US economy in the world. However, if the Clinton Administration manages to reduce the budget deficit, it should be able to benefit gradually from the peace dividend, in anticipation of which former administrations increased their expenditure. The contraction of the budget deficit should provide increased leeway for monetary easing and bring about a downward trend in US long-term interest rates. This should energise investment by civilian industries, which should offset the deflationary impact of cuts in defense expenditure, restore the international competitiveness of American industries, and shrink the nation's current account deficit. Only then can we say that Americans have truly benefited from arms reduction.

Chapter 8

Economists and ECAAR : Our Contribution to Disarmament and Peace

Robert Schwartz

A cursory view of the state of the world today, of Yugoslavia, the Middle East, Somalia, Angola, Cambodia or Haiti, for example, is not very pleasant. If we use this as a base then it would seem that economists can do little to advance the cause of peace, and that the world has not progressed beyond barbarism. However, a further look and we might observe :

1. Almost fifty years of confrontation between the two superpowers has ended.

2. There is some reduction in world military expenditures and the trend is toward further decline.

3. There is a strengthened United Nations with broader international support for peacemaking and peacekeeping than ever before in history. Consideration is also being given to establishing a more democratic and equitable distribution of power in the UN structures.

4. There is recognition of the importance of a greater role for other international organizations in structuring stronger economies, more world trade and less arms expenditures, organizations like the World Bank, The International

Monetary Fund, the World Court, regional financial institutions and trade alliances.

5. There is a growing recognition of the importance of the need for environmental protection and support for a green revolution.

6. There is more optimism in the US that the new leadership in Washington may be less military and more peace oriented.

7. There is greater regional and world multinational cooperation for economic development, environmental protection, human welfare, disarmament, political freedom, peacemaking and peacekeeping, all of which are essential for our survival.

Now having said this we may feel better about our contributions. Economists alone will not resolve world problems but we have an important role which we are increasingly assuming.

ECAAR was conceived during the International Physicians for the Prevention of Nuclear War conference in Moscow in 1987. I thought that if the physicians and health workers could be so organized than so could the economists.

ECAAR has become an [instrumental] organization for American economists concerned with peace and is [now] an international organization through which economists world wide can join to further these objectives. It was founded in December of 1988 at the American Economic Association meeting in New York city after a letter containing the following paragraph by six Nobel Laureates was mailed to members to the AEA.

As economists we have much to contribute to the application to rules of rational behavior for reducing the incidence of armed conflict. We can greatly illuminate the cost of the world-wide arms race, the factors that have enhanced its expansion and the benefits of restricting, indeed ending it. Some of us have been publicly concerned as individuals and we believe that an organization of economists will have a substantially greater impact.

In the four and one half years since its founding ECAAR has established the reality of 'In Union There Is Strength'. With six national affiliates and a number more in near formation, with members in 33 countries and 48 states in the US, ECAAR is a recognized institution. We have interested and active members in many countries.

Recognizing that each country has unique circumstances in which an affiliate must operate, it is ECAAR's policy to encourage each affiliate to set its own program and agenda, as long as it conforms to the organization's stated principles.

In some countries, particularly China, India and Russia, affiliate formation has been slow. I lectured at Fudan University and met with faculty and students at Universities in Beijing. Although China does send delegations to international conferences on peace and disarmament, the government appears to have a policy against its citizens joining any Non Government Organization. A number of Russian economists has participated in our conferences but their country's general disarray has preoccupied economists with other problems, including their personal livelihood. However, we do expect a Russian affiliate to form in the near future. There are active members in India, but no one has come forth who is willing to organize an affiliate. Amartya Sen, an Indian and outstanding world economist, Professor at Harvard University and current President of the AEA, was on one of our ECAAR panels at the annual AEA meeting last year and will be so again this coming January.

The Japanese affiliate is particularly notable for its strong early development in a country with a special economic and political history. Japan is the only country to have experienced the devastation of nuclear bombs. On a personal note, I find it difficult not to tell you that at the time I was on leave from the US Treasury and as a Marine Officer, General Staff, Air Command, Okinawa, I visited Nagasaki soon after the bomb was dropped and observed its destruction. Since then, I have participated every year on Hiroshima Day against nuclear weapons, and this has become a part of my work for disarmament.

Japan has had a history of militarism, but for the last half of the twentieth century it has become a country of peace. Japan's [constitution] limits its military size and military expenditures. Japan probably has a larger percentage of its population engaged in the peace movement than any other country in the world. I proudly welcomed your large delegation (mostly Buddhist) to the June 12, 1982 mass anti war demonstration in New York City.

The United States, by contrast, had a history of limited foreign military involvement until the last half of this century, when it became one of the two military superpowers. It is now the remaining one.

As we approach the end of the twentieth century, the weapons of mass destruction and the build up of military forces and of military philosophy present the most severe problems for planet earth and its people and provide economists with a leading role in resolving them. As our world grows the complications and threats expand exponentially.

In the US, the concentration of resources in, and government support of military production, technology and research, mean consequent negative economic impacts on competitive positions in the civilian economic area in both goods and services. The inroads into the infrastructure and financial strength of the nation have been substantial.

Yet, despite these aspects of a military society, there still remains more democracy and freedom of expression in the US than in most countries. Therefore, there also exists a greater opportunity for economists to speak out against the government's military posture, although at times it is difficult to be heard. On the other hand, the economists in the former communist countries suffered under the totalitarianism of their governments and had little choice but to publicly support their government's policies.

In the US the role for economists striving for disarmament and peace has been larger than in most other countries. Yet, too few of the 16427 American economists registered as members of the American Economic Association are working toward these goals. Those Americans who are engaged in Peace Economics have found that some of the 4260 foreign members of the AEA, including many Japanese, are concerned with these issues and work together in addressing the destructive nature of the military sector.

Each ECAAR member can make his or her contribution individually to disarmament, conversion, and peace but together we have a greater impact. ECAAR members stimulate and support each other. The organization has already sponsored two books by board members, one from our 1990 conference at the University of Notre Dame, *Economic Issues of Disarmament*, edited by Jurgen Brauer and Manas Chatterji, and the other came from our conference at The Hague, *Economics of International Security*, edited by Manas Chatterji, Annemarie Rima and Hank Jager. We have also sponsored a new book edited by Walter Isard and C.H. Anderton, *Economics of Arms Reduction and the Peace Process*. This year for the second time we have had an issue of *Ekonomisuto* in Japan devoted to articles by five of our board member. Under consideration is a quarterly journal on important economic public issues. ECAAR's quarterly newsletter continues to grow in substance and readership. The first mailing of 200 copies has risen to over 1600 today. Besides reaching our membership, copies are sent to foundations, media and selected organizations.

ECAAR, as a Non Governmental Organization (NGO) at the United Nations, has participated in every major NGO Disarmament Conference since its inception. Some of us have formerly addressed conferences at the UN and one of our Board members, Dietrich Fischer, was an official delegate to the 1992 *UN Conference on Environment and Development* in Rio de Janeiro. Our board member, Dorrie Weiss, has gained respect as an NGO leader and is now Chair of the Publications Committee of Disarmament Times. In April she was a delegate at the *Third United Nations Conference on Disarmament Issues* : *"Disarmament and National Security in the Interdependent World"* in Kyoto. We are applying for consultative status with the UN's Economic and Social Council (ECOSOC). ECAAR had a small delegation, including its NGO representative and Executive Director, Alice Slater, visit China in October of last year as guest of the Chinese Association for Peace and Disarmament (CEAPAD). Later, a delegation of CEAPAD members visited the United States as ECAAR's guest to attend the UN April 20-23 international conference, *New Realities : Disarmament, Peace Building and Global Security*, and to participate on the panels.

Another of our special projects is a global register of resource people working on economic aspects of national military sectors which is being funded by the Ford Foundation.

ECAAR began with seven Nobel Laureates in Economics as trustees and is co-chaired by the distinguished Professors Kenneth Arrow and Lawrence Klein, both of whom devote time to formulating policy and working of ECAAR projects. The Dutch-Flemish Association of Economists for Peace held a conference on the *'Economics of International Security. New Realities : Disarmament, Peace Building and Global Security'* at the Peace Palace May 21-23 of last year and the participants included economists from over 25 countries. We are grateful to Professors Jan Tinbergen, the first Nobel Laureate in economics and to that chapter's Chairperson, Professor Annemarie Rima.

Our annual panels at the American Economic Association are now well attended and have sometimes been televised. Some of our senior members from abroad have been included in the major panels presented jointly with the American Economic Association. ECAAR, represented by Professor Klein as the principle speaker, participated in the Canadian Economic Association meeting this month and a Canadian affiliate is now being organized.

Professor William Weida is on leave as Chair of the Economics Department of Colorado University to head a program developed by ECAAR. He is Project Director of the *Community Education Campaign : Local Employment Alternatives for the US Nuclear Weapons Complex.* This is a conversion project with trained economists in 15 states where there are nuclear weapons production facilities. Economists are working with citizens groups, local decision makers, workers and business people and he will tell you about it.

In March 1994, our contest *'Arms Reduction and Global Reconstruction ; A Blueprint for the Year 200'* concluded with over 7000 requests for information on applications and 850 essays were submitted from 49 countries, including 21 from Japan. The contest's 15 judges included nine Nobel Laureates, of whom two, President Arias of Costa Rica and Archbishop Desmond Tutu of South Africa, are Nobel Peace Laureates, plus other distinguished judges representing regional parts of the world.

On May 20 at a Washington press conference Archbishop Tutu, following a visit with President Clinton at the White House, awarded the prizes. The judges determined a tie among the first four essayists and each of these young, bright, serious and enthusiastic persons was awarded $8000. These awards went to David Burress, an economist at the University of Kansas ; Lieutenant Commander Timothy J. Doorey, a US Navy intelligence officer stationed in Virginia ; Boaz Moselle, an English graduate student in economics at Harvard ; and Erik Thompson, a former Peace Corps member and now banker in Milan, Minnesota. While there was no stroke of lightning to resolve world problems, which, as long time scholars and activities in disarmament and peace, we did not expect, there was a serious consideration of global solutions by the participants. Winners of the $1000 prizes were Julie Ajdour, a health care worker in Seattle, Washington ; Frank M. Gren, a U.S. Navy officer and aviator in Tampa, Florida ; Roger E. Ison, a computer scientist in Loveland, Colorado ; Georgi Kirow, a jurist-pensioner in Sofia, Bulgaria ; Karen U. Kwiatkowski, an U.S. Air Force officer stationed in Italy ; Dr. K. Srinath Reddy, a cardiologist in New Delhi, India ; Kevin Sanders, special projects director of the War and Peace Foundation in New York, New York ; and James Wurst, a journalist also from New York. It is interesting to note that three of the winners are in the military. We were very pleased to have Japan represented at the press conference by Professor Akira Hattori whose presence was very welcomed.

As individuals, many of the ECAAR members in the US worked to help elect President Clinton. A number of senior ranking members of the Clinton administration, including one cabinet member, Robert Reich, and the Chairperson of the Council of Economic Advisors, Professor Laura D'Andrea Tyson, are ECAAR members. We maintain our tax exempt, non political status, but may, as individuals, make contributions toward influencing substantial arms reductions and strengthening UN activities, as well as advancing aid to the many distressed areas of the world. ECAAR has completed a study on the cost of the resumption of nuclear testing, which was sent to President Clinton in our effort to stop the US from resuming nuclear testing. We continue to expend great effort to oppose resumption of testing.

While it has become clear that worthwhile projects can be financed, we have some trouble in funding essential administrative costs. In spite of a limited staff (two full time people and a number of regular volunteers) our dues are not sufficient for office management. We are most fortunate that the grant of $50000 a year from the John D. and Katherine T. MacArthur Foundation was renewed for three years, and that the principal essay funding came from the Saison Foundation.

In today's period of world changes, the opportunities for economists to help influence and develop policy to aid in the transition toward a better world are greater than they have ever been. ECAAR is pleased to have a number of the world's leading experts on conversion from military to civilian production among its active members.

ECAAR's purpose is to foster arms reduction and to show how the savings can be channeled to human needs : environmental care, health, housing, education, and infrastructure. To gain these objectives, we jointly sponsor economic studies and projects with universities, foundations, and other non-profit and international organizations. Encouraging economic thinking, discussion, research and writing on matters of disarmament and peace is fundamental to our purpose.

Part II

Arms Production, Employment and Economic Conversion

Chapter 9

Disarmament and Unemployment

*Robert Eisner**

Fearful of damaging the economies of scores of communities nationwide, the Defense Department is preparing to recommend delaying some decisions on base closing until after the 1996 elections, a senior Pentagon official said today...Critics say there is clearly a political motive for President Clinton and some members of Congress to delay or diminish the economic pain of base closings...or at least to limit closing in states like California, Texas and Florida, pivotal battlegrounds in 1996. Page 1, *The New York Times*, May 5, 1994.

The State of California in the United States, the largest state in our Union, with a population (of 30 million) well over half that of France, receives a huge share of the nation's defence dollars. With cutbacks in our military programs, it moved into what its Governor called the worst recession since the Great Depression. To show his concern, and to try to win that state for his party, and, in 1996, for himself, Bill Clinton made 11 trips there in the first 16 months of his presidency.

Some months ago, we had a disastrous earthquake in the populous Southern California area around Los Angeles. Lives were lost and there were many billions of dollars of property damage. Many more billions of dollars poured in to repair roads and viaducts, to sustain those who had lost their homes and jobs, and to get the area back on its feet. And

Southern California experienced, at least briefly, a remarkable economic recovery.

The Great Depression of the 1930s brought unemployment to 25 percent and more in the United States and to comparable levels elsewhere. With World War II, demand from warring nations in Europe and American re-armament expenditures brought huge increases in output and corresponding increases in the effective demand for labor that finally ended the mass unemployment of the Great Depression. With United States entry into the war, government expenditures rose to previously unheard of heights. In 1994, government purchases of goods and services, amounting to $968 billion 1987 dollars, were 65 percent more than the total 1933 Gross Domestic Product of $587 billion. Government was buying 58 percent of total GDP in 1944. The federal budget deficit was enormous, so that the debt-GDP ratio by 1946 reached 114 percent. Unemployment in 1944 was 1.2 percent!

For a numbers of years after World War II, the US economy enjoyed a large reservoir of purchasing power from the large federal debt—financial assets of the private sector. As this was drawn down it was replaced by the military expenditures associated with cold and hot wars. Employment was thus relatively high. Helped along by the Kennedy-Johnson tax cuts which were legislated in 1964, unemployment was down from the 6.4 percent of 1961 to 4.5 percent in 1965 and, with Vietnam war expenditures, to 3.4 percent in 1969.

Over the last several years I have pointed out on several occasions[1] that, despite the end of the Cold War and widely perceived needs for public investment, military expenditures in the United States have been slow to come down. The "Budget Enforcement Act of 1990" ostensibly designed with great pain to reduce Federal Budget deficits, still set 'defense' outlays at $298 billion in fiscal year 1991, $296 billion in 1992 and $293 billion in 1993. To this was to be added some $19 billion for 'international' outlays and untold billions for a variety of secret operations.

The defense outlays themselves were thus to come to 5.25 percent of projected GDP in FY 1991. They were less, as a percent of GDP, than the recent peak ratio of Federal Expenditures for national defense of 6.56 percent in 1986, but substantially more than the 4.84 and 4.86 percent of 1978 and 1979 when the U.S. was only shortly post-Vietnam and the Soviets were plunging into Afghanistan. In calendar 1993, national defense purchases of $303.4 billion were 4.76 percent of GNP (or GDP). In absolute, real terms, the 1993 figure of $242.4 billion (1987 dollars) was still 31 percent greater than the $185.1 billion spent in 1979.

It has now been argued widely that the huge amounts of military expenditures are without defense purpose in the current international situation. President Clinton was elected though on a platform of only modest reduction beyond the modest reductions in defense expenditures that had been contemplated by the Bush Administration. The $292 billion outlays for defense in 1993 were estimated in the fiscal document presented on January 25, 1994 to fall only to $281 billion in 1994 and $271 billion in 1995. Proposals by the Pentagon to eliminate unnecessary weapons systems are lacking. Recommendations to eliminate apparently unnecessary military bases in the United States provoke major lobbying and huge protests from the people's representatives in the Congress.

Why do defence expenditures remain so high compared to those in the years when they were justified as necessary to prevent nuclear annihilation? Why does it it seems so difficult to bring about a drastic reduction in what many see as an unconscionable burden? The reason, I will argue, is similar to a large part of the real reason the expenditures have been so large to begin with. It relates to something those interested in peace and disarmament have often been too reluctant to admit. Simply enough, military spending means not only profits but jobs. A lot of people benefit from military expenditures. And without appropriate macroeconomic policies, unemployment may rise and the economy as a whole may be worse off when they are reduced.

Occasionally a natural disaster like that earthquake in Southern California comes to the rescue. Huge public resources thrown into the effort of reconstruction give people jobs which replace some of those lost in military cutbacks. Perhaps Brest should hope for a typhoon destroy its harbor and much of the city's homes and infrastructure. Jobs would then be created to put the area back together again. But lacking that natural disaster, or the expectation of one, people who lose their jobs in the shipyards suffer, and many more, fearing loss of their jobs, do every thing they can to preserve the *force de frappe* and other components of the military effort.

There are the companies that earn profits on government contracts, the lobbyists and agents who profit from getting them the contracts, the stockholders who receive dividends and capital gains.

But that is only part of it. In the United States, the 3 million Defence Department employees, military and civilian, and over 3 million defense industry employees come to some 4.5 percent of a total labor force of about 130 million. They have an understandable interest in maintqining their jobs—an interest all the more acute if attractive alternatives are not obvious.

And then thene are all those in towns with military bases or defense plants whose livelihood depend on the incomes and expenditures of those directly involved in the defence establishment. As any student of elementary economics can point out further, the multiplier chain can go far beyond that. With cuts in defence expenditures, that multiplier works—in reverse. Unless other forces—from the free market or government policy—intervene, the aggregate loss in output and employment can well exceed the cuts in defence. The peace dividend would be more than offset by these losses. The economic and political issues then become intertwined. Since cuts in defence expenditures *can* hurt many people and indeed the economy as a whole, the political force resisting those cuts becomes formidable.

It is widely asserted by many of good heart that the economy will be better off if a significant portion of the resources now feeding the military could instead be devoted to other production. We are reminded how much the money for one Stealth bomber could buy in the way of child and health care, education, roads or even trips to the seashore. The operative word, however, should be 'could'. Freeing resources from one application does not necessarily mean, in our economy, that they will be devoted to something else. They may simply sit idle.

There are many economists who will argue that the free market will quickly adjust. If jobless former defense workers will only agree to lower wages elsewhere, or rather, accept somewhat lower wages than those hitherto prevailing elsewhere, they will get jobs and employment will increase generally, so that their jobs will not come at the expense of the jobs of others. The lower wages, as production costs are thus decreased, will presumably generate lower prices. With lower prices there will be a lesser demand for any given quantity of money and interest rates will fall. With lower interest rates, investment demand will rise. We will have more production and more employment in the production of new machinery and new factories and additional housing. The lower prices will also increase the real value of the public's assets in the form of government debt, interest-bearing and non-interest-bearing, thus increasing consumption.

This comes down to: (1) faith in some desirable 'natural rate of employment' to which the economy will always tend; and (2) the belief that the economy, left to itself, will quickly get back to this desirable natural rate if something, such as defense expenditure cuts, pushes it off it. I confess to doubt on both counts. The aggregate demand necessary to bring about full or maximum voluntary employment has all too frequently been lacking, and unemployment in the United States has been only occasionally, and then usually in war-time, at its 'full-employment' minimum.

Average annual unemployment as a percentage of the total labour force in the United States, since that 1.2 percent in 1944, has ranged over the past four decades from a low of 2.8 percent in 1953 and 3.4 to 3.5 percent during the height of the Vietnam War, to 9.5 percent in the deep recession years of 1982 and 1983.[2] Indeed, if we look at, monthly figures the range is greater, from less than 3 percent (not seasonally adjusted) in May 1968 and May 1969 to 10.6 percent in December 1982. After averaging 4.9 percent over the 26 years from 1951 to 1976, unemployment rose to a mean of 6.9 percent for the 13 years from 1977 to 1989. It reached 7.7 percent in 1992 and, by the spurns of 1994, had now inched down to 6.4 percent.[3]

Faced with what many now strangely see as the 'natural' or non-accelerating-inflation rate of unemployment (NAIRU), our monetary authority sees fit to raise interest rates in an effort to slow down the economy. And many in our Congress look for further deficit cuts—but not in defense—which would add additional fiscal austerity to this mix.

I see nothing 'natural' about any of these unemployment figures, and nothing acceptable above the minimum reasonably attainable—if not the 2.8 percents of war periods, at least the 4 percent traditionally, in more ambitious times, viewed as our full-employment target. I have remarked that if there is really any 'natural rate' of unemployment,[4] God certainly treats Her children differently in different times and places: 5.7 percent in the United States (by the latest tally for July 1995 !), 2 or 3 percent in France some years ago and 12 percent now, and still 3 percent in Japan!

For whatever reasons, insufficient wage and price flexibility or insufficient responses of investment to lower interest rates or of consumption to greater real wealth in the form of government debt, it would seem clear that the economy has not readily stabilized itself at full employment and has only occasionally, and then usually in wartime, reached that level. Given this kind of variation, generally

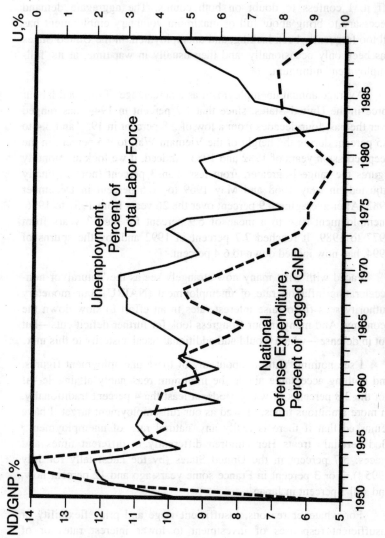

Fig. 9.1 : Unemployment and National Defence Expenditure, 1951-89

downward from what we may hope is our full-employment goal, we may expect that cuts in defense expenditures and the accompanying reductions in defense and defense-related employment will result in increases in aggregate unemployment.

I shall offer some evidence in support of this expectation in the form of a new analysis of data revised and updated from those in a previous paper presented last year in Grenoble.[5] I begin by relating the unemployment rate to the ratio of national defence expenditure to GDP. This relation is graphed in Figure 9.1 and offers a fairly clear picture. The higher the defense expenditure ratio the higher was the employment rate (the lower was the unemployment rate, graphed on an inverted scale).

Table 9.1 : *National defence expenditures and unemployment rates, 1951-93*

$$U = b_0 + b_{01}D1 + b_2 U(-1) + b_3 ND/GNP + b_4 (ND/GNP) (-1)$$

Variable or Parameter	Regression Coefficients and Standard Errors					
	Levels	Diff.	Levels	Levels	Diff.	Levels
C	8.976	0.031	3.409	8.882	—0.022	4.542
	(1.016)	(0.158)	(1.297	(0.832)	(0.159)	(1.186)
D1	—	—	—	2.411	1.129	2.366
				0.696)	(0.739)	(0.673)
U (—1)	—	—	0.634	—	—	0.473
			(0.122)			(0.117)
ND/GDP	—0.407	—0.333	—0.391	—0.412	—0.345	—0.428
	(0.118)	(0.133)	(0.129)	(0.098)	(0.132)	(0.114)
(ND/GDP)(—1)	—	—	0.227	—	—	0.221
			(0.134)			(0.118)
AR (1)	0.636	—	—	0.558	—	—
	(0.121)			(0.133)		
R^2	0.660	0.111	0.652	0.731	0.139	0.730
D-W	1.672	1.957	1.684	1.725	1.929	1.661
n	43	43	43	43	43	43

Notation

C	Constant term
D1	Dummy variable : = 1 for year 1982-1983; = 0 for all other years
AR(1)	First-order autoregressive coefficient
R^2	Adjusted coefficient of determination
D-W	Durbin-Watson coefficient
n	Number of observations
U	Unemployment as percent of total labour force

U(—1)   Unemployment as percent of total labour force, lagged
ND/GDP   National defense expenditures as a percent of previous GDP
(ND/GDP)(—1) National defense expenditures as a percent of previous GDP, lagged

The somewhat more rigorous tests of regression analysis confirm this picture. As shown in Table 9.1, the defence expenditure ratio was related negatively to unemployment over the 1951 to 1993 period. The statistically highly significant coefficient of —0.407, in the "levels" regression corrected for serial correlation shown in column 1, indicates that each percentage point reduction in defence expenditures as a ratio of GDP was associated on the average with four-tenths of one percent increase in the unemployment rate. That the relation is not spurious is supported by the significant coefficient of —0.333 for the defense variable when the estimation is done with first differences, as shown in column 2, and when it is estimated with lagged variables but freed from the first difference specification[6], as reported in column 3. Columns 4, 5 and 6 show the results of regressions including a dummy variable for the exceptionally high unemployment years of 1982 and 1983. They further confirm the basic relation; the defense variable coefficients are, if only trivially so, even more statistically significant[7].

Table 9.2 : *National defense expenditures, adjusted budget deficits, changes in monetary base and unemployment rates, 1962-89*

$$U = b_0 + b_{01}D1 + b_{02}D_2 + b_1 ND/GDP + b_2 PAHED(-1)$$

Variable or Parameter	Regression Coefficients and Standard Errors	
C	10.180 (1.303)	9.680 (1.583)
D1	2.732 (0.713)	2.006 (0.695)
D2	1.624 (0.746)	1.116 (0.825)
ND/GDP	—0.626 (0.187)	—0.515 (0.228)
PAHED(—1)	—	—0.321 (0.162)
AR(1)	0.407 (0.167)	0.594 (0.158)
R2	0.672	0.701
D-W	1.533	1.475
n	36	36

Notation

C	Constant term
D1	Dummy variable: = 1 for years 1982, 1983; = 0 for all other years
D2	Dummy variable: = 1 for years 1958-63; = 0 for all other years
AR(1)	First-order autoregressive coefficient
R^2	Adjusted coefficient of determination
D-W	Durbin-Watson coefficient
n	Number of observations
ND/GDP	National defense expenditures as a percent of previous GNP
PAHED(−1)	Lagged price-adjusted, high-employment budget deficit as percent of GDP

The underlying relation is certainly more complicated. The effect of defense expenditures on employment should depend, at least, on associated fiscal and monetary policy. That this may be so is suggested initially in Table 9.2 where, for the years from the 1958 to 1993 for which observations were available, we introduce into the lagged regressions the price-adjusted high-employment deficit [PAHED(−1)], expressed as ratios of GDP,[8] and a shift dummy for the years 1958-63. We may note then that the relation between unemployment and the defence expenditure ratio is fairly robust.

Without the deficit variable, the national defense coefficient is −0.626, greater in absolute value than the −0.407 coefficient observed in the corresponding regression for the 1951-1993 observations of Table 9.1, without the pre-1964 shift dummy;[9] with the shift dummy, the corresponding national defense variable coefficient for the 1951-93 observations was −0.412. When the deficit variable is included, the defense coefficient remains a substantial −0.515. Its standard error is somewhat higher, however, as there is apparently some multicollinearity with the additional variable picking up part of the explanation of unemployment.

Defence expenditures and unemployment are both slow-moving variables, with substantial autocorrelation. Their interrelations may be better revealed in moving averages. This is confirmed by the results shown in Table 9.3.

In an ARMA(1) formulation, we now note a coefficient of −0.462 for the defense expenditure ratio in a regression of the unemployment rate over the 43 year period from 1951 to 1993. Each percentage point reduction of the ratio of defense outlays to GDP was associated with almost half a percentage point more of unemployment. When the regression was restricted to the years 1958-93 for which observations on the adjusted, 5.5 percent unemployment deficit could be calculated, the coefficient of ND/GDP was −0.510.

Table 9.3 : *National defense expenditures, adjusted budget deficits and unemployment
rates, 1951-93 and 1958-93, moving averages*

$$U = b_0 + b_1 ND/GDP + b_2 PAHED\,(-1)$$

Variable or Parameter	Regression Coefficients and Standard Errors		
	1951-93	1958-93	1958-93
C	9.376	9.727	10.040
	(0.766)	(1.098)	(1.164)
ND/GDP	—0.462	—0.510	—0.535
	(0.091)	(0.149)	(0.158)
PAHED (—1)	—	—	—0.389
			(0.165)
MA(1)	0.439	0.986	0.624
	(0.163)	(0.016)	(0.141)
AR(1)	0.393	0.312	0.455
	(0.132)	(0.127)	(0.129)
R^2	0.678	0.575	0.694
D-W	1.986	2.605	2.082
n	43	36	36

C	Constant term
MA(1)	First-order arithmetic moving average coefficient
AR(1)	First-order autoregressive coefficient
R2	Adjusted coefficient of determination
D-W	Durbin-Watson coefficient
n	Number of observations
U	Unemployment as percent of total labor force
ND/GDP	National defense expenditures as a percent of previous GDP
PAHED(—1)	Lagged price-adjusted, high-employment budget deficit as percent of GDP

Impact of the deficit variable in these moving average regressions
is striking. Its coefficient is —0.389, indicating that each percentage
point more of adjusted deficit as a percent of GDP was associated with
almost four-tenths of a percentage point less of subsequent
unemployment. The coefficient of the defense outlay variable was
virtually unchanged, however, by the introduction of the deficit
variable.

These results suggest that lesser defense expenditures have indeed
been associated with greater unemployment. The results further suggest
that if a 'peace dividend' is used to reduce the structural budget deficit,
the loss in employment will be even greater. In the face of losses of

defense jobs a more appropriate policy, it would appear, would be to *increase* the deficit.[10]

Table 9.4 : *National defense expenditures, adjusted budget deficits, changes in monetary base and unemployment rates, 1962-93, moving averages*

$$U = b_0 + b_1 ND/GNP + b_2 PAHED(-1) + b_3 DMB(-1)$$

Variable or Parameter	Regression Coefficients and Standard Errors	
C	10.717	10.421
	(1.483)	(1.538)
ND/GNP	—0.632	—0.583
	(0.209)	(0.219)
PAHED(—1)	—	—0.316
		(0.201)
DMB(—1)	—2.827	—1.679
	(1.427)	(1.510)
MA(1)	0.376	0.466
	(0.205)	(0.188)
AR(1)	0.477	0.545
	(0.163)	(0.153)
R^2	0.673	0.710
D-W	1.931	1.935
n	32	32

C	Constant term
MA1	First-order arithmetic moving average coefficient
AR(1)	First-order autoregressive coefficient
R^2	Adjusted coefficient of determination
D-W	Durbin-Watson coefficient
n	Number of observations
U	Unemployment as percent of total labor force
ND/GDP	National defense expenditures as a percent previous GDP
PAHED(—1)	Lagged price-adjusted, high-employment budget deficit as percent of GDP
DMB(—1)	Lagged real change in monetary base as percent of previous GDP

Results of introduction of a change-in-monetary base variable to the moving-average relation are shown in Table 9.4. For the years 1962-93 for which observations are now available, the national defense coefficient is —0.632, and the monetary variable coefficient is a significantly negative —2.827. When unemployment is regressed on all three variables—defense outlays, the adjusted deficit and the change in the monetary base—the coefficient of ND/GDP is only slightly reduced. The deficit and monetary base variables retain their negative signs but their levels of statistical significance dip below .05.

The economy is certainly more complicated than the relations depicted in these tables. For one thing, the impact of deficits on unemployment may well depend upon the composition of budget outlays. If a deficit is created by increases in transfer payments or cuts in taxes, the impact on aggregate demand, output and employment may be expected to be less than if the deficit is created by increases in government purchases of good and services. This would suggest that a more effective counter to unemployment created by reductions in government outlays for national defense would be increases in other government outlays for goods and services.[11]

And rather than keep interest rates high to fight inflation or maintain a *franc fort*, monetary authorities would do well to ease credit and lower interest rates. Declines in interest rates might be expected to stimulate housing, construction and business investment generally. They would make it easier for local authorities to borrow to repair and improve public infrastructure. They would also reduce international demand for the local currency and thus, for those countries that lower their rates more than others, reduce the cost of their currency and thus of their goods and raise net exports. By all of these means, jobs would be created.

My empirical relations fit the basic Keynesian-Neoclassical paradigm that plays a large role in most macroeconometric models and serves as a guide for most policy-makers in market economies. They suggest some important conclusions for macroeconomic policy. These are that:

1. reducing defense expenditure in itself may be expected to increase unemployment;

2. using the defense reductions to cut the budget deficit is likely to make matters worse;

3. easier monetary policy will help;

4. cutting taxes would also help but the cut would have to be greater than the cut in defense spending, which would *increase* the deficit over its original amount, to offset fully the deflationary effect of the defense cuts;

5. replacing the cuts in government defense spending with increases in other government spending might fully offset

the effects of the defense cuts in measured constant dollar GDP and reduce unemployment.[12]

If we are serious about reducing military expenditures it would appear then that the U.S. obsession with reducing the Federal deficit militates in the wrong direction. Reducing the deficit, certainly in an economy suffering from too much unemployment, can only add to unemployment. And that makes it all the more difficult to undertake the defence cuts that would otherwise be in order.

Those arguing for deficit reduction insist usually that this is necessary to increase our national saving, which provides for our future. I have indicated elsewhere that even with regard to the narrow conventional measure of saving, the sum of net private domestic investment and net foreign investment, the record does not support this. If anything, bigger deficits have been associated with more consumption and enough more private domestic investment to raise the total of national saving, despite reductions in net foreign investment.[13] With appropriate, broader measures of national saving, including government investment in all manner of infrastructure, the positive relation between the real, structural deficit and provision for the future is all the sharper.

What this signifies is that cuts in defense expenditures should be directed not to reducing the deficit but rather to long overdue investment in the nation's roads and bridges, airports and harbors, water and sewage systems and all of the public capital so vital to growth and our future prosperity. That peace dividend might well go too to investment in human capital—in education, research and health. And some of it might even go to reductions in taxes to encourage private saving and investment. All of this would serve both tomorrow and today, as current employment would be maintained or more quickly restored.

Overall macroeconomic policies directed at maintaining aggregate demand are not likely, however, to be sufficient. Policies must be directed as well to the particular areas and workers affected. There must be specific provision for jobs for those discharged from the armed services and for those for whom that avenue to employment is now lost. The workers in defense plants must be aided by better employment services, retraining, relocation help and adequate unemployment benefits to cushion the transition to non-defense employment. There should be provision for whole communities that will lose from the loss of income of those who have been the recipients of defense expenditures.

I have framed this analysis in terms of the situation in the United States. The nations of Western Europe have devoted lesser proportions of their product to the military and hence have a lesser adjustment to the new international situation to contemplate. I should expect, though, that issues similar to those in the United States must confront them as well. The problems in the states of the former Soviet Union and some other countries of Eastern Europe would appear more severe, as major portions of their economies had been devoted to the military establishment and in some cases, like that of the United States, to sale of arms abroad.

It is all but inevitable that with the best of policies many will suffer from defense cuts, even if in the long run most of us are better off as a result of the transfer of resources to consumption goods we can enjoy promptly and investment that will provide more for later. Policy to minimize the losses of those caught by the cuts is important not only as a measure of justice and equity. It is politically vital if defence expenditures are to be cut successfully in any major way. Without such policy, the potential losers appear likely, as they have so largely done so far, to hold defense cuts to far less than are demanded by any sane evaluation of the international situation and of the needs of their own economy and the world.

Notes

1.   In Eisner (1990, 1991, and 1994b).

2.   See Appendix.

3.   This last figure is by a new method of calculation which yields results possibly some 0.1 to 0.2 percentage points higher as a consequence of correcting for prior misclassification of many women looking for jobs as out of the labor force.

4.   See Eisner (1994c), chapter 8, "The Greatest Misconception of All: Natural Unemployment."

5.   Eisner (1994b).

6.   The long run coefficient of response implied by the this last relation is —0.450, obtained by dividing (—0.39139 + 0.22687) by (1 - 0.63442).

7.   When a linear trend variable was introduced into these regressions its coefficient was generally close to zero and not statistically significant. The coefficient of the

*William R. Kenan Professor of Economics Emeritus, Northwestern University. I am indebted to James H. Gill for important assistance in the statistical work for this paper.

defense variable was affected only trivially. Results with additional shift dummies for the years 1951-1963, 1964-1976 and 1977-1993, not reported in the paper, indicated higher unemployment, after adjusting for defense expenditures, in the first period, and a still more negative coefficient, —0.669, for the defense expenditure variable.

8. The complete series all of the variables used in Figure 1 and Tables 1, 2, 3 and 4 is given in the appendix to the current paper. The original sources and derivations are described in "Data Sources," which follows the appendix. Further discussion of the deficit variable is to be found in Eisner (1986).

9. But about the same as the —0.669 coefficient for the regression with shift dummy variables for the earlier and later periods, observed in note 7 above.

10. It may be noted, as shown in Eisner (1993 and 1994a), that structural budget deficits have also been associated with more investment and more national saving.

11. Support for the results of the single-equation estimates presented in this paper is to be found in simulations of the Wharton (WEFA) Mark 10, multi-equation, quarterly econometric model reported in Eisner (1994b). Our experiments consisted of introducing abrupt reduction of Federal defense expenditures other than compensation in the second quarter of 1993 and comparing the paths of a number of key variables with the "baseline" simulation through the fourth quarter of 1995.

12. These two last conclusions are particularly supported by the WEFA simulations reported in Eisner (1994b).

13. See Eisner (1990, 1993, 1994a and 1994c).

References

Eisner, Robert (1986) How *Real is the Federal Deficit?* (New York : The Free Press).

Eisner, Robert (1990) 'Debunking the Conventional Wisdom in Economic Policy,' *Challenge*, May-June 1990, pp. 4-11.

Eisner, Robert (1990, 1991, 1992) 'Macroeconomic Consequences of Disarmament,' presented at ECAAR-IPPS Conference on Economic Issues of Disarmament, Institute of International Peace Studies, University of Notre Dame, November 23-34, 1990, published in *Challenge*, January-February 1991, pp. 47-50, and in *Economic Issues of Disarmament*, Jurgen Brauer and Manas Chatterji, eds., New York University Press, NY, and by the Macmillan Press outside the US, 1992, pp. 33-41.

Eisner, Robert (1993) 'US National Saving and Budget Deficits,' *Macroeconomic Policy After the Conservative Era*, Gerard Epstein and Herbert Gintis, ed, A Project of the World Institute for Development and Economic Research (WIDER), The United Nations University, Helsinki, to be published by the Cambridge University Press.

Eisner, Robert (1994a) 'National Saving and Budget Deficits,' *The Review of Economics and Statistics*, February, pp. 181-186.

Eisner Robert (1994b) 'Quelques considerations macroeconomiques sur le desarmement, ' *Les Cahiers de l' Espace Europe*, March, 4.

Eisner, Robert (1994c) *The Misunderstood Economy: What Counts and How to Count It* (Cambridge : Harvard Business School Press).

Appendix

Year	U	ND/GDP	PAHED	DMB
1950	5.2	5.51	NA	NA
1951	3.2	11.78	NA	NA
1952	2.9	13.93	NA	NA
1953	2.8	14.01	NA	NA
1954	5.4	11.24	NA	NA
1955	4.3	10.51	NA	NA
1956	4.0	10.07	−1.92	NA
1957	4.2	10.46	−1.79	NA
1958	6.6	10.32	−0.76	NA
1959	5.3	10.20	0.75	NA
1960	5.4	9.17	−1.21	−0.07
1961	6.5	9.33	−0.50	0.04
1962	5.4	9.80	−0.27	0.03
1963	5.5	9.01	−0.13	0.25
1964	5.0	8.36	−0.31	0.22
1965	4.4	7.87	−0.49	0.16
1966	3.7	8.82	0.45	−0.00
1967	3.7	9.53	1.02	0.18
1968	3.5	9.71	2.51	0.10
1969	3.4	8.87	−0.66	0.02
1970	4.8	8.00	−0.92	0.04
1971	5.8	7.33	0.18	0.05
1972	5.5	7.05	0.21	0.22
1973	4.8	6.42	0.18	0.01
1974	5.5	6.12	−1.37	−0.08
1975	8.3	6.14	−0.76	−0.06
1976	7.6	5.89	0.69	0.11
1977	6.9	5.71	−0.24	0.08
1978	6.0	5.52	0.16	0.04
1979	5.8	5.46	−0.69	0.01
1980	7.0	5.73	−0.76	−0.08
1981	7.5	6.19	−1.28	−0.21
1982	9.5	6.39	0.01	0.14

(*Contd.*)

Year	U	ND/GDP	PAHED	DMB
1983	9.5	6.81	1.78	0.27
1984	7.4	6.85	1.88	0.12
1985	7.1	6.85	3.07	0.24
1986	6.9	6.85	3.12	0.36
1987	6.1	6.84	1.16	0.20
1988	5.4	6.51	1.22	0.14
1989	5.2	6.12	0.86	0.00
1990	5.4	5.98	0.98	0.24
1991	6.6	5.81	1.57	0.24
1992	7.3	5.48	2.22	0.42
1993	6.7	5.03	2.23	0.44

U — Unemployment as percent of total labor force

ND/GDP — National defense expenditures as percent of previous GDP

PAHED — Price-adjusted, high-employment (5.5 percent unemployment) fiscal-year budget deficit as percent of GDP

DMB — Real change in monetary base as percent of previous real GDP

Data Sources

Unemployment : *Economic Report of the President, 1994*, Table B-33, p. 306, "All workers."

National Defense Expenditures : 1945-1958, U.S. Department of Commerce, Bureau of Economic Analysis, *National Income and Product Accounts of the United State (NIPA)*, vol. 1, Table 1.1, p. 1 ; 1959-93, *Economic Report of the President, 1994*, Table B-1, p. 269. Taken as percent of gross domestic product.

Gross Domestic Product : 1945-1958, same as above; 1959-93, *Economic Report of the President, 1994*, Table B-1, p. 269, for current dollars, and for 1972-93, Table B-2, p. 270, for 1987 dollars.

Price-Adjusted High-Employment Deficit (PAHED) : From Congressional Budget Office, *The Economic and Budget Outlook* : Fiscal Years 1995-1999, January 1994, Appendix Table E-1, p. 87, Standardized-Employment Deficit (SED), which is calculated for "potential gross domestic product" based on CBO estimates of the NAIRU ("non–accelerating-inflation rate of unemployment"), published in the same source. On the assumption that each percentage point of unemployment adds approximately a point to the deficit as a percent of GDP, which corresponds to both CBO estimates and my own, we adjust the SED to the HED (high employment deficit calculated for a 5.5 percent rate of unemployment) by applying the equation : HED = SED + (5.5 - NAIRU) *GDP/100. We then calculate the price-adjusted high-employment deficit in dollars as PAHED = HED — PE, where PE equals the "price effect," or reduction of real value of debt held by the public as a result of price inflation as measured by the year-to-year changes in the GDP implicit price deflator, second quarter up to 1976, and third quarter from 1977 to 1993. Debt held by the public is taken from *Economic Report of the President, 1994*, Table B-79, p. 362, "Gross Federal debt (end of period), Held by the public." Then, PAHED = 100* PAHEDD/GDPFYD, where GDPFYD is fiscal year actual GDP as reported by CBO in the same source.

Change in Monetary Base (DMB) : Monetary base figures are December daily average, adjusted for changes in reserve requirements, as published in *Economic Report of the President, 1994*, Table B-70, p. 350. These are deflated by the 4th quarter GDP implicit price deflator and their first differences are taken as percents of previous 1987 dollar GDP.

```
┌─────────────────┐
│   Chapter  10   │
└─────────────────┘
```

Lessons from Arms Reduction in OECD Countries

Walter Isard

Introduction

In addressing my topic, I shall begin with a bit of history to see whether expectations on the basis of forecasts from models and their underlying theories were realized. Then I shall consider current thinking.

Historical and General Factors

My observations on lessons from arms reductions in OECD countries go back to the late 1940s. Then the way considered best to handle arms reduction was to conduct a Leontief national input-output study (the input-output technique had been successfully employed for identifying expansions in production directly and indirectly required for the World War II effort in the US). All one had to do was develop a change in the government sector of final demand consistent with an arms reduction scenario, determine which industries would be adversely affected, and then devise government programs to offset adverse effects.[1]

We soon learned that such an approach was inadequate. It did not recognise that an arms industry (inclusive of the several levels of subcontracting) is unevenly distributed over a nation. Adverse impact · and required offsets would need to be identified regionally, both by the industry and region (local area), to attack what many considered the heart of the problem. To handle conversion, we therefore constructed single region input-output, tables like the one on the Los Angeles metropolitan area.[2] But it was quickly realised that subcontracting, at least in the United States, had major interregional linkages which single regional input-output studies fail to capture. Identification of effective cross-regional impacts and offsets require a detailed interregional input-output table. I know of no country that has allocated adequate resources for the construction of such a table, nor do I expect any country to do so. So one lesson we learned from the past is that any program to counter arms reduction will have ignificant shortcomings from inability to identify in full interregional effects.

Another lesson was learned very early. It was recognised that while theoretically a negative impact might be exactly balanced by a positive offset, there are likely to be significant time delays. On average, a machine designed or redesigned to produce a defense item is not likely to be instantaneously retooled to produce a civilian item. More significant, while some entrepreneurs may have both the foresight to anticipate oncoming arms reduction and military demand for their products *and* creativity to enable them to shift the operations of their plants effectively to increased civilian output, a very few do. As a consequence, there is unavoidably a period of decreased employment and effective demand for industrial output. This negative effect may be counterbalanced by pent-up consumer demand that has arisen from curtailed consumer goods production and forced saving in a period of war and intense military build-up; hence the economy can move into a period of major growth with respect to both national income and productive capability via stimulated investment—as was the case of the U.S. economy immediately after World War II. On the other hand, if the arms reduction occurs at a time when no pent-up demand exists, and in particular when consumer demand has been overstimulated by easy credit conditions and encouragement of credit buying, as was the case in the USA in the 80's, then the negative employment and income effect of arms reduction can be multiplied and lead to downward acceleration and major recession.

Consumers, and entrepreneurs are not the only actors in an already advanced economy. So is government. The involvement of government occurs at both the macro and micro level. At the macro level, a key factor has been and continues to be both monetary and fiscal policy. One of the early studies aimed at identifying appropriate macro policy was by Klein and Mori (1973). A Keynesian-type econometric model was employed, and this type of model continues in use today. But it was not long before we recognised that such a model is useful when only small structural changes in an economy and society are taking place. When large structural changes occur, and one necessitated by major arms reduction, the value of such models is questionable. Currently, there is much support for the viewpoint that while these models proved useful for some time subsequent to World War II, currently they often fail miserably in their projections of outcomes of diverse fiscal policies.

Government also acts at the micro level. Its success at attempts to counter arms reduction at this level has been questionable. The empirical evidence since World War II does not indicate widespread 'spinoffs' to the civilian sector from Federally funded programs. Further the very implementation of specific types of arms reduction *per se*, or even passage of legislation to take necessary steps, has frequently been hindered by political figures and the very legislators themselves. For example, Weidenbaum, a former Chairman of the President's Council of Economic Advisers, states : 'I can cite from personal experience the frustration of dealing with members of Congress who, in public, advocate large reductions in military spending and the next day come to the White House in a frantic but private effort to 'save' the weapon system being produced in their districts....Part of the problem is that what passes as benefit/cost analysis in the political sphere is usually done from a local rather than a national perspective. Try closing any unneeded defense base—or reducing the numbers of aircraft or missiles being purchased. The overwhelmingly negative public reaction will quickly demonstrate the point that the political process gives the benefits to the locality far greater weight than the costs borne by the rest of the nation. (Weidenbaum 1990 : 238)

Such reaction of course tempers the extent of realisable arms reduction, and the potential for any decrease in the national deficit and size of any peace dividend.

Over the historical period, particularly since World War II, other problems associated with both the increase and decrease of military production having a bearing on lessons from arms reduction relate to such matters as trade, balance of payments, inflation, and interest rates. Briefly put, export trade stimulation was viewed as a way to maintain employment and income generated by firms adversely affected by cutback—where the exports might be arms for building up the military capability of other nations, or civilian goods to whose production a firm could easily shift. There was of course opposition to any government policy aimed to promote the former type of exports, while civilian goods exports meant entrance into intense competition with firms of other nations. Experience leads us to question seriously the possibility of success in this direction.

Associated with trade were balance of payments considerations. However, Benoit, after pointing to the variety of policy decisions that can significantly affect individual balance of payments items, and the highly interdependent links among these items, neatly summed up possible effects of changes in military expenditures in his statement:

'Since at any one time the number of forces acting on the balance of payments is very large, and since any modifications they make on the balance of payments tend to stimulate offsetting flows,....., it is never possible with any assurance to know what part of the actual measurable change in balance of payments has been produced by one particular set of events'... (Benoit 1973a:212).

He thus concluded that while the Vietnam war expenditures exerted a multi-billion dollar net adverse effect on the balance of payments, arms reduction in Vietnam war expenditures would not necessarily be symmetrical and lead to the elimination of the balance of payments deterioration.

Other considerations related to inflation and interest rates. Boulding (1985) argued that before 1970 it is probably safe to say that inflation was primarily a war-related phenomenon, but that after 1970 inflation become a 'normal characteristic' of society. However, it cannot be denied that military expenditures can contribute to inflation and that arms reduction tempers inflationary pressures. In fact, it can be argued that in a society subject to major inflationary pressures, one lesson from arms reduction is that it does curtail inflation. The current very low rate of inflation in the United States might be pointed to in this connection. Similarly, especially when the need for increased security is espoused and crises conditions prevail, military expenditures can add significantly to national budget deficits and national debt, thereby tending to raise interest rates and have a negative impact upon investment in addition to diverting resources from civilian goods production. Thus, almost by necessity, arms reduction has been considered a force tempering increases in deficits and debt, and to easing interest rates and, in at least some way, fostering national investment and growth of civilian productive capacity.

However, the lessons gained from experience on the impact of arms reduction on a nation's productive capacity relates to deeper questions concerning technological development, the spin-off of military R and D (research and development) as well as other factors, which have been intensively debated in recent years, and which we shall discuss in some detail in the next section.

Specific Aspects of Current Thinking

All the general type of thinking on lessons from arms reduction discussed in the previous section carries over to thinking today. Clearly, fiscal policy and other policy oriented to the behaviour of national macromagnitudes treated in econometric models, by themselves, address only a part of the problem of conversion—of averting unemployment, decline in consumer spending, firm restructuring, and the maintenance of output, employment, and the stimulation of investment. Increasingly, one major lesson from arms reduction relates to the need to concentrate much more than in the past on the microscopic level of behaviour. While input-output studies have helped to point up needed adjustments by specific local areas and specific industries within these areas, the interregional models required to account for interregional (interlocal) effects and the needed adjustments resulting from such effects are too costly to construct.

The problem of time delays encountered in conversion or reconversion, the play of political leaders in opposing back-yard reductions, the inabilities of bureaucratic government to implement programs still remain and their associated historical lessons on arms reduction still hold. *But we have learned still other very important lessons—lessons that stem from the mounting increase in the last decade or two in the specialisation of firms, fixed capital, labor, entrepreneurship, management, marketing, R and D efforts and in government administration and policy making.*

a. Some General Remarks on the Current Background

Today's choice of national industrial policy—how allocate resources among military and civilian production and what kinds of specialisation a country should pursue—takes place in an expanding internationalisation of the economy. (In this regard see Gold [1994] on the internationalization of military production.) The conversion problem has thus become severe for economies of Western Europe and North America that not only have a large military industry but also for those which have increased their specialisation in part because of the increasing internationalisation of markets for military goods. It is generally regarded that the militarisation of these economies has led to the weakening of a number of important civilian industries and losses in their productivity, capability to meet competition, and creativeness. The resulting tendency towards subsidisation (on political grounds) and protection obviously excaberates the tendencies for industrial decline.

While it is recognized that at the *aggregate* level the disruptions caused by major arms cutbacks may sometimes not be momentous, and be of no greater scale than the responses to significant shifts in consumer demand and the basic technology changes an economy often confronts, nonetheless at the micro level drastic distortions may be experienced.

b. Problems Associated with Management and Entrepreneurship

As Weidenbaum has stated (1990) :

'It is not hard to understand why defense company managers are now reluctant to move from fields they have mastered into lines of business alien to them. They lack knowledge of products, production methods, advertising and distribution. They are ignorant, too, of financial arrangements, funding of research and development, contracting firms, and the very nature of the private customer's demands.

The type of company that can successfully design and build a new multimillion-dollar ICBM network or space exploration system has a capability very different from that of the soap, steel, toy or other typical low-technology company operating in the commercial economy.

On the other hand, the large defense firms possess resources. They possess great capability to perform research extending the state-of-the-art, as well as preparing complex engineering designs. Related to that attribute is a management capable of managing the development, production and integration of large, complex systems. However, the demand for that valuable array of talent is limited.'

Let us elaborate on this statement. The market environments faced by military-oriented and civilian-oriented firms are miles apart. Prime defence contractors have, in practice, only one customer, the government, whereas civilian-oriented firms sell their product in the market place. The former typically operate under cost-plus contracting, and their administrative structure has been formed to respond to the unique reporting and control requirements of the government consumer. Moreover, they as well as their dependent subcontractors do not have to make fine calculations of revenues as do those projecting the tastes and needs of civilian consumers. Their capital equipment may be highly inflexible, and their marketing forces are specialized in military work and focus on the performance quality of product. These firms have limited experience in producing at high volumes and low unit cost. In sum, they have developed a subculture geared to the military market. As all cultures, they resist, are obdurate, or indifferent to change. Thus they continue, as they have in the past, to exercise and even further build up their political persuasion capability now aimed at the preclusion of cancellations. *One basic lesson is that these firms, as part of the larger military-industrial complex, a strong political subculture, have and can be expected to be, as is characteristic of all cultures, resistant, and slow to change*—as is all too clear in the recent history of NATO countries.

Nonetheless, cultures wanting to survive do change when under pressure, and defense firms have, but not too successfully, altered their practices and operations in several directions. Here we can profit much from the findings of Udis in his recent Peace Science paper 'Adjustments to Reduce Domestic Defense Spending in Western Europe: Scale, Scope, and Government' (1992). (Udis is a long-time scholar of conversion theory and experience). He has edited a pathbreaking volume on 'The Economic Consequences of Reduced Military Spending' (1973).

In discussing the drive to preserve or expand scale economies as a firm responds to military cutbacks, Udis states:

'The search for scale economies takes two forms: an effort to expand exports, and/or an effort to fashion collaborative ventures with firms in other countries which are designed to yield a significantly larger product run than would be possible for single country/firm. In the case of the smaller European states the collaborative approach often is combined with an effort to carve out a special niche in which comparative advantage can be developed. This effort, in some cases, also involves a downsizing of staff associated with the abandonment of a wider product line and a movement toward specialisation.' (1992, p. 3). Often firms sell off their non-main product divisions. With regard to the collaborative propensity. Udis later states : 'Clearly, all future aircraft development in Western Europe will be, to some extent, collaborative (1992, p. 8.)'...reflecting a preference to maintain traditional lines of business, a manifestation of the aforementioned lesson.

Reppy (1993) in her extensive study reports that in contrast with the U.S. experience, 'joint ventures have been more common in Europe, where small national weapon buys must be combined to achieve production economies' (p. 3). The typical goal of European joint venture is to create an entity that will approach the size of U.S. company in the same submarket, such as helicopters, in order to compete effectively in the European market and export markets.

Note that the formation of consortia and teaming arrangements broadly defined tends, at least somewhat, to limit competition among the circle of countries with capability in a given military product (for example an aerospace item). Such provides an alternative scale economies route (via meeting the combined needs of the countries involved) to increasing undesirable arms exports to third world and other countries. (See Senesen [1994] for data on arms exports from industrialized countries to the developing nation of Turkey.)

Another type of response is to seek *economies of scope*. Such should not be interpreted to mean diversification. For we have learned that efforts at diversification have been, with a few exceptions, unsuccessful if not devastating—and the lesson is: diversification *per se* is not the way for firms to avoid the negative impacts of cutbacks in arms orders. What is appropriate is a redefinition of a firm's potential to achieve comparative advantage, one based on its superior capabilities. This then limits the direction of diversification which a firm should pursue, the specific civilian goods in whose sale it should enter competition. For example, it may be based on the specialisation of its scientific staff and the specific know-how of its engineers. Such permits spread of critical fixed and other costs over many units, the total of the units of all products the firm produces—in effect economies of scope from what may be designated 'efficient borders of the firm'. There of course have been cases where diversification with little regard to scope economies has been successful—but it is generally concluded by scholars of conversion that these have been few and far between.

Still another response to arms cutbacks has been acquisition of one firm by another, specifically designed to eliminate competitors. (As already noted, acquisition for diversification *per se* has not been successful.) The record has again been bad. There has been much experience with acquisition in the US. See Weidenbaum (1992) and

Markusen (1992). But, few if any, successful ventures in this direction can be reported.

c. Problems Associated with Labor

The general 'conversion problem' associated with labor has always been extremely difficult. One obstacle is the very large number of people involved. Others relate to problems of retaining and relocation of both skilled and unskilled labor, reorientation of managers, provision of placement services and of financial assistance during interim periods of unemployment and for relocation. These are familiar problems and need no discussion here. However, they have rarely ever been managed successfully ; and the lesson is that the scale of the problem in the context of the political structures, union organisation and social welfare programs in the OECD countries (which contribute to resistance to change and inertia within the labor force) is likely always to remain a major one.

d. Problems Associated with Community Restructuring

Typically, but not always, a community faced with unemployment from the closing down of a base or cutback in military contracts to local industry will indicate a desire to find alternative employment opportunities. At times they will take an initiative on their own or pressure the state or national government to undertake this task and subsidize replacement activity. Frequently, the community will lack resources to plan for and identify such activity, and at times this will be so for the state or national government when involved. The community may face resistance from a firm's management, and even its labor force, to shift from a mainstream military operation to a civilian one or to mainstream operation involving a combination of military and civilian goods production. Often the community will not be able to amass sufficient financial funds required for the reuse of sites and plants and other economic adjustment. Even if a community does take the initiative, obtain the support of the state/national government and exploit programmes of an Office of Economic Adjustment, put together the necessary funds, secure top engineering, economic and planning expertise, it does not follow that replacement activities with sound economic basis can be found for that community. It may not have any foundation for economic survival—no matter how much infrastructure investment may be made, no matter how close the interaction of the local social structure and cooperation of local authorities, producers' organizations and trade unions. Recall that often military installations including bases are located at sites (which then grow into communities) which are not useful for economic and other development, for example the Hanford site in the eastern part of Washington state. Also, bear in mind that engineering and other consulting firms are under pressure to identify new activities for a declining community and often make recommendations which turn out to be sour and uneconomic, in part because of failure to consider or inability to appreciate the dynamics of a given state of technological development,

Thus, one more lesson from arms reduction and other experience is that a number of communities cannot survive arms cutbacks, and unless other very strong social welfare or political considerations are present, should be allowed to decline and even die. Such communities can, of course, provide fertile ground for special interest groups to press for the preclusion of desirable military production decisions.

e: Problems Associated with Government

In its operation the government (federal, state, provincial and the like) imposes numerous obstacles to adjustment of firms and local communities to military cutbacks. In addition to red tape, excessive bureaucracy, fickleness and other well-known negative features of government which interfere with market operation, there is politicisation in the government's implementation of various subsidisation and other policies aimed at softening the blows of military reductions and in the required allocation of resources. For example, the government is not good at making technological choices when large-scale direct subsidies are to be made to high-tech civilian enterprises in connection with research and development. The market place performs much better. Government's role is best restricted to the support of basic research. Moreover, government's mounting array of statutory and administrative requirements, let alone the existing array, are roadblocks that hit hardest at new enterprise, old enterprise often being exempt from such. But it is the new enterprise to effect economic readjustments from arms cuts that most needs to be encouraged. Of course this is not to deny that government programmes should not be made available. Numerous ones differing from country to country and among regions have been suggested. But the lesson is that as far as possible decisions should be left to the market place in areas where the market place does not break down ; and when statutory and administrative regulations are imposed, they should be carefully designed not to discriminate against new enterprise.

In addition to the direct support of military-oriented industry, government itself may engage in military production. As with industry a basic lesson that seems to be learned by governments when they must face up to reduced domestic military spending is to seek collaboration. By and large, governments must give up the self-reliance goal of a domestic weapons capability across a wide range of items in order to concentrate on selected weaponry with a volume of production (based on sales to country partners) sufficient to achieve scale economies and to purchase the sacrificed items from the country partners. This is a direction in which the historical highly self-reliant neutral nation, Sweden, is moving. As already noted, such a policy has the additional side effect of tempering the pressure for scale economy attainment to export arms to third world countries and regions of high current or potential instability.

As already indicated, there are innumerable ways in which government is, either necessarily or can be involved, in restructuring firms, industries, and communities. Recently in the US there has been much thought given to national legislation requiring prenotification of plant closings, worker adjustment assistance, alternative use planning by firms for plant, equipment and labor force. Putting together the various suggested roles of government over the set of OECD countries would constitute a laundry list of almost infinite length. There is, however, one important aspect which is most inadequately appreciated *and* addressed—namely, the interdependence among the many policies (which are still mounting in number) which governments must consider. With regard to just defence sector goals, Reppy (1994) has stated that the US government has particular policy goals for the defence sector, not all of which are necessarily mutually consistent, but each one of which depends on the cooperation of the defence industry. 'These goals include maintenance of a defence industrial base, promotion of the national technology base, elimination of costs associated with excess capacity in the defence industry, and conversion to civilian employment of the resources being released from the defence sector. The changes in the structure of the defence industry affects the ability of the government to carry out its policies ; as the networks are reconfigured some actors will disappear and others will, by virtue of their new relationships, especially their international links, be less willing to cooperate in the government's policies than in the past.' (p. 4).

'The policy goals listed above cannot be viewed in isolation, since each has implications for the others. Thus, the elimination of excess capacity, which is being accomplished through the actions of the defence firms with very little input form the US government, has profound implications for the maintenance of the defence industrial base and national technology base, as well as for conversion. DOD costs will fall, as overhead charges associated with maintaining excess production lines are eliminated. Competition will clearly be reduced, since the number of companies in each of the sub-markets is shrinking. Supplier networks are collapsing along with the shutdown of production lines, and some highly specialized and essential subcontractors may disappear. International links carry with them both increased access to other sources of supply and markets and the danger of unwanted technology transfer abroad and foreign dependency.' (pp. 4-5).

In the context of the current US situation I would go on to consider the interrelation of all the above with environmental policy, infrastructure investment and employment policy, tax policy, deficit reduction policy, energy and other resource development policy, trade and arms export policy, and foreign policy with respect to the Middle East, North Korea, Bosnia and other parts of the world. Of course, no human mind, with or without teams of scholars armed with high-speed modern computers, can treat all possible policy interrelations. But clearly policies associated with arms reduction phenomena must and can consider some of the major policy interconnections—more than is typical of scholars researching such policy. After all, we form a specific scholar subculture with our own specific form of blindness. We are more blind than need be.

f. Issues Associated with R and D and Technological Progress

A hotly debated issue is : does large-scale R and D spending and technology progress in the military sector occur at the expense of R and D and technology development in the civilian sector ? Over the years a number of scholars have pointed to the many spillovers from military research and development programs to civilian life : eradication of yellow fever, chlorination of water, blood plasma substitutes, modern aircraft, electronics, and so on. There can be no question of the great significance of these spillovers. Yet has all this been at the expense of even greater contributions that might have been achieved by an equivalent resource allocation directly to civilian R&D and technology development ... a basic question today.

In 1989 after careful study Reppy (1989) argued that : 'there are technological and institutional features of the defence market that promote and limit opportunities for transference of military R&D ; and we must turn to empirical evidence to evaluate the relative importance of these opposing tendencies—aggregate productivity studies and patent studies suggest little or no benefit to the civilian economy from military R&D. Preliminary data from international trade competitiveness studies, however, show a stronger US performance in military, related high technology product groups than in product groups based on civilian technology. Reppy suggest that general conclusions about military R&D and innovation are hard to make because each technology and industry is potentially a special case. Nevertheless, Reppy suggests that military R&D can benefit the civilian economy (Isard and Anderton, 1992, p. 33). It can to the extent that : 'it is aimed at developing generic technologies with wide applicability ; ... institutional barriers such as secrecy and specialized accounting requirements are minimized ; and ... the military customer values low cost and producibility as well as high performance.' However the 'record for reform of military R&D and procurement is not good, and we are entitled to be skeptical as to whether the current interest in using military R&D as the vehicle for improved commercial competitiveness is likely to succeed.' (Reppy, 1989, pp. 7-8).

Since 1989, the record of reform has not significantly improved in the United States nor has there been any evidence of greater efforts at and implementation of the transfer of military R and D to civilian uses, though there has been much talk of such potential. Obstacles lie in the military market orientation of defence industry entrepreneurship and the play of forces within the political subculture of the military industrial complex. Nor am I aware of any change of the situation from the late 80s within other OECD countries. The lesson is that while there may be much talk of the potential for transfer of military R and D, the evidence does not suggest that such R and D does not retard civilian technological development were equivalent resources currently allocated directly to civilian R and D.

Summary

1. Input-output models, while helpful in the task of identifying impacts of and offsets to arms reductions, fail to capture important interregional effects, and suffer in a major way from inability to consider nonlinearities and other aspects of the aggregate and disaggregate economies affected.

2. Econometric models, while also helpful for short-run analysis, are unable to take into account structural change that is concomitant with arms reduction policies ; and today, with increasing complexity of

the economy and its internationalization, the forecasts of these models are being found to be lacking and increasingly questioned.

3. Computable general equilibrium (CGE) models, while adding another dimension to the analysis of the national economy, are much too aggregative and fail to get down to specific regions (communities) and industries.

4. Sound arms reduction analysis must fall back upon different uses of existing models, singly or in combination, depending on the specific impact or problem to be attacked and policy (measure) to be taken.

5. When pent-up demand exists at or during a period of arms reduction, the necessary and inevitable time delays in initiating and implementing diverse conversion (offset) programs need not intensify the problem of maintaining a healthy economy. However, when an economy is already in a recession period, or when there has previously been overstimulation of consumer demands and overinvestment by private and governmental units, arms reduction may initiate or intensify a recession.

6. Arms reduction during inflationary periods can contribute to cooling down the economy, help reduce national budget deficits and national debt and temper rising interest rates.

7. There seems to be no effective legislative or other way to curb politicization of conversion effort. While the presence of lobby and special interest groups has become increasingly recognised and efforts to curb their activities taken, the pressures and skill at politicization of these groups have steadily mounted. We seem to be no better off today in eliminating politicization than in previous years.

8. The mounting increase in specialisation of firms, fixed capital, labor, entrepreneurship, management, marketing, R & D efforts and government administration and policy making has steadily increased, perhaps exponentially, the difficulties in conversion and development of offset programs.

9. At times the heart of the arms reduction problem is at the macrolevel, and at other times at the microlevel. Often the problem involves both levels, and necessitates adjustments all along the line.

10. Arms reduction is severely hampered in some OECD countries by past heavy militarization. Such occurred at the expense of attaining and/or maintaining top civilian productivity. This now

reduces the competitiveness of reconverted (let alone converted) plant in markets which have become increasingly internationalised and penetrated by new, highly productive economies (for example Japan). Pressure for subsidisation, protection and resort to arms exports have accordingly increased in these OECD countries.

11. Management of military-oriented plant is more than. ever indifferent and resistant to conversion and slower to change. This reflects the increased specialisation of its production, the resultant decrease in its knowledge of and ability at civilian goods production, marketing, promotion and distribution, its unfamiliarity with effecting contractual and financial arrangements for such—all a result of having been oriented to the ways and practices of serving a single consumer, the government, on a cost-plus basis, and in general its membership in the 'military-industrial complex' political subculture. Nonetheless, the need for economic viability has forced it to change.

12. In a number of cases maintenance of and/or the search for scale economies has led to formation of collaborative and teaming arrangements (in which governments may be involved as producers). Each military producer in one or more countries becomes further specialized in a niche in which it has comparative advantage (non-niche production being relinquished to others) and where the combined needs of the customers formerly served by all the team members yields a sufficient volume for the niche product producer to achieve low unit cost. To some extent this effort has been successful and has helped reduce the pressure for and the use of arms exports to developing countries to achieve scale economies.

13. In other cases, there has been a search for *scope economies*—economics associated with a firm's potential to achieve comparative advantage based on its superior capabilities and a delineation of its 'efficient borders'. Diversification per se, with regard to these capabilities, has in general been a dismal failure. (Economies of scope permits the spreading of critical fixed and other costs over the combined total of the units of all kinds of products produced).

14. Acquisition of one firm by another to eliminate competition has its merits and demerits, but does induce economic viability. Acquisition for diversification purposes alone has rarely, if at all, been successful.

15. Labor force adjustment programs involving retraining, relocation, reeducation of managers and administrative personnel, and

financial assistance of all sorts have rarely been successful, partly because of the large number and diversity of individuals involved, union-oriented obstacles, and the existence of anti-incentive social welfare programs.

16. Community efforts also have rarely been successful, partly because they have failed to obtain the full range of contributions necessary for successful adaptation—contributions of vigorous entrepreneurship and management, union support, government and state financial aid, competent engineering and planning personnel and partly because the community may have no sound economic basis for any activity.

17. Governments unwaringly have imposed obstacles to industrial restructuring and community adjustment. Aside from the inevitable red tape and bureaucracy, government tends to impose a larger array of statutory and administrative requirements on new establishments than on old when often it is new enterprise that a community and its labor force are encouraged to seek. Also while government support of basic research is essential, because of the intervention of politics government cannot make wise decisions on directions for technological development to be supported.

18. With respect to the reduction of its defence establishment, the government confronts many complex tissues, which together require a set of policies, necessarily interdependent. Frequently, however, we find inadequate recognition of and attention given to the interdependence of this set with key policies in the non-defence sectors. Such often gives rise to poor policy formulation in all these sectors.

Positive Suggestions from Experience

1. For individual firms, scope economies rather than diversification *per se* is to be encouraged.

2. Relinquishment of national goals for sufficiency in arms production, combined with collaborative and teaming arrangements among industries in one or more countries (inclusive of government military production) which permit military producers to specialize in a niche and obtain volume sales at low unit costs can temper their efforts to achieve scale economies by exporting arms to developing areas and unstable regions of the world that have militarization objectives.

3. Joint efforts of community, unions (labor) and management are essential for effective industrial restructuring of communities when

such can be economically justified. This requires engineering and planning expertise and other technical assistance (for example, to conduct new product assessment, and marketing and financial feasibility studies), perhaps early mandatory alternative use planning by management, education at the union, community and management levels, and financial support by local, state and national governments. All the entailed costs are to be evaluated against community and other gains, particularly in avoiding social welfare costs which would otherwise be involved.

Notes

1. See Leontief, et. al. (1953).

2. See Isard and Schooler (1964), and also the 500 sector study of the Philadelphia region designed to reduce errors from insufficient disaggregation of industry operations (Isard and Langford, 1971).

References

Benoit, Emile (1973) *Defence and Economic Growth in Developing Countries* (Lexington, MA : Lexington Books).

Dumas, Lloyd J. (1986) *The Overburdened Economy* (Berkeley, CA : University of California Press).

Gold, David (1994) 'The Internationalisation of Military Production,' *Peace Economics, Peace Science and Public Policy*, Vol 1, no 3, Spring.

Isard, Walter and Eugene W. Scholar (1964) 'An Economic Analysis of Local and Regional Impacts of Reduction of Military Expenditures,' *Papers*, Peace Science Society, International, Vol 1, pp. 15-44.

Isard, Walter and Thomas Langford (1971) *Regional Input-Output Study* (Cambridge, MA : MIT Press).

Isard, Walter and Charles H. Anderton (eds.) (1992) *Economics of Arms Reduction and the Peace Process* (Amsterdam : North Holland).

Klein, Lawrence and K. Mori (1973) 'The Impact of Disarmament of Aggregate Economic Activity : An Econometric Analysis,' in B. Udis (ed.) *The Economic Consequences of Reduced Military Spending* (Lexington, MA : Lexington Books).

Leontief, Wassily, et. al. (1953) *Studies in Structure of the American Economy* (New York : Oxford University Press).

Markusen, Anne and Catherine Hill (1992) *Converting the Cold War Economy* (Washington, DC : Economic Policy Institute).

Pianta, Mario (1992) 'Industrial and Technological Aspects of a Strategy for Converting Military Industry,' *MOST* (Economic Journal on Eastern Europe and Russia), May 2.

Reppy, Judith (1989) 'Technology Flows Between the Military and Civilian Sectors,' paper presented at the meetings at the American Association for the Advancement of Science, January.

Reppy, Judith (1994) 'Defence Companies' Strategies in a Declining Market : Implications for Government Policy,' *Peace Economics, Peace Science and Public Policy*, Vol. 1, no. 2, Winter.

Senesen, Gulay, Gunluk (1994) 'Spillover from the Developed Countries in the Arming of the Developing Countries,' *Peace Economics, Peace Science and Public Policy*, Vol. 1, no. 3, Spring.

Udis, Bernard (ed.) (1973) *The Economic Consequences of Reduced Military Spending* (Lexington, MA : Lexington Books).

Udis, Bernard (1992) 'Adjustments to Reduced Domestic Defence Spending in Western Europe : Scale, Scope and Government,' paper presented to the Peace Science Society, International, Rotterdam, May 1992.

Weidenbaum, Murray (1990) 'Defence Firms Should Fight, Not Switch,' *Newsday*, March 26, 1990.

Weidenbaum, Murray (1992) *Small Wars, Big Defence* (New York : Oxford University Press).

Chapter 11

The Internationalisation of Military Production

David Gold[*]

Introduction

In recent years, the internationalisation of military production and military procurement has become an issue of concern with respect to defence planning and policies regarding the domestic defence industrial base (Moran, 1990 ; US Congress, OTA, 1990). Governments and private firms have engaged in international trade in military goods and services for some time, in recognition of the fact that even the largest and most well endowed countries cannot be the best producers of everything. Smaller and less well endowed countries utilise arms purchases as a significant component of their national security planning. At the same time, governments seek as much autonomy as possible in military procurement and zealously protect their domestic capabilities. Indeed, national security arguments are often successful in maintaining trade barriers and rebuffing direct investment, including buyouts, by foreign firms when the targets are defence-sensitive industries.

The recent concern with internationalisation goes beyond the purchase and sale of weapons systems and major components. Internationalisation has pervaded the vast networks of suppliers and subcontractors that are needed to design, test, manufacture and assemble a modern weapons system. This phenomena is partly due to the growth of international trade ; the new element, however, is the growth of international investment and other means of direct linkages across firms located in different countries. The growth of these international linkages, defined below as transnationalisation, has been less visible in the United States defence sector than in Western Europe and other countries, in part because the United States government has been more restrictive regarding foreign investment at the prime contractor level then have the governments of other countries. In fact, many cross-border mergers and collaborative agreements among defence contractors in Europe have been encouraged by their respective governments, while the Japanese government has aided its defence firms in seeking international collaborative arrangements. In the United States, transnationalization in defence production may be less visible, but it is not necessarily less important.

This paper suggests that the increasing dependence of the United States military on foreign controlled resources is the result of three interrelated phenomena, the rising cost of military goods and services, the shift of innovative primacy from military to civilian firms, and the growth of transnationalisation within those civilian sectors most closely linked to defence production.

The Division of Labour is (Usually) Limited by the Extent of the Market

During the first decade of the aerospace age, a state-of-the-art aircraft should be designed, developed, constructed and assembled at a single site, with a research and development staff consisting largely of those who also manufactured the aircraft, some components, such as engines, purchased externally and modified on location, materials purchased from suppliers and designs fabricated on-site or with the aid of local machine shops (Biddle, 1991).

By the 1970s, an aircraft or missile had become the centerpiece of a weapons system or platform, with a vast network of prime contractors, subcontractors and suppliers, and separate, albeit linked, contractual arrangements covering research and design, testing, manufacturing and assembly. Over time, the growing technological sophistication of the components of weapon systems increased the need for specialized skills. Many of these skills were developed, or absorbed, by the large firms, but it was not possible, or efficient, for such firms to be sufficiently specialized in everything that is required. In the early to mid-1970s, while still in the testing and development stages, the B-1 bomber had a single prime contractor and over 5000 subcontractors and suppliers spread over the continental 48 states plus the District of Columbia (Adams, 1976). Such a spread was partly for political reasons, to build Congressional and public support for the system, but it also illustrates the growth and complexity of the networks needed for major aerospace projects.

A decade later, the Air Force itself acted as prime contractor, or systems integrator, for the MX missile system, and established a network of 37 associate contractors in order to subdivide responsibility for key elements of the system. These associate contractors then had their own networks of suppliers and subcontractors (Gold, Paine and Shields, 1981). By the end of 1980s, the process of extending the division of labor was clearly and visibly involving foreign as well as domestic companies (US Congress, OTA, 1990).

Accompanying the extension of the division of labour in defence production and systems management was a rapid increase in the costs of developing and producing major systems, as well as a significant growth in the time elapsed from conception to delivery. The rapid cost growth has largely been the result of mushrooming R & D costs to achieve state-of-the-art sophistication, with gold-plating, procurement inefficiencies and firm-level inefficiencies also contributing. While some of this cost growth has been offset by performance improvements (Gansler, 1989), problems of managing defence projects in a high cost environment have been compounded by a stagnant domestic market, albeit one exhibiting periods of sharp growth and decline. Thus, from 1956 to 1988, at the peak of the Reagan build-up, real military spending grew at an estimated compound rate of 0.2 per cent per year, while the military share of gross output declined by 2.8 per cent per year over the same period.[1]

This stagnation has contributed to substantial excess capacity in military production. Capacity problems have been compounded by the fact that successive buildups have tended to emphasize changes in military strategies and shifts in the mix of weaponry demanded—aircraft, missiles, electronic warfare, strategic defence, and so on. Thus, new capacity has been required even as existing capacity becomes redundant (Markusen and Yudkin, 1992). Moreover, the Department of Defence has been slow to permit the rationalization and consolidation needed to reduce problems of overcapacity, as the DoD has tried to preserve as much domestic autonomy as possible in the production of weaponry, often awarding contracts to firms with the objective of maintaining capacity in a specific weapon or technology.

Some similar patterns have emerged in production for civilian markets in aerospace and electronics. Development costs in particular have risen rapidly in such industries as computers, semi-conductors, telecommunications equipment, air frames and engines, and even automobiles. Civilian technology has also been subject to increasing specialization. In computers, IBM was once dominant in everything but now is under attack from different sets of competitors in each product category. At the same time, some technologies require the joining together of other, specialized technologies. For example, modern telecommunications requires combining telephones, computers, microelectronics, new materials (fiber optics), space technology (satellites), and so on. Production increasingly involves far-flung networks of subcontractors and suppliers, many of them foreign own and based.

However, unlike the military market, many civilian high technology products have dropped in price and/or increased markedly in quality and performance. One need only look at the price and performance curves for personal computers to see perhaps the most dramatic illustration of these phenomena. At the same time, product and process life cycles appear to be shortening, and competitive pressures are forcing firms to reduce the time involved in product development.

Both the military and civilian segments of the aerospace and electronics industries have been subject to an extension of the division of labour, an increase in the cost of development and production and an increase in the complexity of the networks required to generate final output. Yet this has occurred in a stagnant military market and a growing civilian market, leading to rising unit costs in the former and generally declining units costs in the latter. In the military market, requirements established by the monopsonist–the Department of Defence—replace the stimulus to productive and organizational innovation that the market mechanism provides for civilian output.

From Spin-off to Spin-on

The computer industry is largely a creation of the United States military, although its present status has evolved far beyond what would be required for the needs of the DoD. In the years during and immediately after World War II, the lions' share of funding and direction for computer research was provided by various Pentagon agencies and the military services, allowing the fledgling industry to simultaneously pursue a variety of technological options (Flamm,

1988). In addition, the Pentagon was the largest buyer and military purchases of computers allowed firms to move along their learning and cost curves and be in a better position to seek civilian markets. During the 1950s, markets began to grow for civilian applications of computer technology, and by the mid-1960s, the bulk of new research and development funding was being raised privately. By the 1980s, the dominant technological initiatives were towards civilian applications and the overall directions for computer research were being established by global entrepreneurial initiatives.

During and after World War II, military technology was a leading force in technological development, in such areas as aircraft, computers, and electronics. Gradually, and especially since the 1960s, civilian sectors have assumed more, if not most of this role (Mowery and Rosenberg, 1989). Thus, although there are important exceptions, the military has looked increasingly to the civilian sector for technological input, and has been subject to the impact of the economic, organizational and technological pressures that have shaped civilian industries.

In addition, firms from Europe, Japan, Canada and even some developing countries have challenged United States firms in many industries on world markets. In part, this is due to the ability of these countries to commercialize existing technologies and develop new technologies. New technologies, especially in electronics, biotechnology and new materials, and heightened competition have shortened product life cycles while simultaneously raising the costs and risks of R&D. This combination of increased, and increasingly costly, specialization with cross-fertilisation has led to various forms of inter-·firm, inter-industry and international collaboration.

Thus, as the United States military looks to the civilian market for its products and, increasingly, its technology, it is confronted with the growth of cross-border arrangements, and with the presence of desirable products and producers outside of the United States.

Transnational Corporations and the Organisation of International Production

A transnational corporation (TNC) is, most simply, a firm that controls productive assets (that is, value-adding activities) outside its home country, while the production that occurs outside of the home country under the governance of TNCs is defined as international ·production. Transnationalisation, then, refers to the expansion of international production, as well as the expansion of resources under the common governance of TNCs. The most visible means of expanding international production is through the purchase or creation of productive assets via foreign direct investment (FDI). In recent decades, FDI, measured either as annual flows or accumulated stocks, has grown more rapidly than other measures of aggregate economic activity, providing an indication of the growth of the transnationalisation of the world economy (UNCTC, 1988 ; UNCTAD, 1993). For the United States, real inflows of FDI have grown at a estimated compound rate of 14.3 per cent from 1960 through 1991, using annual data, compared with growth rates of 5.9 per cent for merchandise imports, 3.6 per cent for business fixed investment and 2.8 per cent for gross domestic product.[2]

Transnationalisation is a product of imperfect competition. Firms possessing a proprietary asset (that is, technology, organization, brand name, etc.) utilize advantages within host countries (that is, large markets, low production costs, etc.) while gaining cost savings from internalizing transactions that would otherwise occur on an arms-length basis (Dunning, 1993). Transnationalization does not have to occur only through FDI. Governmental restrictions have prevented or restricted FDI in many industries and countries, a condition especially applicable in industries deemed vital for national security. Indeed, the decline in restrictions and the liberalisation and privatisation trends in such industries as telecommunications, finance and air transport helps explain the rapid surge in FDI in the late 1980s (UNCTAD, 1993, Ch IV).

When FDI is restricted, firms will seek alternative means, including joint ventures, licensing, franchising and subcontracting, to obtain the benefits from a host country presence while retaining the advantages of common .governance. In addition, the rising costs of research and development, and the growing risks of R&D in a world economy that is increasingly competitive and where technological change is advancing at a rapid rate, has stimulated many high technology firms to enter into joint ventures, strategic alliances and other sharing arrangements with firms both at home and abroad. These arrangements frequently involve firms that cooperate in certain activities or product lines, while competing in others. This also involve firms seeking synergies in technologies, as with the links between computer and telecommunications firms, or the recent surge in links between telephone and television companies.

Firms increasingly seek to expand in all large markets, either via exporting or international production. This is partly to expand output, and partly to challenge their own competitors wherever possible, given the nature of oligopolistic competition. Global expansion often involves cross-border mergers and acquisitions, mostly in older industries with established technologies, such as petroleum, chemicals, consumer goods, retailing and publishing, or where established firms seek to purchase new technologies rather than develop it themselves, as in automobiles and pharmaceuticals. Firms have also sought to pool risks and capital costs ; in high tech, this has often involved various types of alliances among companies, such as forming joint ventures, joining research consortia, and so on, and these have increasingly involved cross-border arrangements.

Transnationalisation has proceeded at different rates and utilised different forms across industries, partly due to differences in products and production conditions and partly the result of policy differences. In air frames and engines, the industry has gravitated towards a small number of producers with extensive networks of suppliers and subcontractors and reliance upon joint ventures and consortia, frequently involving foreign firms (Mowery, 1988b). In micro-electronics and computers, learning economies, niche markets and growing technical sophistication have led to a proliferation of producers and greater specialization, forcing United States firms to rely upon foreign suppliers in many areas or even to cede effective control of some product markets to foreign firms. There have been a number of mergers, minority share holdings, joint ventures and research consortia involving United States and foreign companies.

Government attitudes and actions have also differed. West European governments have encouraged cross-border linkages among defence producers in order to rationalize capacity, achieve budgetary savings and compete more effectively against larger United States firms for foreign military sales (Skons, 1993 ; Bitzinger, 1993 ; Kolodziej, 1987). European governments have also encouraged collaboration among firms in civilian high technology industries. Japan has pursued a dual-use strategy, with the government supporting extensive military-civilian cross-fertilization and encouraging collaboration between Japanese and foreign firms in military and military-related products (Drifte, 1985 ; Ikegami-Andersson, 1993). One result is that Japan is the largest licensee of major United States weapons systems. Some developing countries have utilized international linkages as a means of obtaining technology to build up domestic civilian and military industries (Franko-Jones, 1992).

In the United States, the approach to cross-national collaboration has been less accommodating. On the one hand, there has been encouragement of cross-national collaborative arrangements within alliances, and the negotiation of offset agreements as a means of securing foreign military sales. However, the United States has frequently opposed the specific involvement of foreign firms in defence producing industries, successfully preventing some mergers and acquisitions (Graham and Krugman, 1991). For the United States, the spread of transnationalisation has largely occurred among firms whose primary activities are oriented towards civilian markets. As a result, it is in the subcontracting and supplier networks that the internationalisation of military production is proceeding most rapidly.

National Security Implications

Issues surrounding the internationalisation of military production have usually been raised in the context of their implications for national security planning (Moran, 1990). The analysis presented above implies that the trend towards internationalisation is likely to continue, so that attempts to maintain nationally-based production systems are likely to be costly, and ultimately ineffective.

This does not mean, however that a country in the position of the United States must be cast in the thankless role of a King Canute, expanding its resources to hold back an inevitable tide. The increasing internationalisation on the production side may need to be accompanied by greater internationalisation on the policy side.

One example can be found in the extent of military commitments to be undertaken. So far, the United States has altered its commitments in recognition of changed strategic realities, but has not necessarily reduced the resources needed to carry them out, setting up a significant budgetary conflict. There have also been attempts to improve multi-lateral linkages, either through a strengthened United Nations or on an *ad hoc* basis as in the military operation against Iraq in 1990-91. While multi-lateralism has many difficulties, the short run costs—both political and economic—of constructing an effective multi-lateral peace-making as well as peace-keeping operation are likely to have significant long-term payoffs for the United States, in terms of reducing defence requirements and making foreign dependence less of an issue.

A second and related example is the long-standing desire on the part of defence planners to maintain an industrial base capable of meeting long-term defence needs and short-term surge requirements. The latter is an increasingly difficult objective, given the costs and time involved in defence production, while the former implies a growing reliance upon an international production base. Defence planners may be increasingly forced to support civilian technological initiatives while adapting civilian organizational patterns in the management of defence projects.

A third example is the problem of technological dissemination and its link to weapons proliferation. The growing civilian roots of military technology has made it more difficult to control or even monitor the spread of military goods and services. In the waning days of the Cold War, NATO allies found themselves in the position of restricting exports of such relatively inexpensive off-the-shelf items as Apple computers on the grounds of potential military applications, thereby denying themselves export markets and denying useful products to countries that would benefit from their civilian applications. The combination of civilian primacy in research and internationalization of production makes it likely that proliferation via dual-use output will be an increasingly serious problem. The solution may be in a reviving and widening of the arms control process, bringing in more countries and more products, to create an international arena that has capabilities with some equivalence to the international arena on the production side.

Notes

*   The view expressed in this paper are personal and do not necessarily represent the views of the United Nations or any of its subsidiary bodies.

1.   Growth rates are calculated from the following regression equation : $\ln X = a + b\text{TIME} + u$, where X is, respectively, annual military purchases of goods and services deflated by the GNP deflator for government purchases, and real military purchases as a share of real GNP, all data taken from the national income and product accounts, (NIPA), TIME is a time trend and u is a random error term.

2.   See note 1. The dependent variables are annual inflows of FDI to the United States (from the IMF balance of payments data tape), deflated by the GDP deflator for business fixed investment, real merchandise imports, real business fixed investment and real GDP, all taken from the NIPA.

References

Adams, Gordon (1976) *The B-1 Bomber : An Analysis of Its Strategic Utility, Cost, Constituencies and Economic Impact* (New York : Council on Economic Priorities).

Biddle, Wayne (1991) *Barons of the Sky : From Early Flight to Strategic Warfare* : The Story of the America Aerospace Industry (New York : Simon and Schuster).

Bitzinger, Richard A. (1993) *The Globalisation of Arms Production : Defence Markets in Transition* (Washington, DC : Defence Budget Project).

Drifte, Reinhard (1986) *Arms Production in Japan : The Military Applications of Civilian Technology* (Boulder and London : Westview Press).

Dunning, John H. (1993) *Multinational Enterprises and the Global Economy* (Wokingham, England : Addision-Wesley Publishing).

Flamm, Kenneth (1988) *Creating the Computers Government, Industry and High Technology* (Washington, DC : Brookings Institution).

Franko-Jones, Patrice (1992) *The Brazilian Defence Industry* (Boulder and Oxford : Westview Press),.

Gansler, Jacques S. (1989) *Affording Defence* (Cambridge, Mass : The MIT Press).

Gold, David, Christopher Paine and Gail Shields (1981) *Misguided Expenditure : An Analysis of the Proposed MX Missile System* (New York : Council on Economic Priorities).

Graham, Edward and Paul R. Krugman (1991) *Foreign Direct Investment in the United States* (Washington DC : Institute for International Economics).

Ikegami-Anderson, Masako (1993) 'Japan : A Latent but Large Supplier of Dual-Use Technology,' in Wulf.

Kolodziej, Edward A. (1987) *Making and Marketing Arms : The French Experience and Its Implications for the International System* (Princeton : Princeton University Press).

Markusen, Ann and Joel Yudkin (1992) *Dismantling the Cold War Economy* (New York : Basic Books).

Moran, Theodore H. (1990) 'The Globalisation of America's Defence Industries : Managing the Threat of Foreign Dependence,' *International Security*, 15.

Mowery, David C. (ed) (1988a) *International Collaborative Ventures in US Manufacturing* (Cambridge, Mass : Balliner Publishing Company).

Mowery, David C. (1988b) 'Joint Ventures in the US Commercial Aircraft Industry' in *International Collaborative Ventures in US Manufacturing* (Cambridge, Mass : Balliner Publishing Company).

Mowery, David C. and Nathan Rosenberg (1989) *Technology and the Pursuit of Economic Growth* (Cambridge : Cambridge University Press).

United Nations Centre on Transnational Corporations (UNCTC) (1988) *Transnational Corporations in World Developments : Trends and Prospects* (New York : United Nations).

United Nations Conference on Trade and Development (UNCTAD), Programme on Transnational Corporations (1993) *World Investment Report 1993 : Transnational Corporations and Integrated International Production* (New York : United Nations).

US Congress, Office of Technology Assessment (OTA) (1990) *Arming Our Allies : Cooperation and Competition in Defence Technology OTA-ISC-449* (Washington DC : US Government Printing Office).

Wulf, Hubert, editor (1993) *Arms Industry Limited* (Stockholm and Oxford : Stockholm International Peace Research Institute and Oxford University Press).

| Chapter 12 |

Division of Labour in the Non-Soviet Warsaw Pact Arms Industry, 1945-89

Jurgen Brauer and Hubert van Tuyil

Introduction

With the conclusion of the Second World War, the Soviet Union gradually brought into its political, economic, and cultural orbit six states collectively known as the Non-Soviet Warsaw Pact (NSWP) nations, namely East Germany, Poland, Czechoslovakia (the 'northern tier'), and Hungary, Romania, and Bulgaria (the 'southern tier'). The Council for Mutual Economic Assistance (CMEA) was formed in 1945 and the Warsaw Treaty Organisation (WTO), or Warsaw Pact, was established in 1955.

In the aftermath of the Second World War, the national security, particularly the territorial integrity of the Soviet Union became a primary objective of Soviet leaders. The NSWP nations and their resources were used to assist in achieving that objective. One such resource was the provisions of machines of war—that is, the arms industry both in the Soviet Union and in those of the NSWP nations.

Eighty or ninety percent of all WTO weaponry was produced in the Soviet Union (C.D. Jones, 1986 : 144 [80 percent] ; Zukrowska, 1991 : 6, [90 percent])[1]. Much, if not most, of the remainder was licensed production in NSWP nations of weaponry of Soviet design.

Among the NSWP nations, Poland and Czechoslovakia were the prime weapons producers, while East Germany produced the least. Bulgaria, Hungary, and Romania also produced relatively little by way of weaponry.

This division of labour cannot be explained by referring to industrial capacity existing prior to 1945 alone. For example, East Germany, despite its reasonably advanced and growing industrial capacities, was a minor WTO arms producer. In contrast, Poland, despite its deteriorating economic capacity, was a major WTO arms producer. Relatively poor Romania began a sustained 'go-it-alone' arms production effort in 1969, renouncing WTO division of labour and arms production specialisation (Alexiev 1979). While pre-1945 capacity does help to explain some aspects of the division of labour, other factors must be considered ; most notably, the development of industrial capacity during the communist era of arms production, and, more important, the overt 'visible hand' of Soviet politico-military considerations that steered the arms production development of the NSWP nations.

In the context of the present volume, this chapter is a contribution from economic history. The chapter can be read as an illustration of how arms spending (here in the form of arms production directed by the Soviet Union) stifled independent economic development of the Soviet satellite states and led to reduced security within the Soviet bloc. By extension, the chapter also illustrates the twisted Soviet-era economic structures from which the nations of East-Central Europe must now convert to post-Cold War economies of peace.

The Development of the NSWP arms Industry between 1945-1989

A prime reason for the Soviet's desire to form CMEA (sometimes referred to as CEMA or Comecon) and WTO was to shore up its defence efforts. The politico-military, economic, and socio-cultural structures of Moscow's satellite states to the west and southwest were formed in the Soviet image. 'Priority was given to those industrial branches that were important for creating the foundations of defense industries' (Checinski 1987 : 15). Consequently, the economies of the six NSWP countries in Eastern Europe were designed to accommodate primarily Soviet defense needs. 'Indeed,' writes Checinski (1987 : 16), 'one could reasonably argue that defence tasks are the basis for the main areas of economic cooperation with CEMA, even though this is never explicitly stated' (see also, Brus 1986 : 244).

For several reasons, **East Germany**'s role in the arms industry's division of labour was held largely to the production of militarily relevant education and training, research and development, and support services, rather than arms production *per se*. First, historically—and exceptions such as the famous Jena optical works notwithstanding—much of Germany's arms production expertise lay in its western, not in its eastern, part. Second, Germans and Soviets had a long history of joint military education and training ventures. Third, East Germany and Soviet Union were World War II enemies, limiting Soviet desires to endow its new-found ally with military production capability, although immediately following the war—a CIA-report of 9 December 1949 (U.S. Government, CIA, 1982) suggests—part of East German industry was dismantled and shipped to the USSR, the remainder being 'employed to capacity' to make radar, aircraft and tank parts, military vehicles, anti-aircraft guns, and ammunition, apparently for the USSR.

A fourth reason for restricting East Germany's role in actual weapons production may well have been that (East) Germany was a former enemy not only of the USSR, along with Hungary and Romania, but also of Czechoslovakia and Poland. For example, Martin (1983 : 3) writes that 'units or individuals who fought either on the German side or the Allied side in World War II ... were excluded from service with the postwar East European national armies.' Indeed, the *Nationale Volksarmee* (NVA) was founded only in 1956, one year after the formation of the WTO. But whereas it was also carefully structured and equipped for a first-echelon role as a *component* of the Group of Soviet Forces Germany,' (our emphasis), Martin writes,

'Soviet units in East Germany are receiving the newest tanks, infantry combat vehicles, self-propelled artillery, and air defence missiles at a much faster rate than the NVA. It appears that although the NVA enjoys a higher priority than other non-Soviet Warsaw Pact armies and its wartime role would be more significant, it does not enjoy the same [equipment] priority as Soviet forces' do (Martin 1983 : 9).

Fifth, East Germany (and West Germany, too) was assumed to form the frontline of confrontation between the U.S. and Soviet ideological systems and it would not have been wise strategically to place crucial weapons manufacturing plants at the front of the expected war zone which, in the 1950s, still was geared more toward ground forces than toward air and missile force.[2]

The apparent decision to keep Germany out of actual weapons manufacturing, coupled with the southern tier's relative lack of arms production history and industrial prowess, left only Poland and Czechoslovakia as possible large-scale arms production partners. And it became East Germany's role in the WTO arms production scheme to specialize in military scientific-technical R&D as well as in associated education[3] and training tasks, the design and building of support systems and services, and the production of robotics, optics, computers, electronics, radars, communications gear, and so on.[4] The GDR was also an important conduit for western technology (Korbonski 1983 : *xix*).[5] The only country to compete with East Germany in high-tech arms R&D was Czechoslovakia (Wheeler Soper in Burant 1988 : 379 ; Korbonski, 1983 : xx ; Popper, October 1988 ; but also see Checinski 1981 : 13 and 1987 : 16).

Ironically, it is perhaps precisely because of being directed to concentrate on education and research and pushing the frontiers of military technology and absorbing higher-level technology through West Germany, that East Germany and the ČSSR in particular in some areas become more advanced than the Soviet Union itself (Popper, October 1988). If so, a tension develops : the more scientifically advanced the satellites become, the more the USSR wishes to put their research findings in practice elsewhere in the Soviet bloc—consequently, the higher the potential losses to the economies of the GDR and ČSSR. The dilemma for the USSR then is that, on the one hand, by extracting technology knowledge out of these two countries at increasing rates, it also affronts them increasingly. On the other hand, the more the GDR and ČSSR are permitted to retain these technologies for the advancement of their own people, the more the pressure grows to spin away from Soviet control.

While East Germany's role in the WTO was specialized and limited, it had an important 'representative' role in the Third World, supplying training, technology, and management expertise (Korbonski 1983 : *xix* : Gilberg 1983 : 83 ; Wheeler Soper in appendix to Gawdiak 1989 [same appendix as in Burant 1988)]). This contrasts with Polish and Czechoslovakian arms production which appears more specifically geared toward meeting Warsaw Pact needs. 'Hungary and Poland have confined their Third World involvement to commercial assistance,' writes Gawdiak (1989 : 338). This is corroborated by a study by Araszkiewics who writes : 'Usually, the pattern of sectoral involvement of East European states in technical assistance for developing Africa reflected the specializations of their own economies—for example,

Poland used to share its technology for creating coal-mines, chemical, power generating and fish processing plants' (1985 : 157).

The division of labour in arms production did not necessarily affect the country's own level of military equipment. East Germany (like Poland) actually received more advanced weaponry and in larger quantities in the 1970s and 1980s than the southern tier, which 'received used equipment that was being replaced in Soviet or Northern Tier forces' (Wheeler Soper in Burant 1988 : 378). He further writes :

'[The Soviet Union] has organized an efficient division of labour among the NSWP countries ... The Northern Tier countries produce some Soviet heavy weapons, including older tank, artillery, and infantry combat vehicles on license. However, the Soviet Union generally restricts its allies to the production of a relatively narrow range of military equipment, including small arms, munitions,' communications, radar, optical, and other precision instruments and various components and parts for larger Soviet-designed weapons systems' (Wheeler Soper, in Burant 1988 : 379).

The case of East Germany in particular raises the most important question of the economic cost of defense-related non-arms production. It is apparent that East Germany was designated to invest massively in education and training, in infrastructure related to military and projected battlefield needs, and to sundry support equipment and services, about which we seem to possess very little knowledge in general, let alone of its cost.[6] These costs were probably substantial, because East Germany appeared to receive the largest amount of Soviet economic subsidies of all six NSWP countries (Korbonski 1983, *passim* ; similar conclusion in Crane, May 1986).

Poland's arms industry experienced a period of industrial rebuilding (late 1940s-1950s), followed by modernization (1960s), increasing dependence on the Soviet Union for more complex weaponry (1970s), and, finally, industrial decline, if not devastation (1980s). Poland's 'industry, build from the ashes after World War II, is also the most 'Sovietised' (Checinski 1981 : v). The military-industrial connection was modeled on Soviet notions, that is, a full integration of all things industrial toward military needs (Checinski 1981 : *viii*). In stark contrast to East Germany, immediately after World War II 'ammunition factories, armories, and airplane-repair installations were reconstructed' (Checinski 1981 : 3). A two percent limitation on investment toward the military sector (1950) was shelved when the

Korean War began (Checinski 1981 : 4). Immediately, the best workers, managers, technicians, and engineers were transferred to military-production duties, leading to adjustment problems in the economy at large (Checinski 1981 : 4). Thus,

'in the early 1950s Poland undertook the manufacture of tanks, military airplanes, radar, and communications equipment. Old artillery and ammunition factories were modernized, and new ones were build. This also led to the modernisation of nonmilitary industries that operated in conjunction with the military sector—for instance the steel, rubber, electronics, and machine-building industries. The military representatives were sent to those nonmilitary installations that were collaborating with that sector' (Checinski 1981 : 5).

Along with industrial rebuilding came civilian-military co-production.

[i]n Poland, ... plants which manufacture military equipment also account for about 50 per cent of the national output of motor cycles and scooters, 80 per cent of the sewing machines, 70 per cent of the washing machines and 30 per cent of the refrigerators produced in the country' (United Nations 1962 : 30).

As production surged in the 1960s and 1970s, Poland (and Czechoslovakia) began to adapt Soviet equipment and to develop their own designs. For example,

'In 1959, the Czechoslovaks completed the initial set of plans for their transporter, which was to be built around the Tatra 928 engine. It was called SKOT [or OT-64]... The prototype was completed in 1961 and, that year, Poland expressed interest in participating in the test and development programme. Joint production was eventually undertaken, and the new vehicle was first publicly displayed in Warsaw in 1964 at the 22 July Liberation Day Parade' (Zaloga 1979 : 38).

Poland and Czechoslovakia developed their own trucked transporter and anti-aircraft artillery (Zaloga, 1979 : 40, 76). Poland came close to developing a truly comprehensive arms industry :

'Polish industry produces complete tanks, armored personnel carriers, trucks, artillery, and small arms. Components, spare parts, and subassemblies for major weapons, communication and electronic equipment, and miscellaneous items of engineering and chemical equipment are also produced within the country' (Keefe 1973 : 302).

Tank production was especially important as WTO forces relied heavily on massive armored formations. Poland and Czechoslovakia jointly built the T-72. Both countries built the T-34, Poland building an improved version until 1956 (Zaloga, 1979 : 8). Production continued with the T-54 (Zaloga, 1979 : 11, 15). 'The Poles and Czechoslovaks make no secret of the fact that they consider their locally produced [T-54] models to be distinctly superior to the Ural original' (Zaloga 1979 : 19). Mechanised artillery was also produced.

In addition, Poland produced warships and planes, but cannot be considered a leader in these areas : Its shipyards built two classes of landing ships (Jordan 1982 : 128-131), and whereas its aviation production was more substantial, 'all combat aircraft are of Soviet design ; the more modern aircraft are manufactured in the Soviet Union' (Keefe 1973 : 293). Crane (1987 : 21) refers to Poland as producing military aircraft, 'most notably helicopters,' but the only specific helicopter reference we could find is to the WSK-PZL Swidnik MI02 and MI2B, and attack helicopter of Soviet design, Polish built, first flown in 1961 (Taylor 1990). Attempts at 'further development of advanced combat aircraft [were abandoned] in 1969' (Johnson, Dean, and Alexiev, 1980).

Cooperation with allies also extended into the area of research and manufacturing process (Keefe 1973 : 301-302). In the 1960s and 1970s, Poland increasingly become dependent on the Soviet Union for 'the larger and more complex weapons and heavy equipment' (Keefe 1973 : 301), leaving Poland with the production of a wide range of equipment and items of somewhat lesser nature and technical challenge.

... East Central European countries ... did not have the needed R&D potential nor the necessary resources which would allow them to carry their own works [arms production] on a wider scale. Moreover, in the name of standardization and unification of the utilized equipment most of the countries in East Central Europe had to stop their own weapon production in the post war period. Such a solution was not only supported by economic and strategic reasoning but also by political arguments' (Zukrowska 1991 : 7).

Still, even the Soviet Union could not always impose its decisions at will. A Comecon attempt to end Polish aircraft production was decisively blocked by the Poles (Keefe 1973 : 188). This limited independence may stem from the fact that Poland was a raw materials exporter (Korbonski 1983 : *xx-xxi*) of tremendous importance to the Soviet Union.

Politics, as Zukrowska suggests, may have been the most significant constraint on NSWP arms production. Following the 1953 (East Germany), 1956 (Hungary), and 1968 (Czechoslovakia) rebellions, the Soviet Union appears to have hesitated even more in its modernization plans for NSWP armies. Polish, East German, and Czechoslovak troops themselves participated reluctantly in the 1968 invasion of Czechoslovakia, and the Soviets must have asked themselves the value of equipping their 'allies' with modern weaponry, especially since the Soviet army itself needed constant modernization funding.[7] Martin writes : 'Funds for military modernization will likely remain in short supply, and with the possible exception of indigenously produced equipment such as Czech and Polish T72s and OT64s, there is no reason to expect an influx of large quantities of new equipment into the national forces in the foreseeable future' (Martin 1983 : 10).

In sum, not only did part of Poland's arms production complex become increasingly obsolete as a result of the imposed division of labour, but its present-day conversion has been complicated—as elsewhere—by a decline in military-civilian co-production. By the 1960s and 1970s more specialized methods and equipment were used, more clearly separating military from civilian production.

> An example illustrating this problem can be differences in construction visible in the trucks used in the civil and military sphere. The cars commissioned by the MND [Ministry of National Defence] are generally of a different construction, designed to overcome ground obstacles, and are equipped with special mechanical and electronic equipment' (Wieczorek 1991 : 11 and, in a similar vein, Zukrowska 1991 : 7).

By 1989, toward the end of the communist era,

> 'there were 128 industrial, service and trade enterprises with the status of enterprise of the defence industry. Of this number, 84 enterprises were under the Ministry of Industry, 36 under the Ministry of National Defence, 3 under the Ministry of Internal Affairs, and the remaining 5 functioned as part of other departments. The average index of the share of arms (special) production in the entire production of enterprises included in the

arms industry was 38 percent. The share of arms production in the global industrial production of the country did not exceed 3 percent (Wieczorek 1991 : 4).

Czechoslovakia's post-war arms production story is one of few upheavals and drastic changes. Not even the Soviet invasion of the country in 1968 appears to have greatly influenced the Czechoslovak arms industry. Even though some data suggest a reduction in Czechoslovakia's arms production efforts after 1968, other pieces of information point to a continued and successful Czech insistence on indigenous weapons work. It appears that the events of 1968 had more impact on the *other* NSWP countries (particularly Romania), than on Czechoslovakia itself.[8]

Victory in 1945 and the Gottwald coup of 1948 gave the Soviet Union control of an arms industry with a proud history, including pre-war links with the USSR (Scott and Scott 1984 : 304).[9] After World War II, even after it 'was forced to adapt to the needs of the Soviet Union' (Zorach 1977 : 87), Czechoslovakia continued the design, development, and modification of a wide array of weaponry, supplying about one-half of its output to fellow WTO member states (Adelman and Augustine 1992 : 38). For example, Czechoslovakia produced and modified the T-34, the T-54, the T-55 (Zaloga 1979 : *passim*), and the T-72 main battle tanks of Soviet design (Gawdiak 1989 : 237 ; Albrecht and Nikutta 1989 : 258).[10] Between 1976 and 1980, Poland and Czechoslovakia together produced 800 tanks per year ; some 600 tanks in 1982 (500 T-55s, 100 T-72s), 550 in 1983, 450 in 1984, 700 in 1985, and again about 700 in 1986 (Albrecht and Nikutta 1989 : 258-259).

Czechoslovakia designed and produced the BTR-60, 'one of the rare occasions when the Russians consented to the development of an armoured vehicle outside the Soviet Union by a Warsaw Pact country' (Zaloga 1979 : 38). Czechoslovakia produced 'small-caliber weapons and various models of guns, howitzers, rocket launchers, grenade launchers, antiaircraft guns, and armored personnel carriers' (Gawdiak 1989 : 237) as well as scout cars (Burant 1988 : 251), various aircraft from trainers (Albrecht and Nikutta 1989 : 259) to MiG jet fighters (Zorach 1977 : 87).

Technical superiority of Czech weaponry has been observed in both indigenous Czech designs and modifications of Soviet-derivative armaments (for example, Wheeler Soper in Burant 1988 : 379 ; Zaloga 1979, *passim* ; Checinski 1981 : 13 ; Korbonski 1983 : *xx* ; and Popper

1988, *passim*). The technical superiority of the Czech trainer aircraft L-29 and L-39 was such that even the Soviet air forces accepted them as the standard model for the entire Warsaw Pact (Albrecht and Nikutta 1989 : 259). The Soviet Union was not always happy about NWSP technical achievements. At one point, 'Czechoslovakia's representatives introduced a modern radar apparatus that was superior to one designed by the Soviets. The Soviets, however, succeeded in winning approval for the production of their own radar equipment by playing the embittered Poles against the Czech' (Checinski 1981 : 15).

From the limited information available it would appear that Czechoslovakia's arms industry made up a major part of its industry before 1968. Keith Crane (1987 : 26) computed an estimate of 13.2 percent of military-durables production out of the entire Czech machinery output in 1967, a figure that drops comfortably below ten percent after 1968. Crane estimates a similar pattern for Czech arms exports as a percentage of total machinery exports. Before 1968, the figures are at 20 percent and above ; after 1968, the figures drop to between 10 percent and 16 percent.[11]

The post-World War II era saw a deliberate shift of arms production sites to Slovakia. Federal Ministry of Economic Development official Joseph Fucik wrote that 'in the first half of the 1950's an armament industry was built in Slovakia on the basis of political decisions' (Fucik 1992 : 9). He provides some interesting index numbers. If 1950 is set equal to 100 percent, then arms production reached an index of '198 percent in 1951 and 375 percent in 1953. The share of the industry went from 4 percent of total engineering production in 1950 to 27 percent in 1953' (Fucik 1992 : 9). Fucik (1992 : 10) reports a territorial breakdown of 60 percent Slovak and 40 percent Czech in 1987,[12]

The communist era wrought a distorting structural impact on industrial and infrastructure development in Czechoslovakia as well as in other NSWP nations (Checinski 1987 : 15-16 ; Albrecht and Nikutta 1989 : 261-262). In the case of Poland and Czechoslovakia—and also, to a more limited extent, Romania—emphasis was placed on heavy weapons and related research and development work as well as on military-related infrastructure. In addition, Soviet pressure to produce arms free of charge for some poor Third World allies (Cuba, Vietnam, Egypt) encouraged NSWP arms production. But some Third World (semi-) allies were able to pay for the weaponry—Libya, Syria, Iraq, and so on—thus also affording renumerated weapons export opportunities.

Fucik specified 1987 as the culmination of Czech arms production efforts with 'more than 100 enterprises,' in 36 of which arms production constituted more than 20 percent of total production. Altogether, some 73000 workers were employed in the manufacturing of weapons. Some 30 percent of sales were for the Czech armed forces, the other 70 percent for export. Of the exports, 85-90 percent went to other NSWP countries, the remainder into the Third World (Fucik 1992 : 10).

In sum, Czechoslovakia's larger motor vehicle industry (see Skoller and Crane 1988 : 49-52) may be the best candidate for successful post-Cold War conversion. But others, like the large post-war tank factories in Slovakia, are likely to be a burden, while the Czech Republic is facing internal political pressure to maintain its arms manufacturing and export capacity.[13]

As in the case of all other NSWP countries, civilian production in **Hungary** was subordinated to military needs (for example, Crane and Yeh 1991 : 40). Hungary's actual arms production appears limited, similar to East Germany's role. Hungary also shared East Germany's 'representative' role :

> 'cases of coordinated military sales and support services can be documented, such as the agreement between Hungary and Mozambique of November 1978, according to which the former was to provide military equipment, including tanks and aircraft (most likely on behalf of the USSR, since Hungary does not produce such equipment), as well as military advisors and instructors to train Mozambican troops in the use of the weapons' (Kanet 1984 : 5).

Checinski (1981 : 13) refers to the growing technological role of the NSWP countries, including Hungary, that induced some changes in Soviet-dominated military production coordination, apparently involving shifts in responsibilities between the WTO committees and Comecon's Military-Industrial Commission, MIC. And Korbonski (1983 : *xxi*) writes of Hungary's elevated status within the Soviet realm during the 1970s after Janos Kadar's economic reforms proved reasonably successful. That success may have provided Hungary with some measure of independence, even to the extent of being 'released' from Warsaw Pact arms production duties.

In Hungary military production absorbed perhaps one-third to one-half of one percent of GNP (Crane and Yeh 1991 : 79). Crane (1987 : 26, Table 8) estimates Hungary's production of military durables at less than one percent of 'global machinery output' between 1970 and 1979.

Under the Soviet security-umbrella, Hungary was the NSWP least-encumbered state with respect to military-production. It is ironic to note that post-Cold War Hungary is thinking seriously about constructing a somewhat independent arms industry to fend off potential threats from its former Soviet-bloc allies (Slovakia, Romania, Serbia).[14]

Romania's indigenous arms industry reflected its status as a reluctant ally of the Soviet Union (Alexiev 1979). By 1963, the Soviet Union was already annoyed at the extent of Rumanian industrialization and Romania's apparent effort to sell agricultural products and raw materials to Western countries in exchange for desired Western products (J.F. Kennedy National Security Files, 1987 : May 11, 1963). Romania was careful to build an arms industry geographically located almost entirely in originally Rumanian territory. Whereas Romania's desire and drive for independence was thus visible in policy and implementation even before the Warsaw Pact invasion of Czechoslovakia in 1968—which Romania refused to join—it became exceedingly pronounced thereafter. Romania accelerated its efforts of weapons independence, often by co-producing with western and neutral nations. It pursued independent policies and design/production strategies not necessarily approved by the WTO bodies responsible for coordination and planning.

'Romania is the only Warsaw Pact country that has escaped Soviet military-technical domination. In the late 1960s, Romania recognized the danger of depending on the Soviet Union as its sole source of military equipment and weapons. As a result, Romania initiated heavy domestic weapons production of relatively low-technology infantry weapons and has produced British transport aircraft, Chinese fast-attack boats, and French helicopters under various coproduction and licensing arrangements. Romania has also produced a fighter-bomber jointly with Yugoslavia' (Gawdiak 1989 : 355).

The latter arrangement was a disappointment to the Yugoslavs, who found Rumanian manufacture to be substandard—and deliveries late (Banzie 1989 : 275).

'Romania still remains backward in its military technology because both the Soviet Union and Western countries are reluctant to transfer their most modern weapons to it. Each side must assume that any technology given to Romania could end up in enemy hands' (Gawdiak 1989 : 355).

The decision to develop a truly independent arms industry was apparently made in April of 1968, in order to meet domestic needs and to obtain high-tech weapons via license agreements or joint ventures (Alexiev 1979 : 10). The Rumanians simply refused to work with the WTO Technical Committee (established in 1969), and their relationship with the Military Scientific-Technical Council was unclear (C.D. Jones 1986 : 143). Nor was Romania pleased with all the Soviet weaponry it received, being so disappointed with the T-54 that 'it approached several West German firms for bids to completely rework the existing vehicles ...' (Zaloga 1979 : 19). But domestic production has not been an unqualified success. 'Romania's armed forces are considered the most poorly equipped in the Warsaw Pact' (Banzie 1989 : 270).

Romania's limited military spending[15] did not limit the diversity of the arms industry's production, partly due to the 1968 decision to remove the industry from WTO limitations. Romania initiated agreements with Britain, China, and France, for the production of transport aircraft, fast-attack boats, and helicopter, respectively, as well as the joint construction of a fighter-bomber with Yugoslavia (Gawdiak 1989 : 355). This aircraft, the IAR-93, led to the development of the IAR-99 and IAR-317 (Albert and Nikutta 1989 : 258). As of 1990, it was in use only in the two countries that developed it (Taylor 1990). Romania manufactured 15 types of aircraft by 1988 (Banzie 1989 : 275) and a miscellany of equipment in virtually every category (except tanks), including assault rifles, antiaircraft guns, various vehicles including rocket trucks, and small arms. (Wiener 1981 : *passim* ; Taylor 1990).

The Rumanian export trade was substantial but not impressive when compared to its NSWP compatriots. Its 1985 estimated export total of $430 million, while making it a respectable number 13 on the world list (of 39 exporters), still left it next to last among NSWP countries (ahead only of Hungary). Romania only imported $30 million worth of arms in 1985, however (United States Government, ACDA 1988 : 30-31). In a word, the strategy set in 1968 of relying on a domestic industry to minimise arms imports seems to have worked.

Arms made up 3.53 percent of total exports, about average for NSWP countries, but only 0.29 percent of imports (United States Government, ACDA 1988 : 36-37). Arms exports peaked at an estimated $800 million (current dollars) and 6.9 percent of exports in 1982, but this figure must be regarded as an aberration (United States Government, ACDA 1980 : 148 ; ACDA 1988 : 116).

The long-term future of the Rumanian arms industry in uncertain. Between internal unrest and continuing crises to the country's southwest and northeast, it would seem unlikely that Romania would significantly reduce its military forces. But, in fact, this is what has been happening. In 1989 Romania announced a five percent reduction in defense spending (*Jane's Defence Weekly* 1989 : 207). This was followed by a 1990 announcement that the army was eliminating half of its ground-based tactical missiles, 47 percent of its tanks, 40 percent of its field artillery, 40 percent of its combat aircraft, and 20 percent of its air defence missiles (*Jane's Defence Weekly* 1990 : 187). These reductions can perhaps be explained, despite the regional unrest, by the disintegration of the country's two largest neighbors, the Soviet Union and Yugoslavia.

As is the case for the other southern tier NSWP nations, information on **Bulgaria** is scarce. Not only is the economic impact of Bulgaria's arms industry unclear, so are the reasons for its existence. Unsurprisingly, a substantial portion of its production is licensed. Even so, in the mid-1970s, Bulgarian defence spending as a percentage of GNP was greater than any other NSWP member (United States Congress 1977 : 270).[16] Another analysis of WTO defence spending (in 1980) suggests that Bulgarian defence spending was in fact quite typical of the region : Bulgaria spent 3.2 percent of its GNP on defence, compared to the GDR's 5.0 percent and Romania's 1.6 percent (Brown 1988 : 500). Only Hungary actually spent less. But Bulgaria's defence budget provided a rather mediocre home market for the Bulgarian arms industry. This was underscored by a Bulgarian announcement in 1989 that it would scrap 200 tanks, an equal number of artillery pieces, 20 aircraft, and five warships (*Jane's Defence Weekly* 1989 : 207).

The scant published information on Bulgarian arms production reveals that Bulgaria produces the ubiquitous Kalashnikov assault rifle (Lewis 1982 : 300) and trucks, tanks, and anti-aircraft guns, none of indigeneous design (Wiener 1981, *passim*). Apparently some other miscellaneous weapons were also produced, including a mine detector (Owen 1979 : 13). As for aircraft, as of 1979 Bulgaria had 'no aerospace industry as such, but its developing electronics industry is improving its capacity for the production of aerospace electronic equipment' (E.G. Jones 1979 : 16). The possibility that Bulgaria could compete internationally in this high-technology market, or even meet its domestic needs for military aircraft, must be considered remote. But Checinski (1987 : 16) notes that Bulgaria specialized on military electronics. Noting that Bulgaria also featured a relatively expansive motor vehicle industry, it is perhaps not far-fetched to suggest that Bulgaria's assigned role in specific arms production contributions to the Warsaw Pact lay in component production, such as in the areas of electronics and motor vehicles.

Despite its limited role in weapons production, Bulgaria was an arms exporter, although primarily to Third World allies, functioning as a Soviet 'proxy' in this (Gawdiak 1989 : 338). In 1985, Bulgaria ranked 12th in exports among 39 exporting countries, with exports totalling $400 million, exceeding those of Romania, Hungary, and even Yugoslavia. Arms imports, however, were even greater ($675 million) (United States Government, ACDA 1988 : 30). Arms exports constituted 3.68 percent of total exports, a figure close to the NSWP average. The peak year was 1985, when estimated exports reached $625 million (current dollars) and 5.0 percent of exports (United States Government, ACDA 1988 : 36, 93). Only in 1982 and 1984 were arms exports greater than imports (United States Government, ACDA 1988 : 93). Arms imports typically constituted three to five percent of all imports (United States Government, ACDA 1988 : 93). Imports grew sharply after 1971, while exports did not grow until 1978, with a substantial increase in 1981 (United States Government, ACDA 1980 : 125 ; United States Government, ACDA 1988 : 93).

Summary and Conclusions

We conclude with an attempt to summarise the vast NSWP arms production history with a series of 'one-liners', an explanatory sentence or two, and a concluding statement.

East Germany : the 'supportive ally'. For the most part barred from actual arms production for historical and strategic reasons, East Germany's role was to specialize in military-related research and development, education and training, and various other support services (to WTO and its Third World friends), even though it proved its productive abilities in non-military areas.

Poland : the *'miserable ally'*. Initially an important, perhaps even favored, ally, Poland received much industrial and military rebuilding and modernisation assistance up until the late 1960s. Thereafter, a variety of factors combined to leave Poland's arms production ambitions lurching and unsupported, its economy shattered, and its future in disarray.

Czechoslovakia : the *'stand-tall ally'*. Despite Soviet occupation of Czechoslovakia in 1968, the nation clung to its proud history of arms production achievements, and insisted upon and was granted arms production privileges denied other NSWP members. For strategic reasons, almost all of the armored vehicle production was located in Slovak lands, a development that both Czechs and Slovaks may find painful as they go their separate post-Cold War ways.

Hungary : the *'unperturbed ally'*. Hungary, as did East Germany, served as an important entre-pôt of Western technology into the Eastern bloc. Moreover, Hungary's economic experiments of the 1970s proved reasonably attractive and successful so that even the Soviet Union permitted Hungary to become a somewhat distanced ally, even to the point of being 'released' from WTO arms production duties (in part also because of Hungary's limited arms production background and limited strategic importance).

Romania : the *'unwilling ally'*. Displaying a fiercely nationalist independence stance, based upon a non-Slavic culture, Romania decided upon a 'go-it-alone' approach of heavy-industry industrialization with military application, especially following the events of 1968. This implied broad-based weaponry production, denying Romania specialization and economies-of-scale advantages.

Bulgaria : the *'immaterial ally'*. Bulgaria's arms production ventures were immaterial to the Warsaw Pact. The real question is : why did Bulgaria produce 'any weaponry or components at all ? We would have expected to be able to retrieve more information on Bulgaria as a Black Sea nation and therefore on dock-and harbor support services but have not noted any relevant material.

All in all, we conclude that the Soviet Union did impose on its allies a measure of arms production 'division of labour'. Precisely because of its imposition, the division did not result in mutual gains, predictable tensions among trading partners developed, and the division was not 'efficient' (Wheeler Soper in Burant 1988 : 379). To the contrary, an efficient sharing of arms production duties within the

Warsaw Pact would have left Bulgaria out of arms production altogether, except perhaps for a program of maintenance of sea-going vessels and as an arms export shipping point, would have utilized better Hungary's access to Western markets, would have directly involved East German production knowledge, would have utilized East German, Polish, and Bulgarian shipbuilding expertise, would have better allocated Polish and Czech arms manufacturing expertise (instead of often playing one against the other politically), and certainly would have required the Soviets to give up much of the control over arms production that they actually assumed. But that might have provided its 'protectorate allies' with more freedom and responsibility that the Soviet Union was willing to grant them.

Notes

We acknowledge and appreciate research assistance rendered by Ms. Yang Thang. A longer, more detailed and documented version of this paper was read at the annual meetings of the Allied Social Science Associations in Anaheim, CA, January 1993, and is available from the authors upon request.

1. The literature is unclear to just what the figures of 80 percent and 90 percent refer to. Do they refer to the value of the produced weaponry (and if so, how was that value computed), or do those figures refer to quantities (and if so, how were quantities compared, weighted, and added), or do they refer to some other measure ?

2. Of course, Czechoslovakia shared a common border with West Germany, but about 75 % of Czechoslovakian weapons production was located in the Slovakian republic, geographically sheltered by neutral Austria, rather than in the Czech republic (Adelman and Augustine 1992 : 38).

3. Actual information on this education function is minimal. But Banzie (1989 : 186) mentions the Friedrich Engels Military Academy in Dresden, 'which offers advanced training to field officers and functions as the research center for the military sciences in the GDR.'

4. For example, see Burant 1988 : 251 ; Checinski 1987 : 161; Dawisha 1988 : 84 ; C.D. Jones 1986 : 144 ; and Keefe 1973 : 301-302.

5. Another technology influx point, of course, was the economically more freewheeling Hungary, at least since the early 1970s (see Popper, August 1988).

6. Cost, incidentally, also includes environmental and health damage attributable to military-related production. For example, after the U.S. and Canada, East Germany apparently was the world's third largest uranium miner to fuel the Soviet military-nuclear industry (see Aeppel 1992).

7. The Middle East war of 1967 provided the Soviets and their East European allies with an opportunity to produce arms for hard-currency exports, particularly to Egypt, Syria, Libya, and Iraq who imported more modern Soviet equipment than the NSWP countries were permitted to introduce.

8. Note that the discussion refers to arms production, not to armed forces *per se*. Banzie, for example, writes (Banzie 1989) : 'The Czechoslovak People's Army (Czeskoslovenska Lidova Armada) was, until 1968. considered one of the better equipped and trained of the Warsaw Pact armies. Following the 1968 Soviet intervention, the Czechoslovak Army ceased to function as an effective force, as it underwent purges and reorganisation, and was reduced in size from nine to five motor-rifle divisions by 1972. On 26 August 1968 Czechoslovakia signed an agreement with the Soviet Union allowing for the stationing of Soviet forces on Czechoslovak territory, and this agreement was confirmed by the 'Status of Forces' Agreement signed in October of the same year. There are 80000 Soviet troops (Soviet Central Group of Forces) of the buffer zone between NATO countries and the Soviet Union in the Central Region' (p. 163).

    Nonetheless, '[t]he Czechoslovak army is now the second largest of the non-Soviet Warsaw pact armies, and is relatively well equipped' (p. 163).

9. The Czech 'exhibit considerable pride in their arms industry ... and it is no mere coincidence that our English words 'pistol' and 'howitzer' are of Czech origin dating back to the Hussite wars' (Zorach 1977 : 87).

10. Remarkably, we have not noticed any reference to NSWP production of the T-62 tank, Ray Bonds (1981 : 22) explicitly notes that the T-64 was *not* produced in Czechoslovakia, nor in Poland.

11. Interestingly, a telegram from Thomas L. Hughes to the Acting U.S. Secretary of State of 23 September 1963 claims that an economic crisis in Czechoslovakia was partially caused by the cancellation of a weapons export order by China (J.F. Kennedy National Security Files, 1987 : 23 Sept. 1963).

12. A figure of 75 percent for Slovakia has been suggested by Adelman and Augustine (1992 : 38).

13. For example, shortly after independene, Czech president Havel authorized—to avoid unemployment, but against his conscience—a $200 million arms export deal with Syria. The U.S. Arms Control and Disarmament Agency (ACDA) lists Czechoslovakia as the world's nineth-largest arms exporter in 1991.

14. Personal communication with Hungarian scholars.

15. Even at its peak, Rumanian military spending was modest, being last in the Soviet bloc when measured as a percentage of GNP (United States Congress 1977 : 270). This means that the Rumanian arms industry was producing in very limited quantities, or at relatively low cost.

16. The United Nations has suggested that disarmament measures could result in as much as a ten to twelve percent increase in Bulgarian investment, were it not for its military outlays (United Nations 1962 : 30-31).

References

Adelman, Kenneth and Norman Augustine (1992) 'Defence Conversion : Bulldozing the Management,' *Foreign Affairs* 71, pp. 26-47.

Aeppel, Timothy (1992) 'East's Uranium Sites Leave Costly Legacy in Germany,' *The Wall Street Journal* Tuesday, 18 August 1992, p. A13.

Albrecht, Ulrich and Randolph Nikutta (1989) *Die Sowjetische Rüstungsindustrie* (Opladen : Westdeutscher Verlag).

Alexiev, Alex (1979) *Romania and the Warsaw Pact : The Defence Policy of a Reluctant Ally* (St. Monica, CA : RAND).

Araszkiewics, Halina (1985) 'The Economic and Scientific-Technological Cooperation Between East-European and African States,' *Afrika Spectrum* 2, pp. 153-166.

Banzie, S.E. de ; et al. (compilers and translators) *The RUSI Soviet-Warsaw Pact Yearbook 1989* (Coulsdon, Surrey, UK : Jane's Defence Data).

Bonds, Ray (ed.) (1981) *Weapons of the Modern Soviet Ground Forces : Major Equipment of Today's Red Army* (New York : Arco).

Brown, J.F. (1988) *Eastern Europe and Communist Rule* (Durham, NC : Duke University Press).

Brus, W. (1986) '1966 to 1975 : Normalization and Conflict,' pp. 139-250 in M.C. Kaser (ed.) *The Economic History of Europe, 1919-1975. Vol. 3 : Institutional Change Within A Planned Economy* (New York : Clarendon Press of Oxford University Press).

Burant, Stephen (ed.) (1988) *East Germany : A Country Study* Area Handbook Series, 3rd edition. (Washington, DC : Library of Congress, Federal Research Division.)

Checinski, Michael (1987) 'Warsaw Pact/CEMA Military-Economic Trends', *Problems of Communism*, March-April, pp. 15-28.

Checinski, Michael (1981) *A Comparison of the Polish and Soviet Armaments Decisionmaking Systems* (Santa Monica, CA : RAND Corp.).

Crane, Keith (1987) *Western Leverage and Eastern European Military Spending* (St. Monica, CA : RAND).

Crane, Keith (1986) *The Soviet Economic Dilemma of Eastern Europe* (St. Monica, CA : RAND).

Crane, Keith and K.C. Yeh (1991) *Economic Reform and the Military in Poland, Hungary, and China* (St. Monica, CA : RAND).

Dawisha, Karen (1988) *Eastern Europe, Gorbachev and Reform : The Great Challenge.* (Cambridge, MA : Cambridge University Press).

Fucik, Joseph 'Defence Conversion and Armament Production in the Czech and Slovak Federal Republic,' Background paper to the NATO—C&EE Defence Conversion Seminar, Brussels, 20-22 May 1992.

Gawdiak, Ihor (ed.) (1989) *Czechoslovakia : A Country Study.* Area Handbook Series, 3rd edition (Washington, DC : Library of Congress, Federal Research Division).

Gilberg, Trond (1983) 'Eastern European Military Assistance to the Third World.' pp. 72-95 in Copper, John F. and Daniel S. Papp (eds.) *Communist Nations' Military Assistance* (Boulder, CO : Westview Press).

[Jane's Defence Weekly.] (1990) 'Major Force Cuts Planned,' *Jane's Defence Weekly* 3 February.

[Jane's Defence Weekly.] (1989) 'Two More Liaison Pact Nations Cut Arms,' *Jane's Defence Weekly* 11 February.

Johnson, A. Ross; Robert W. Dean ; and Alexander Alexiev (1980) 'East European Military Establishments : The Warsaw Pact Northern Tier,' (St. Monica, CA : RAND, December 1980 R-2417/1-AF/FF).

Jordan, John (1982) *An Illustrated Guide to the Soviet Navy* (New York : Arco).

Jones, Christopher D (1986) 'Agencies of Alliance : Multinational in Form, Bilateral in Content,' pp. 127-172 in Jeffrey Simon and Trond Gilberg, eds., *Security Implications of Nationalism in Eastern Europe* (Boulder, CO : Westview Press).

Jones, E.G. (ed.) (1979) *Guide to Science and Technology in Eastern Europe* (Harlow, Essex, UK : Longman Group, Francis Hodgson).

Kanet, Roger (1984) 'NATO-Warsaw Pact Rivalry in the Third World Arms Market,' Paper Prepared for the Center for Naval Warfare Studies, Naval War College, Newport, RI 02841.

Keefe, Eugene K. et al. (1973) *Area Handbook for Poland* (Washington, DC : GPO).

[Kennedy, J.F.] (1987) *The John F. Kennedy National Security Files : USSR and Eastern Europe—National Security Files, 1961-1963* (Frederick, MD : University Publications of America).

Korbonski, Andrzej (1983) 'Foreword,' pp. xi to xxvi in Marrese, Michael ; and Jan Vanous *Soviet Subsidization of Trade with Eastern Europe : A Soviet Perspective* (Berkeley, CA : Institute of International Studies, University of California).

Lewis, William J. (1982) *The Warsaw Pact : Arms, Doctrine, and Strategy* New York : McGraw Hill and Institute for Foreign Policy Analysis).

Martin, Richard C. (1983) 'Warsaw Pact Force Modernisation : A Closer Look,' *Paramameters* Vol. 15, No.2, pp. 3-11.

Owen, John (ed.) (1979) *Brassey's Infantry Weapons of the Warsaw Pact Armies* (London : Brassey's).

Popper, Steven W. (1988) 'Conflicts in CMEA Science and Technology Integration' (St. Monica, CA : RAND, P-7491.)

Popper, Steven W (1988) 'East European Reliance on Technology Imports from the West.' (St. Monica, CA : RAND, R-3632-USDP.)

Scott, Harriet Fast ; and William F. Scott (1984) *The Armed Forces of the USSR.* 3rd ed., rev. and updated (Boulder, CO : Westview Press).

Skoller, Deborah ; and Keith Crane (1988) 'Specialization and Cooperation Agreements Within the Motor Vehicle Industry of the Council for Mutual Economic Assistance' (St. Monica, CA : RAND, N-2575).

Taylor, Michael J. H. (ed.) (1990) *Soviet and East European Major Combat Aircraft : Including the World's Non-Aligned Nations* (London : Tri-Service Press Ltd).

United Nations (1962) *Economic and Social Consequences of Disarmament* (New York : United Nations Department of Economic and Social Affairs).

United States Congress. Joint Economic Committee. (1977) 'East European Economies Post-Helsinki : A Compendium of Papers Submitted to the Joint Economic Committee, Congress of the United States, August 25, 1977.' Washington, DC : GPO, 1977.

United States Government. Central Intelligence Agency (CIA) (1982) 'CIA Research Reports : Europe, 1946-1976. Germany.' (Frederick, MD : University Publications of America).

United States Government. Arms Control and Disarmament Agency (ACDA) (1988) *World Military Expenditures and Arms Transfers 1987* (Washington, DC : GPO).

United States Government. Arms Control and Disarmament Agency (ACDA) (1980) *World Military Expenditures and Arms Transfers 1969-1978* (Washington, DC : GPO).

Wieczorek, Pawel (1991) 'The Polish Arms Industry in the New Political and Economic Reality,' Warsaw : Polish Institute of International Affairs, Occasional Papers No. 23.

Wiener, Friedrich (1981) *The Armies of the Warsaw Pact Nations : Organization, Concept of War, Weapons, and Equipment* (Vienna : Ueberreuther).

Wheeler Soper, Karl (1988). 'Appendix C,' pp.. 376-380 in Burant, Stephen (ed.) *East Germany : A Country Study.* Area Handbook Series, 3rd edition. (Washington, DC : Library of Congress, Federal Research Division) ; also as 'Appendix C,' pp. 352-356 in Gawdiak, Ihor (ed.) (1989) *Czechoslovakia : A Country Study* Area Handbook Series, 3rd edition (Washington, DC : Library of Congress, Federal Research Division).

Zaloga, Steven J. (1979) *Modern Soviet Armor : Combat Vehicles of the USSR and Warsaw Pact Today* (Englewood Cliffs, N.J. : Prentice-Hall).

Zorach, Jonathan (1977) 'Recent Studies on Czechoslovak Arms and Armament Works.' *East Central Europe/L'Europe Du Centre-Est* Vol. 4, No. 1, pp 86-92.

Zukrowska, Katarzyna (1991) 'From Adjustments to Conversion of the Military Industry in East Central Europe,' Warsaw : Polish Institute of International Affairs, Occasional Papers No. 27, 1991.

Chapter 13

From Theory to Practice : Policies for Enabling Effective Conversion

Lloyd J. Dumas

Introduction

Psychologists tell us that transitions are among the most stressful of life processes. Even when a transition is to a better situation, there is nearly always some difficulty in readjusting. It is not difficult to understand why. Habit and routine carry us through much of day-to-day life almost automatically. Without requiring much effort or attention, many of the things that need to be done, get done. But during transition the rules change, disrupting routines and requiring attention to be focused on doing even ordinary things differently. The more dramatic the transition, the fewer things can continue to be effectively accomplished by relying on already established patterns. Eventually, new habits and routines are formed and the amount of effort and attention needed to carry out ordinary activities settles back to more normal levels. But until that happens, until we become more used to the new situation, day-to-day life literally becomes more complicated, more involving and thus more stressful.

Because they are inherently stressful, most people tend to resist most transitions almost reflexively. This is especially true when the new situation seems risky or when the benefits of the new situation are not at all clear to them as individuals. So it is not surprising that, around the world, so many people who have been deeply involved in military-oriented activities are so resistant to converting to the very different and (to them) unfamiliar world of productive civilian activity.

Again and again, managers of military-industrial enterprises insist that their companies cannot successfully convert their managements, workforces and facilities to profitable, civilian oriented activity. They do not know how to operate in that alien world. Since the 1960's, the record of attempts by large American military-industrial firms to use their military-oriented workforces and facilities to produce civilian commercial products is not encouraging. Virtually all of these attempts have been disastrous for both the companies and their customers, leading the Chair of the Board of one of America's largest military contractors to comment that the record of industrial conversion in the U.S. is 'unblemished by success'.

While one can appreciate the cleaver irony of that remark, the fact is that none of these firms ever attempted anything that deserves to be called conversion. What they did was to try to produce civilian products without doing any of the retraining, reorientation and restructuring critical to the successful transition of workforces and facilities to the very different world of productive civilian commercial activity. I know of no serious analyst of conversion who has ever suggested that military industrial enterprises had the slightest chance of becoming successful, efficient civilian producers without paying close attention to these critical transition activities. If the record of conversion of large American military contractors is "unblemished by success", that is because it has never really been tried.

Rather than playing the game of clever rhetoric or engaging in ideological debate, the time has come to get down to cases and find ways of taking conversion from theory to practice. In that spirit, this analysis focuses on how to make conversion work.

The Importance of Decentralisation

There are both technical and social psychological reasons why it is important that the planning and implementation of conversion be highly decentralised. From a technical point of view, conversion plans must be tailored to the specifics of the workforce, facilities and surrounding community. Since no one is more familiar with those specifics than local managers and workers, it is only logical that they should have primary responsibility for blueprinting their own conversion. If additional expertise or other assistance is required in either planning conversion or ultimately implementing the plan, they can bring in whatever consultants or advisors might be helpful.

The first step in effective conversion planning is always to carefully assess the capabilities of the workforce, equipment and facilities involved. The skills, talent and experience o the workers must be analyzed, along with the quantity, condition and location of each type of machinery and equipment in the facility and the character of the facility itself. With this information in hand, it is much easier to evaluate the range of civilian products that the enterprise is technically capable of producing. Having a clear picture of the capabilities of the enterprise is also key to designing efficient programs for retraining the laborforce, retooling the equipment and restructuring the organization once alternative civilian products have been selected.

It is very important to emphasize that technical feasibility is a necessary but not sufficient criterion for choosing the range of civilian products to be targeted. In the end, the only way the converted enterprise will be successful is if there is strong demand for the new products it has chosen and if they can be produced at low enough cost. Economic viability is thus broader and more encompassing than technical feasibility. It should be the ultimate criterion for choosing the civilian lines of work to which the enterprise will be converted.

From the view point of social psychology, decentralisation has the considerable advantage of simplifying the process of directly involving the workforce in conversion planning. That is not merely a nice thing to do, it will greatly increase the likelihood of success. Aside from the valuable knowledge and experience workers bring to the table, there is by now a great deal of evidence that real participation in decision making substantially improves worker motivation. People tend to be much more determined to make programs succeed if they feel strongly that they have played an important role in shaping them. Even if the final program design is different from the approach they most favored, believing that their views were taken seriously can give them an important feeling of ownership in the project. A highly motivated workforce determined to make conversion succeed is a very valuable, perhaps even critical asset.

Why is Retraining and Re-Orientation so Critical ?

Long years of training and experience in the peculiar performance-driven, cost-insensitive world of the military sector creates a highly paid laborforce with a trained incapacity to function effectively in a highly cost-sensitive civilian-commercial environment. With the possible exception of wage expectations, this is not much of a problem for most factory floor workers or lower level administrative and clerical personnel. But the skills and orientation of managers, engineers and scientists need to be substantially reshaped. Without special attention to this problem, they are unlikely to be successfully converted themselves, and will almost certainly drag the rest of the conversion process down with them. They need to learn some new skills (retraining), and just as important, they must be taught to look at what they do from the entirely different perspective of a cost-sensitive operating environment (re-orientation).

Military enterprise managers are used to operating in an essentially one-customer world. Even weapons systems that are not directly sold to the government typically require government approval to be sold abroad. One way or another, managers of this high priority sector are guaranteed both access to needed resources and profits. Critical management skills in such an environment are maintaining capability with little regard to cost, meeting the bureaucratic requirements of the government procurement system and knowing how to maneuver within the political system to secure a continuing flow of contracts and arms export licenses. This has little in common with the critical civilian commercial sector skills of minimizing cost, serving many and varied customers and navigating in the constantly changing commercial marketplace where nothing is guaranteed.

Military sector engineers and scientists are pushed hard to squeeze every possible ounce of performance out of products they must develop for use in extremely hostile operating environments. They are able to use the most exotic equipment and materials, and must specify the most exacting tolerances to meet those stringent military performance requirements. For the most part, military sector engineers are not even aware of just how dramatically overly tight tolerances and exotic materials can increase the cost of manufacturing, operating and maintaining the final product. They do not know because they have not had to know. But unless engineers, scientists and managers come to understand the cost implications of their work, they will be unable to function effectively in a civilian commercial environment.

Internal vs. External Conversion

"External" conversion involves the transition of workers, equipment and facilities that are released by a military-oriented enterprise to re-employment elsewhere in the civilian sector. "Internal" conversion involves the transition of the existing workforce, equipment and facilities of a military-oriented enterprise that is shifting to serving a civilian market. Some argue that external conversion is preferable because it is easier to get off on the right foot in a new enterprise than to completely reshape patterns within an existing company. There is something to be said for this point of view. It is more difficult to change patterns of behavior when much of the surrounding physical and sociological environment remains the same.

Which structure will be most effective for any organisation depends strongly on the characteristics of the environment within which that organisation functions and the nature of the goals the organisation has been set up to accomplish. It therefore makes sense that the structure that is best for a military enterprise will tend to be different from the structure that is best for a civilian producer. External conversion has the advantage of moving labor and capital into an organization that has already been set up for effective civilian sector operation. Any serious attempt at internal conversion must pay attention to restructuring the organization so that it is better attuned to civilian operation. Furthermore, every enterprise tends to develop its own organizational culture, and it also makes sense that the organisational cultures of military-industrial firms often fall outside the range of cultures compatible with efficient operation in the civilian commercial world. Any attempt at internal conversion must come to grips with this problem as well.

On the other hand, internal conversion has much to recommend it. It minimises disruption of the lives of workers and their families, since it aims at keeping as much of the workforce intact as is economically sensible. It minimises disruption of the surrounding community as well by maintaining the tax base and the geographic patterns of living, spending and commuting. And working within the familiar context of an existing firm and workplace has its advantages as well. There is less on-the-job adjustment, and thus less transition stress on the workforce. In general, internal conversion is likely to be a lower cost transition strategy.

One of the great technical advantages of internal conversion is that job retraining programs are much more effective if they are targeted to specific job opportunities. With internal conversion, the nature of any job the worker will be doing after conversion is known nearly as well as the nature of their present job. With this information in hand, it is much easier to design and operate a successful program of retraining and re-orientation. In fact, knowing what position the trainee will be filling is such an advantage that it must also be taken seriously in external conversion. Mechanisms need to be developed for connecting workers to future employment before retraining begins. For example, a government might sign retraining contracts with potential civilian employers. The government would pay for or heavily subsidize a program of retraining tailor-made to the needs of the employer, provided the employer agrees to hire (and retain for a specified minimum period) the trainee upon successful completion of the program.

Even if all military contractors were committed to planning for internal conversion, some external conversion would be unavoidable. The engineering intensity and management staff size common in military industry are simply insupportable in any economically viable, unsubsidised civilian firm. One striking anecdotal illustration is the B1 bomber plant in southern California. At the height of its operation, it had 14,000 workers—5,000 production workers, 5,000 engineers and 4,000 managers. What civilian manufacturer would need or could support one engineer and almost one manager per production worker? Effective conversion almost inevitable requires some paring of the workforce. External conversion would therefore be needed to reshape the engineers, scientists and managers laid off from military enterprises, and to reconnect them with civilian employment elsewhere.

Furthermore, no matter how carefully internal conversion is planned some of the plans will not work well. There are many reasons for this, some of which have more to do with the uncertainties of life in the world of business than with the quality of planning. External conversion will be required to find new civilian uses for the workers (and perhaps the facilities and equipment as well) associated with the plans that failed. Finally, if so many large military contractors continue to resist any form of internal conversion planning, there will be no socially or economically acceptable alternative but to convert their labour and capital resources externally.

The Role of the Market

Even in market economies, the military-industrial sector often bears more resemblance to a centrally planned economy than to a market-driven system. The sole or primary customer is the central government which tells the firms what they should produce to satisfy its security needs. To be sure, it is an iterative process, and the firms clearly do influence the specific form the final product takes.

Nevertheless, they are not like independent firms operating in an ordinary market. Prices are not determined by the free play of supply and demand forces, but are set through negotiation. The government itself may actually own a considerable fraction of the production facilities and equipment. Similarly, it is the central government that decides how much of which products of the military sector may be sold to other customers (primarily foreign governments), not the managements or private owners of the military firms. None of this has much to do with the operation of a market economy.

While it is likely that some significant fraction of converted military enterprises will remain in the service of the civilian activities of the central government, it is very unlikely that all, or even most, of the transferred resources of the military sector can or should be converted to that purpose. Whether in established market economies or economies in transition, the market-based private sector will be critical to successful conversion.

Having said that, it is important to understand what role the market can and cannot play effectively in the conversion process. Many economists argue that the entire conversion process can be left to the market, that there is no need for special attention to be paid to this particular problem of economic transition. Those labor and capital resources that are no longer needed in the military sector should simply be released, as they should be in any declining industry. The market will see to it that they are quickly and efficiently reabsorbed in the parts of the civilian economy that are expanding. In short, according to the ideology of free market capitalism, external conversion is automatic.

This is simply wrong. Because they have become specialised to the peculiar world of the military sector (as described earlier), much of the capital and especially the labor resources released from employment there will not be capable of functioning efficiently in civilian commercial enterprises. There is, in effect a wall between the military and civilian sectors. Without special attention to taking that wall down (by reshaping the resources), much of the labor and capital released from the military sector will either be severely underemployed or completely unemployed. Eventually, some part of the managers, engineers and scientists, in particular, will find ways of retraining and reorienting themselves, but it will be an extremely inefficient, stumbling, and long delayed process. They will go through a great deal of unnecessary personal suffering, and the society will suffer a prolonged loss of the important economic contributions they are capable of making. In the end, a generation of skilled workers may be sacrificed before the market readjusts. That is unnecessary and unacceptable.

The role the market can and should play in conversion is critically connected to the issue of economic viability raised earlier. The choice of civilian products to be produced by the converted enterprise should be market-driven. It is a common mistake for military enterprises looking for alternative civilian products to simply try to repackage the products or components they have already developed in some form they believe will perform a function that would serve some specified civilian purpose. In the first place, technologies or products developed for the military tend to be very high cost. Even if there is some direct civilian use for them, they may well be too expensive for a cost-sensitive civilian customer to be interested in buying. In other words, what they do may be useful, but it may not be useful enough to justify their cost in the eyes of potential civilian customers. Such products will not be economically viable because they will not have a market unless their cost can be drastically reduced.

Secondly, military sector technologies and products may be too complex and thus too difficult and/or expensive for ordinary civilian customers to operate and maintain. Performance-driven, lavishly funded militaries can afford to have scads of highly trained technicians working with expensive state-of-the-art equipment to constantly readjust and maintain complex systems. Most civilian customers, especially ordinary household consumers, want products that are simple to use and need no more than occasional attention and maintenance. Even if they are quite useful, there will be strong customer resistance to products that are not reliable and easy to use and maintain. For this reason too, such products will not be economically viable even though they are technically feasible for the firms to produce and can technically be applied to civilian use.

Instead of merely trying to repackage essentially military technologies and products for civilian use, converting military enterprises should evaluate their core competencies, that is, evaluate the categories of expertise and capability in which they have an advantage. For example, a military industrial enterprise may have special capabilities in microminiaturization, in precision manufacturing or in dealing with toxic or otherwise dangerous materials. The company should then consider what sorts of civilian products require or would be substantially enhanced by those capabilities. The characteristics of the market for each product on that list should then be carefully assessed, including what prices would be competitive and what other product characteristics are necessary to elicit a strong market demand. It must then carefully evaluate its ability to produce, at reasonable cost, versions of these products with the characteristics that will make them marketable, and sell them at an attractive price.

The Role of Government

The public sector clearly has an important role to play in converting the economy. The central government in particular, must be heavily involved. It is, after all, the customer of military industry as well as the locus of the military system itself. But while the coordination and assistance of the central government is very helpful, it is vital to keep in mind that conversion must be highly decentralised or it will not work well. There are things the government should not attempt to do. Blue printing the conversion of military industrial enterprises is perhaps the most important one of them.

The search for and choice of productive and profitable civilian activities for a given military serving facility and workforce must be tailor-made and guided by the market, as discussed earlier. It is best left in the hands of the organization's management and workers. So should the development of plans for reshaping capital and labor to supply these new civilian markets. Even close oversight of microlevel conversion by the federal government is a poor idea. It's unlikely to improve and the effectiveness of conversion plans and very likely to be expensive and inefficient. This is simply not a fruitful area for government intervention.

A more appropriate role for the central government is to use its leverage as customer to pressure military industry to begin planning for internal conversion without further delay. One low cost approach is to require military contractors to set up independently funded labor-management "alternative use committees" as a condition of continued eligibility for any future government contract, military or civilian. There can be no airtight guarantee that this planning will be taken seriously and done well. But with independent funding and the combination of competing and common interests of labor and management at work, the probability that this will trigger effective internal conversion can be considerable increased.

During World War II, much of American civilian industry converted to supply the military needs of the war effort. After a few years, the majority of this industrial capacity was reconverted to its accustomed civilian function. This "reconversion" was remarkably successful, with about 30 percent of the U.S. economy shifting relatively smoothly from military to civilian production within a year. The prevailing myth is that this reconversion succeeded purely as the result of the ordinary operation of market forces driven by the pent-up demand created by shortages of civilian products during the war years. In fact, beginning two years before the war ended, there was a great deal of planning by both the private and public sectors for this reconversion. All levels of government—federal, state and local—played a crucial role. They took care of education and training and planned public works projects to create productive jobs building the nation's infrastructure. But the private sector did all the microlevel company planning.

The problem today is considerably different. Over the half century of the Cold War, industrial research and production systems have been set up in a number of nations around the world whose central purpose and experience has been in military-oriented work. Generations of engineers, scientists and managers have never done anything else. They have no experience with the civilian sector. They do not face 'reconversion' (going back to something familiar); they are confronted by the need to enter an entirely new and unfamiliar world. This kind of transition is much more complicated. Nevertheless, I believe the basic division of labor that worked so well in reconverting American industry at the end of World War II is still a good general model to follow. In short, the proper role of government is to facilitate conversion, to plan major civilian projects that address the pressing needs of society in those areas that are legitimate responsibilities of government, and to help in reshaping the military sector laborforce to the brave new world of civilian activity.

Designing in Feedback Loops

The relative handful of my colleagues and I who have made conversion a central focus of academic work over the past twenty-five years or so have, by now, come to understand a great deal about the nature of conversion. While I cannot claim that our analysis of conversion from planning to implementation is comprehensive, at least the economic and engineering dimensions of the process are conceptually quite well understood. Nevertheless, there are bound to be surprises when these ideas begin to be applied.

It is a basic principle of engineering that it is not wise to go directly from the drawing board to full-scale production. (Although the U.S. Department of Defense has tried it a few times, with predictably poor results). No matter how well the principles underlying a design appear to have been thought through, no matter how carefully the design itself has been crafted, things are unlikely to work exactly as expected when it is actually built and put into operation. Consequently, it is wise to design feedback loops into the process.

Feedback loops are a concept central to the field of cybernetics, the study of control and communications systems in human beings and machines. A feedback loop returns part of the system's (or subsystem's) output as input so that any disparity between the way the system is functioning and the way it should be functioning can be detected and corrected. In other words, the system is designed so that it monitors its own performance and adjusts itself when it is not doing what it should be doing.

When conversion planning is under way and plans are beginning to be implemented, there is little doubt that some things will be easier and work better than expected, and other things will be much more of a problem than had been supposed. Building in a mechanism for detecting problems and correcting them before they get too far out of hand makes a great deal of sense. The details may need to be different in each particular case, but there are certain general principles to follow. Probably the most important is maximizing the enthusiasm and involvement of those who are both the subject of conversion and the means for carrying it out—the workforce.

By the very nature of who they are, all of the employees of a converting enterprise, from the top management to the least skilled workers are directly involved in the changes that are being made. Each of them, from his/her own vantage point within the organization, has a somewhat different angle from which to see the details of what is happening as the conversion process unfolds. And each has a different role to play in determining the success or failure of the process. To the extent that they feel that their input has been taken seriously in planning and implementing conversion, they are that much more likely to do every thing they can to ensure that the process succeeds.

Employees that are motivated and involved will pay attention when something begins to go wrong. If they have the capacity and flexibility to correct it themselves, they will. If not, they will bring it to the attention of those who can correct the problem, provided an atmosphere that encourages the reporting of developing problems has been established. A specific and deliberate effort must be made to create and maintain such an atmosphere. Although employees may well report some kinds of problems without great encouragement, it is well established that they are ordinarily very reluctant to point up problems that resulted from errors in their own judgement or decision making or that of their superiors. Yet these are exactly the kind of errors that will sabotage a complex decision making process like conversion. They must be uncovered and corrected as soon as possible. Employees, must be convinced that reporting "bad news" will not only be accepted, but rewarded.

Since real conversion has never been tried by large military contractors (and so rarely been tried by others), it would be a very good idea to set up a series of demonstration projects around the world to help debug the conversion process. These projects would play the role of prototypes for the design of conversion, a crucial intermediate stage between the drawing board and full-scale implementation. They would be treated as case studies, perhaps with special assistance given to the converting enterprise as a *quid pro quo* for opening the project to expert scrutiny and permitting the results to be widely disseminated.

Conclusions

The Cold War is over. Military budgets have declined and should continue to decline. Hundreds of thousands of talented and highly skilled military sector workers have already been laid off around thee world, and the personal economic futures of millions more are at risk. We are in danger of wasting the huge social investment that has been made in their education and training, an investment that has given them the capability to make an enormous, positive contribution to human economic wellbeing.

We are also in danger of sidetracking the process of international demilitarisation—one of the greatest potential benefits of the end of the Cold War. In the absence of well designed and implemented programs of conversion, politicians in the military producing countries may find it easier to encourage international arms sales and maintain domestic procurement levels than to face rising unemployment. There had been more than enough of that already.

It is time to stop resisting the complex and difficult transition from military to civilian-oriented activity. We must stop making excuses for our reluctance to move into the future. Continuing to complain that conversion can't possibly work no longer makes sense. The transition is here and more is coming.

These are times of truly historic significance. We are poised on the brink of a new era in international relations. What we do now will set the pattern for the next century, perhaps for the next millennium. Human beings have long dreamed of a more peaceful, less militarized world. Now we stand at the threshold of that dream. Instead of anguishing over all the reasons why the dream might fail, it is up to us to see that it does not.

Chapter 14

Conversion in Russia : Problems and Prospects

Serge V. Malakhov and Vladimir Bondarev

During the past few years, the problem of disarmament has become interesting for economists, who have neglected for year theoretical and practical issues of conversion. Fontanel (1994) notices that 'there are many economic theories, which are trying to analyse problems of defence, but they could hardly go beyond general principles.'

The crisis in Russia revealed limits of the theoretic base of economics and disarmament. The failure of conversion programs in the USSR and Russia afterwards proved the weakness of 'pure economies', which has treated the peace as an external and given factor. But that failure has been already previewed years before (Geress 1991). While economic thought payed attention to correlations between disarmament, growth, and unemployment, the problem of the military industrial complex's (MIC) behaviour under the transition to market economy attracted almost no attention. And the evolution of the Russian military industry is crucial not only for Russia itself, but for the world economy and global peace strategy. The importance of this problem is something, which sounds louder than the simplistic dilemma 'disarmament-unemployment' does.

The Russian conversion has been studied profoundly (Despress 1993, Fontanel 1993, Hummerl 1993. Menshikov 1993, Yaromeno 1992). The authors have analysed different aspects of the Russian conversion, and have proposed rather definitive conclusions. However, being separate by different economic approaches, the analysis rested very fragmented, while it needed more flexible ties between theoretical and practical tools, micro-and microanalysis economic and social phenomena.

Russian conversion ; macroeconomic aspects

The problem of the Russian conversion has attracted attention of economist and political scientists during the last six years. And it seems from an external point of view, drastic changes in the Russian economy have had almost no influence on conversion itself. It is so to a large extent. After many conversion programs have been adopted, top managers of military enterprises continue to speak about the lack of funds, and promise, once the centralised financing would be expanded, it would fill the market with consumer goods. Today, these is no well conceptually and financially developed conversion strategy in Russia.

After the first announcement of conversion in 1988, based on administrative methodology, many proposals and programs have been envisaged and launched, which have resulted in effects far from moderate. The most well known conversion program was presented on March 20, 1992 within the framework of the "Law on conversion of defence industry of Russian Federation". Viewed as a long-term program, it consisted of general principles and appeals only without any practice steps. And this law has had very limited influence on real process of conversion in Russia.

During the last few years the Russian economy has been involved in the process of transition, which has changed it drastically. The economic crisis of the last few years has deteriorated industrial mechanism and cooperation, while the inflation has reduced the capacity of the federal budget to sustain military industry.

Probably, the Russian conversion is the best case, when one has to keep in mind the Eisner (1994) recommendation to treat any macroeconomic model of disarmament with prudence. The Russian economic crisis resulted not only in a decrease in production, but in the quality of statistical data also. The quantative macroeconomic analysis of conversion seems to be unproductive. For example, while the correlation between disarmament and unemployment is theoretically and to a large extent practically significant, there is no evidence in Russia today, that this correlation really exists.

· · None the less the unemployment rate is inferior to the rate of military expenses reduction. The total decrease of labour in the Russian industry was a little more than 7 percent in 1993. Of course, if one takes into account to politics of 'temporary vacancies for employees', effectuated widely in Russia, the unemployment potential in Russia could be estimated as extremely high. But today we speak only about the weak influence of disarmament on unemployment in Russia.

The expenditures of the federal budget on conversion in 1993 was 0.16 percent of the Russian GDP, while military expenditures decreased to 4.45 percent of GDP. But the federal budget is not the only source of military and conversion financing. It is the ancient Soviet tradition to finance the MIC by different means. Today, the MIC is provided by means of the federal budget itself as well as by local budgets. And there are too many federal extrabudgetary funds, which are available for military and conversion purposes, such as the technology development fund, the fund of industrial investments, and the fund of new products and researchers. In other words, although there is no exact figures on the Russian MIC finance, one would not exaggerate to say, that it is supported on a rather broader base, than the federal budget. According to our estimations the total conversion financing, including all indirect sources, was around 2.3 percent of the 1993 GDP.

While the current situation of the Russian military industry is very difficult, it hasn't lost its possibility for self-financing. The total of its self-financing capacity was 162 billions of rubles in 1993. And the significant part of profits and depreciation funds were invested in conversion projects, such as telecommunications, and medical technology and so on.

It is very easy to find the official data, which proves the success of conversion in Russia. The part of civilian production augmented to 75 percent of the total output of military industries. More than 1500 of enterprises have completed their conversion programs to 1993. The decrease of civilian production in the military sector was only 15 percent in 1993—not much more, than total decrease in industrial output. The increase in unemployment is insignificant. But at the same time there are very strong countertendencies, which bear witness to the failure.

Among 500000 retired in military industry in 1993, 160000 retired from civilian production departments. Between those 1500 reorganized enterprises, 1000 of them were selling their production with losses. The decrease in production of certain consumer durables, such as tape recorders and TV sets, resulted in 25-55 percent in 1993. And the present share of civilian production within the MIC of 75 percent reflects not only the conversion itself, but the process of structural changes with a decreasing tendency of production.

The federal finance policy affected negatively the MIC in two ways : it hasn't only reduced financing, but also has delayed budget expenditures. At the end of 1993 the federal debt to the military industry amounted to 1.2 trillions rubles. The special purpose conversion loan of 301 billion rubles compensated only partly the lack of funds.

The preliminary analysis provides no grounds for definite conclusions. The statistical data are so inconsistent and full of contradictions, creating barriers for analysts, which could hardly be overridden in the nearest future. Nevertheless, it is possible to describe certain tendencies, many of which are of latent nature.

The general trends within the Russian MIC are common for the whole Russian industry. The .unemployment, today at least, is comparable with the whole process in Russia, and it is not elastic to decrease in military expenditures. The backward technological base cannot challenge demand constraints, while depreciation funds give too weak support for renovation. So, the decrease in production is inevitable, and the impact of backwardness to it is much more, than of conversion itself, which only accelerates the process, but not initiates it. It means, that macroeconomic analysis could hardly draw out conversion as a complete and integral process from general economic trends. At least, it has very low predictive capacity today, and it is very difficult to observe prospects for conversion in Russia by macroeconomic tools.

The industrial dimension.

The electronic industry was one of the most promising in Gorbachev and other Perestroika leaders' strategy of conversion. Both politicians and economists were sure, that industry could realize in few years its advantages of high-tech. The alternative to increased production of consumer electronics looked very attractive in comparison with cloudy prospects of other branches of the military industry. The conversion of military electronics seemed to be easy and fruitful because of low conversion costs and extremely high levels of demand on durables. This was enabled by the concentration of management. The electronics were produced traditionally within the framework of the four ministries of the Soviet Union—Minaviaprom, Minelectroprom, Minsvyazy, and Minradioprom. The share of military electronics equaled 35-40 percent of the total electronic production, but the MIC controlled, either directly or indirectly, from 8ɔ to 100 percent of the latter. Military orders on semiconducters amounted to 50-70 percent of the value of the total electronic production in 1990, which was much more than MC 8 percent in the United States.

The Soviet electronics industry is considered to be competitive, and the first conversion projects had definitive goals to meet both internal and external demand. But the following years, we discovered its inability to work competitively. After the Soviet foreign trade had been liberalised, the imports of electronics became one of the most attractive commercial investment. In two years the Soviet electronics lost a significant part of its domestic market. The price liberalisation of the beginning of 1992 almost completed that process—it discovered the archaic cost structure of the Russian electronics. The burden of the enormous fixed costs demanded an increase in sales, but the market had been already filled with imports. In 1989 alone the Soviet industry spent about 10 billion dollars on electronic imports. Next year inventories of domestically produced computers doubled. Finally, the decrease of internal demand for domestic electronics was 500 percent !

Today, the prospects of the Russian electronics industry looks very pessimistic. The equipment is old-fashioned with its mean usage term of 20 years. The production of such consumer goods as TV sets or tape recorders now is less than a half of the output before Perestroika. Only a few enterprises found their part of the market, especially in collaboration with telecommunications and aerospace industry. Military orders couldn't absorb the total production capacity of the electronic industry, which seems to be useless at a half.

The other example is the nuclear industry. It is more successful than electronics[5] in two senses - it would increase its domestic sales during last two years without significant losses on external markets, and it has launched few rather efficient conversion programs. Different case studies demonstrate the more flexible politics of production and sales of conversion projects within the nuclear industry, than in electronics. During the last two years managers of nuclear enterprises have developed a diversification approach. They invested in transportation, machinery, medical technology, and, finally, brewery. There are too many other examples of that kind to cite them. But, generally, their approach differs from conversion behavior in electronics. Rather than try on highly competitive markets, they are searching for niches in production and gaps in demand. In 1993 the nuclear industry increased their fixed assets by 91 billion roubles only within conversion programs.

Regional dimension

Nobody could deny the importance of the Siberian region for Soviet industry in general, and for the military sector, in particular. Today the share of Siberia in total Russian industrial fixed assets is more than 20 per cent with a population of 16 per cent. About 11 percent of the MIC employees work in Siberia. A half of them are scientists, engineers, and technicians. Military production amounts to 30 percent of the total machinery regional output.

At the end of 1990 the military production decreased to 46 percent of the total output, while the part of the civilian production increased to 54 percent, 20 percent of the latter were consumer goods. In 1992 the total output of the Siberian MIC decreased by 20 percent relatively to the previous year. But the part of the military production diminished more rapidly - at a rate of 70 percent in comparison with the level of 1991. But at the same time the regional MIC personnel was reduced only 5.6 percent and by 17 percent in military production itself.

The regional MIC consists of 85 enterprises, which are concentrated around large cities like Krasnoyarsk, Novosibirks. Omsk and Irkutsk, whereas 70 percent of the regional MIC personnel works in Novosibiosk, Omsk and Karasnoyarsk. The most advanced industries are machinery, radioelectronics, and aerospace. Radioelectronics has leading positions, while there are many other types of military equipment produced in the region.

Radioelectronics is leading in conversion also. About 35 percent of conversion programs in Siberia deal with different electronic and radio technologies (13 percent for transportation programs and 7 percent for chemistry).

The conversion in Siberia seems to by partially successful. During 1990-91 the output of civilian production amounted modestly to 0.9 percent and the production of consumer goods increased 2.3 percent. And the military production declined at the same time. But the change in employment rates provoke serious assumptions. During the same period the employment in civilian production increased 5.6 percent and in consumer goods production 11.5 percent respectively. It means that Siberia choose a way of labor-oriented conversion programs. It could be either within projects with a dominant social factor, or conversion projects need more unqualified labour than military production substituted by civilian ones. The second reason looks more realistic, because many local experts pay attention to the tendency of losing technologically advanced production within conversion programs.

While radioelectronics rests the most popular sector for conversion ideas, it gives almost the worst results. According to data from different regional analytical surveys, the financial positions of radioelectronic enterprises under conversion is the most dangerous, while nuclear enterprises seem to survive.

The exhaustion of fixed assets creates another barrier on the way to conversion. More than 40 percent of the equipment has been utilized more that 10 years. And in Tomask there are examples of 75 percent exhaustion of equipment (the world-by-word translation gives us a notion of depreciation, but in developed countries 75 percent rate of depreciation could prove evidence of efficient and accelerated depreciations policy, while in the case of Russian accountancy principles one could find equipment, which has been used 20 years and has 50 percent depreciation rate. So, and exhaustion explains the situation better than a depreciation.

Finally, the conversion is complicated by the giant social infrastructure of military complexes. The high level of concentration puts the social network in a dependent position from production. Today 60-80 percent of net income of military enterprises is spent on social needs. In the situation of economic crisis it is more than a burden. The inflation reduces significantly both depreciation efficiency and long-term borrowing. So, net income now is almost a unique source of development and conversion, and it should be used as more efficiently as possible. Unfortunately, the substitution of military production by civilian one in most cases doesn't increase efficiency and rates of return.

However, in spite of the many difficulties, regional authorities and local entrepreneurship are trying to accelerate the conversion process. There are few regional and local conversion programs, which try to deliberate from radioelectronic dominance, and to develop other industries, like transportation, machinery, and chemistry. As regional experts notice, Siberian radioelectronics hasn't become the conversion locomotive for the regional military complex.

Behaviour of enterprises

During the last two years the conversion process has become more decentralised. After the weakness of the State and its inability to support conversion efficiently had been realized by military enterprises, they began to look for their own ways and opportunities. The privatisation and the liberalisation of foreign trade enable that tendency. Certain enterprises choose the way of strengthening military exportation, the other plunged into civilian markets. Today three general models of behaviour can be distinguished:

— the development of military production, especially export-orient;

— the reorganisation of modes of production in favor of civilian products on the same technological base towards markets already known and compatible with technological base;

— the total conversion of production on a new technological base towards new markets

The first model, which could be described as 'development of military production,' is very common in aerospace industry, which seeks primarily opportunities abroad. The second model, which might be called 'well-known market' strategy, is trying to realise advantages of import substitution, is common for radioelectronics and electrotechnics. The third model ('unknown market' strategy) is a new one, and its seeks gaps in demand, especially internal one, in any possible market. This strategy tries to gain from the historically deficit nature of demand in the former Soviet Union, underdevelopment of ones markets and inexistense of other markets.

Although all three models or strategies are quite different, three are many common traits among them.

Firstly, all models come across with rather new effects for them, like budget constraints and demand constraints. The management almost has no skills to operate in such conditions on an open market. Financial management and marketing are quite uncommon techniques to be used widely and efficiently now. The accommodation to new rules of business takes time, which often might be enough not to survive, but to die and go bankrupt. While the privatisation makes the idea of cost of capital vital for enterprises (the concept neglected in the planned economy by definition), the transition to market means for military enterprises the draw-out from seller's markets and plunge in buyer's markets. It needs more flexibility, more knowledge, and, sometimes, more courage.

Today the adaptation goes very slowly. The exportation of aircrafts is a good example to clarify this. Prices for Russian aircrafts sometimes are inferior to world competitive prices from 40 to 50 per cent. The art of bargaining turns against Russian exporters, who topple down prices in a struggle among themselves.

As it was mentioned above, the 'well-known market' strategy dominated at the initial stage of conversion. It was supported by officials and by enterprises also. That strategy seemed easier than any other alternative with conversion costs either modest, or minimal. Today it is rather evident, that that strategy is doomed to failure. If we speak about electronics, we could observe this tendency not only in Russia, but to a large extent in the United States. Of course, direct comparison could be viewed as incorrect, but there are many common reasons for conversion failure in Russian and American electronics. The cut of federal budget support presumes the increase of weighted average cost of capital by definition, because financial markets require rate of return higher, than government financing. After that, military production traditionally is more bureaucratised with very sophisticated hierarchical organisational structure. It means, that factory overhead and administrative expenses are high. In such situations when fixed costs are higher, than competitors on a civilian market have, enterprises face new costs to be incurred—marketing costs. It means, that, at least, at the initial stage, production is not competitive.

However, it was and it is very difficult to avoid temptation to choose the simplest way of conversion. The 'well-known marked' strategy needs comparatively low costs and its benefits are quite obvious and predictable. On MC controversy, 'unknown market' strategy needs more financing, and future inflows are unpredictable. Back-of-the-envelope calculations, based on the discounted cash flow analysis could prove that idea.

But the conversion is not a bond, holding by a passive investor, to be described by a simplistic DCF (discounted cash flow) approach. The idea of a conversion itself is optional by its nature.

And it is optional just from the very beginning.

Suppose a military enterprise, which decides to convert its production or not. The decision to convert, that is, to invest into conversion program with abandonment of military production profits immediately, or to defer the conversion means whether to exercise the call option immediately or to wait. If forecasted cash inflows are rather small, an enterprise would postpone the conversion, even the DCF analysis gives the positive NPV of conversion investment. High uncertainty of conversion and time needed for it only increases the value of an option to postpone the investment.

Of course, in many cases we have no option to convert or not. The decision to convert implies more than calculations of profitability. We have to take a wide range of considerations, like world politics, institutional changes, and so on. Sometimes the conversion is not optional, and the military production should be stopped.

But, if so, we are not going beyond the option pricing model. An enterprise under conversion has the other option: to entry well-known markets on the same technological base, only substituting the military profile of output by its civilian correspondent production. Or to plunge in a market, unknown for the enterprise, on a new technological base.

If we take an enterprise which has produced military electronics and now has to convert its production. It could enter the market of civilian electronics, which, to its turn is rather volatile, but well-known to the enterprise. Usually markets of that kind require high rates of return. But the enterprise could predict more or less precisely its future share of the market. But if the enterprise plunges in the unknown market, its share could not be predicted, although the market itself requires modest rates of return. It means that in the case of unknown market future benefits for the enterprise could be more uncertain, than in the case of volatile, but well-known and predictable market. And it would be so especially in the case, when the enterprise plunge in an unknown market with a new product or with the idea to fill the niche.

Of course, a new technological base and new market need more investments and more time to succeed. It seems too inattractive within the DCF approach. But it could be the right choice, viewing through OPM (option pricing model). High uncertainty and a long payback period could be estimated as advantages.

Suppose a Russian electronic enterprise envisages two conversion programs with the same discounted forecasted cash inflows of $2 mln.

But the first program needs only $2.5 mln. of investment and only two years to be realized on a well-known market. The second program needs two times more both investments and time on an unknown market. Four years and five million are not competitive with the first program in general, but in this particular case they could be. Suppose, that future benefits from the second program will be more uncertain and more volatile. It means, that the standard deviation of returns on the second program is higher, than of returns on the first program. If we take for illustrative purposes the standard deviation of returns on the first program as 35 percent and of returns on the second program of 55 percent, the picture changes significantly (volatility of this kind is a common thing in the economy in transition like the Russian one). We could estimate the value of each strategy today. Using the Black-Scholes option pricing model, we have $0.236 mln. for the choice of the first strategy, and $0.360 mln. for the second. Being compared today, the second program looks more attractive.

Anyway, the choice between 'well-known market' strategy and 'unknown market' strategy could be described in the following manner.

	COSTS	BENEFITS
WELL-KNOWN MARKET	LOW	LOW
UNKNOWN MARKET	HIGH	HIGH

Using OPM to estimate 'peace dividends'

The conversion process is very sophisticated by its nature. One could say, that is a, rather unique process in an economic world. Its means and ends are very complex, and the mixture of economic, political, and social factors doesn't enable its analysis. One of the most important features of conversion is its behavioral pattern. Agencies and institutions, involved in conversion, are not passive investors, they have different options, and they could manage the process. The structure of means and ends of conversion is not fully tangible, and it needs more sophisticated approaches, than the DCF model.

Any conversion program, being implemented either in Russia, or in the United States, has its costs and its benefits. But these costs and benefits go far beyond direct cash outflows and inflows of a conversion project. If a conversion program is optional, that is, if an enterprise has a choice to continue military production, conversion costs should include opportunity costs of military production. If not, these opportunity costs become a sort of sunk costs and they should not be accounted. But there are other indirect costs like reduction of employment and therefore demand. Costs of this kind are not compatible with the DCF model, but they could be estimated within framework of social accounting. The OPM is also not a useless tool in this case. We could take direct costs like conversion investments as an option exercise price, while indirect costs could be estimated as an option premium.

Future benefits of conversion are also of sophisticated nature. They include not only positive cash flows from a civilian production, but employment also. Finally, it includes also irrevocability of conversion. If we speak about 'peace dividends', the price of conversion with no way back might be the best substance for this notion. And new jobs, created by conversion program, are included here, and serve also as a guarantee for irrevocability. So, discounted cash inflows of a conversion program could be seen as an asset's value in the OPM. And, finally, 'peace dividends' could be estimated directly as an option value itself.

The OPM illustrates the idea, why the 'unknown market' strategy could be more efficient:

The option W of a 'well-known market' strategy has low conversion costs CCw, while the 'unknown market' strategy U has high conversion costs CCu. The both options have the same upper bound of their values, expressed by the dashed line. The straight lines W and U express lower bounds of values of options W and U respectively. But the behaviour of options' values, that is, peace dividends' between upper and lower bounds are quite different. The behaviour of two values is expressed by W' and U' curves respectively. Because of low uncertainty, W' curve is very close to W. It means, that 'peace dividends' of the 'well-known market' strategy usually are very close to their lower bound. At the same time, high volatility of the 'unknown market" strategy presumes the intermediate position of a U' value between its upper and lower bounds.

Both options wouldn't be exercised that is, the conversion wouldn't begin, if its benefits will be less than its costs. But it doesn't mean, that 'peace dividends' will be zero in this case. The chance to be profitable gives to conversion itself positive 'peace dividends'.

We can see, that at a point A the 'unknown market' option has more 'peace dividends' than the 'well-known market' option. Really, after U' intersects W at a point B (more precisely, a little bit earlier, after U' meets W') the "well-known market" option becomes more valuable, than the 'unknown market' option. But such situation has very low probability. The W' curve becomes parallel to the W line long before point B. It means, that the exercise of this option becomes a virtual certainty before it becomes more valuable, than the 'unknown market' option. In the conversion process in means the following: if the analysis of the 'well-known market' strategy proves evidence of conversion project's profitability, whatever it would be low, the project will be implemented. Even modest results are in favor of 'well-known market' strategy, if we do not take into consideration high 'peace dividends' of uncertain 'unknown market' strategy.

Of course, it will be great exaggeration to manifest absolutely the 'unknown market' strategy. Sometimes it could be less valuable, and has less 'peace dividends'. The only purpose of consideration presented here is to expose the possible way to evaluate conversion decisions.

Conclusion

The conversion in Russia takes time and place during grandiose economic transformation on a large territory. The transition to market facilitates, on one hand, the conversion, and imposes barriers, on the other. Today conversion in Russia has more negative results, than positive. Partly it explains by specific features of the Russian economy, but to a large extent this failure has common grounds with common difficulties of conversion process all over the world. Economists should pay more attention to disarmament and should use different tools to analyse its sophisticated nature. The future will require more social accounting technics for conversion analysis. But even today the use of the Option Pricing Model could enable the evaluation of conversion alternatives.

Chapter 15

The Role of Local Plants in Conversion Strategy in the Arms Industry: the Case of Brest, France

Roland de Pennaros, Thierry Sauvin, and Marie Noelle
Le Nouail

The arms industry is today confronted by a changed environment characterised by economic uncertainty, shrinking markets, increased international competition and shifts in the nature of military requirements.

Because of this state of affairs, companies now attach increasing importance to the definition of new strategies. The solution may lie in diversification or even in conversion of activities.[1] However, apart from the fact that this strategy may not necessarily be desired, it is an option which is not open to all companies. Given their history[2] they have path constraints which limit their choice of strategies.

Beyond the various trajectories taken by industrial groups, it would seem that they have a common objective: the preservation of strategic coherence, a coherence underlined by the tendency to refocus on the trade. These two concepts, coherence and trade are central to our analysis.

Coherence implies that there exists a substantial degree of proximity between the activities of the firm. 'A firm may be considered coherent when its activities are linked up, in the sense that they have certain common characteristics' (Dozi, Teece, Winter 1990). And the trade may be defined not by the firm's main activity, but by its ability to combine it various fields of know-how so as to be able to supply various demands and to incite a need for them (Batsch 1993). In other words, a firm is organised around the trade, that is around 'a combination of competence needed to satisfy a number or complex and differentiated requirements' (de Montmorillon, 1986). These competences are possessed by the employees and may only be developed provided that the firm is able to preserve them and even to renew them. In this respect, maintenance of the workforce and the qualitative development of its skills may be considered as a strategic necessity for firms in the arms sector.

We define a coherent strategy as a social construction involving the different entities of the productive system, notably a top management (head office, or even the parent company), of local plants and communities. In this way, coherence can never be apprehended uniquely at the level of the top management. It is also present at two other levels : on the one hand in relations between top management and local plant, and on the other hand, in the links between the plant and local economic agents (firms, employees, unions, University, local authorities . . . .). These three levels coordinate, and do not merely run alongside each other. The socio-economic coherence of the whole depends on the dialectic which establishes itself between the top management, the plant and local economic agents.

Because of its territorial implantation, the plant, the executive agent which carries out the instructions of the industrial group, leaves its mark on the local socio-economic environment which, in return, may suggest it counter proposals to put to the group. The plant, therefore, presents itself as being the best place for the confrontation of two logics : industrial profitability and regional development (cf Fig 15.1).

A mere executive agent or a power for proposal, what is the role of the plant in drawing up conversion strategy for arms industry groups ? To what extent can it become an active partner in regional development policy ?

In an attempt to answer these questions we shall consider a case study : that of the arms sector industrial plants in the region of Brest.

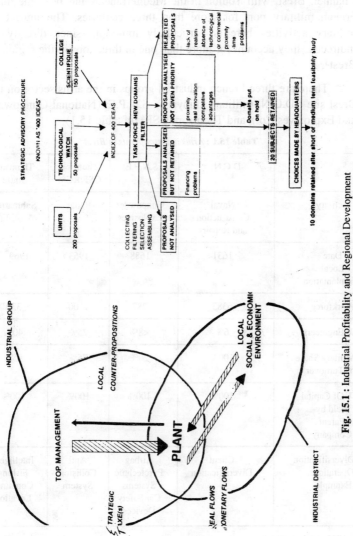

Fig. 15.1 : Industrial Profitability and Regional Development

Situated at the western tip of Europe, at the mouth of the English Channel, Brest, with Toulon in the Mediterranean, has been the main French military port for more than three centuries. The impact of military activities here is particularly important since directly or indirectly, they account for more than one in three jobs in the region of Brest[3].

There are three French industrial groups in the arms sector on the Brest site : D.C.N. (military shipyard), S.N.P.E. (National Gunpowder and Explosives Co.), and Thomson -CSF[4] (cf table 15.1)

Table 15.1 : *Arms industry plants in Brest*

	D.C.N.	S.N.P.E.	Thomson CSF/RCM	Thomson Sintra
Activity	Naval Construction and Repairs	Gunpowder Explosives	Radar	Submarine Acoustics
Date Local Implantation	1631	1688	1963	1969[**]
Workforce[*]	6383	287	1100	330
% Engineers	6%	<8%	25%	40%
Military Share of Turnover	100%	20%	75%	95%
% Of Capital Held by Parent Company	100%	100%	100%	100%
Diversification Operations Examples	Client Diversification	Airbag Pyrotechnic Systems Car Safety Devices	Anti Collision System	Intelligent Fishing Container Location

[*] 1993, 1st January

[**] submarine action division of Thomson-CSF/RCM until 1985.

With a production value of about 3.5 billion French francs and more than 6000 employees, the D.C.N plant in Brest is by far the largest local industry. It is essentially a production centre, working almost exclusively for the French Navy. Specialized in the construction of large surface-vessels (nuclear aircraft carrier 'Charles-de-Gaulle'), the repair and maintenance of the F.O.S.T nuclear submarines of the F.O.S.T. (oceanic strategic force) based in the bay of Brest, D.C.N. Brest is, by its 7 million production hours per year and workforce, the largest of the eight industrial D.C.N. centres.

The Pont-de-Buis gunpowder factory, one of the 9 S.N.P.E. plants, is much smaller (fewer than 200 employees and a turnover of 140 million French francs) and its link with Defence activities is now much reduced. The military activity of the plant (military gunpowders and pyrotechnics, products for the maintenance of law and order) barely represents more than 20 percent of its turnover. The production of warfare explosives has been overtaken by the production of gunpower for hunting.

Thomson-CSF, and with it Defence electronics, are represented by two plants. The first, specialized in the construction of radars and other air navigation systems comes under the direct control of the group's parent company. Attached to its Radar and Counter Measures (R.C.M.) division, it employs 1100 people. The second, specialized in the manufacture of sonars for warships is linked to a 100 percent affiliated company of the group : Thomson-Sintra A.S.M. it employs 330 people and has a turnover of 280 million French francs.

These plants, which play an important role in local economy are undergoing the effects of the crisis currently hitting the industrial groups from which they depend (cf Table 15.2). For these reasons they are particularly exposed to the confrontation of the two logics, industrial and regional, which we have already mentioned. The study of their role in drawing up group strategy is thus of even greater importance.

Table 15.2 : *Employees in Arms Industries in Brest, France*

Groups and their plants in Brest	Number of employees			Evolution 93/83 (82)
	1st January			
	1982	1983	1993	
Direction des Constructions Navales	31251		25942	—17%
D.C.N-Brest	6884		6363	—7.6%
Societe Nationale des Poudres et Explosifs		6843	4961	—27.5%
S.N.P.E. Pont-de-Buis		516	287	—44.4%
Thomson-CSF		64200[**]	42350	-34%
Brest Plants[*]		2096	1430	-31.8%

*one plant in 1983 : Centre Electronique Brestois (CEB).

two plants in 1993 : the submarine action division of the CEB became in 1985 an industrial establishment of a new affiliated company of the group-Thomson-Sintra.

**in 1984

Sources : D.C.N.-Boucheron report 1985 and "Info DGA", octobre 93.

S.N.P.E.-annual reports

Thomson-CSF-annual reports, C.E.B. statement of accounts, local press.

The study is structured round three points.

The first will underline the importance of the strategic power wielded by top management and exercised on local plants. The second will examine the capacity of local plants to reorientate group strategy, that is move from the role of executive agent to that of strategy maker. In the final point we will consider the impact which the local environment may have on the coherence of the plants and the industrial groups. The question of the insertion of the plant within the regional socio-economic fabric is posed here. Is it a means to increase the negotiating power of the plant and, thus, to further reorientate the orders given by the central office or parent company ?

The top management and its strategic power

An industrial group is generally defined as a group of firms subject to the control of a parent company. In other words, the firms belonging to a group submit to a dominating effect and thus lose a large part of their autonomy of decision. Therefore they are forced to integrate their action within a much larger group with its own specific logic to be respected (Morvan, p. 279, 1991). Not all the firms in the armaments sector form groups in the strict sense of the word and not all local plants have a legal personality. Nevertheless, given their size and the specialization of the plants, it is possible to establish a relative assimilation : the question deals with a group of economic units subject to one decision making unit assumed by a headquarters (top management), and to a controlling power.

Three characteristics of the group deserve our attention.

A group is an integrated number of firms. This characteristic poses the problem of belonging to an industrial group that is the place of each entity within the group's system of management (structure, information and decision system).

The group is subject to the strategic control of a top management. The central power to a certain extent makes it possible to overcome the heterogeneousness of the activities of the various entities within the group. The top management has a homogenising function. However, respect for the heterogeneous is necessary to the definition of new strategies.

The group is a structure which guarantees a certain unity. The top management supervises the coordination of action between firms which may be engaged in very different activities. The French industrial groups in the armaments sector represented in the region conform all the more to these characteristics since the top management in question (central offices, parent companies) has complete control over their affiliated companies and plants of Table 15.1.

The local plants, therefore, find themselves in a situation of financial and strategic dependence. Officially, their autonomy of decision may only result from a decentralization of power decided on by the top management. For strategic reasons (seeking economies of scale and effects of organisational synergy) as well as for cultural reasons (tradition of hierarchical organization linked to an identification relative to the military 'nature' of their customers), these industrial groups have centralized strategic decision-making procedures. It is always up to the top management to make the definitive choice of strategic projects to be developed in accordance with its own objectives (refocussing on the trade, seeking short term profitability). Local plants may nonetheless be called upon during the strategy defining process. This happens all the more frequently when research and development structures are decentralized.

In this respect the case of the Thomson-CSF group is interesting. The reflection focussing on the possibilities of diversification towards the civilian sector was given concrete expression in 1991 with the setting up of 'New Domains Task-Force' (TF-DN), a structure placed under the immediate supervision of the chairman and given the task of reflecting on new domains of activity. The TF-DN set up a consultation procedure concerning all the units of the Thomson-CSF group, technological watch centres, and the 'College Scientifique et Technique'[5].

This procedure, called '400 ideas' (of aneex) was made necessary because of the structure of the group. Indeed, technology, which is the cultural base of this firm, is managed at a local level, within departments which function autonomously (existence of an trading account). Technological coherence is the responsibility of the Techniques and research management unit which organises the six main poles of professional electronics competences[6]. At the organisational level, the projects are followed up by a Project Manager belonging to the operational structure. The projects are financed half by the industrial group, a third by local self-financing and the rest coming from outside resources.

Thanks to this decentralised research structure the local plants of the Thomson CSF group thus have the opportunity and the competences required to make proposals for strategy and to develop them.

This is hardly the case of the other two groups where there is no such decentralisation : for the main part, research is concentrated in the Paris region[7] close to the head office. On this subject, the recent creation of a diversification cell in the DCN Brest should not delude people : without the scientific, technological and human resources necessary to accomplish the task it has been assigned, it will hardly get beyond the stage of a simple suggestion box. Moreover, it should be noted that R&D has not been much developed in the DCN as a whole[8].

Local plants - do they have a role to play in strategy ?

The analysis of the local situation shows that some plants may be a real bargaining force within groups. They appear to be able to integrate strategy projects of local coherence into the global strategy of the group. From a number of achievements, let us take the following examples : Intelligent fishing at Thomson Sintra, the Thomson RCM anti-collision system, and the Airbag and safety belt pre-tension device[9] at the SNPE Point-de-Buis.

To understand this mechanism, we should return to the definition of strategy which can be analysed as a number of instructions, of rules defined by the governing body with the aim of achieving the objectives assigned to the organization : this is **control regulation.** At the same time and for the same subjects, local actors design their own strategies conforming to their own interest : this is **autonomous regulation.** 'The hierarchical relationship is not the only one where autonomy and control are in confrontation. Functional relations do the same (. . . .). The control-autonomy relationship is therefore general and very diversified. It arises every time an individual or a group has the capacity to intervene in the running, the organization and the activity of another group (. . .). Thus the opposition between autonomy and control is indeed the opposition of two collective strategies' [J.D. Reynaud, 1993].

It is thus appropriate to study the conditions of the emergence of this autonomous regulation, then to analyse its consequences on the strategic coherence of industrial groups.

Autonomous regulation emerges to the extent that local actors have the necessary strategic resources. Indeed, the existence of a social group capable of a will for action within an organization is not something to be taken for granted. A strategy group is a group with opportunities to seize and able to develop offensive behavior [Crozier, Friedbers, 1977]. For this, the group must have a relative degree of autonomy officially granted by the organization or obtained by its action. Gaining autonomy means gaining access to mastering the main sources of uncertainty within organization :

— those concerning specific competences, particularly technological competences. The possession of a competence or a functional specialisation which the organization would find difficult to replace increases the degree of autonomy of the possessor.

— those concerning the relationships between the firm and its environment. This is referred to as the power of the 'marginal secant', that is of an actor who is a party to several interlocking systems of action and who can act as an interface between different, even contradictory logics of actions.

— those linked to mastering communication and information. The possession of information affects the capacity for action and within the organization becomes the object of negotiation and bargaining.

— those resulting from the existence of general organizational rules. Rules initially intended to reduce sources of environmental uncertainly create further uncertainties. Their respect is not guaranteed but gives rise to more or less formal negotiation between the superior officer and subordinates.

The objective existence of a zone of uncertainty is a necessary condition to autonomy, though not sufficient in itself - the actors also need to be able to exploit it.

Hence, not all social groups are on equal footing within the organization. Only those with expertise and specific competence which cannot easily be replaced may become strategic actors.

Finally it should be noted that the logic of the controlling actor (top management) cannot amount to a logic of efficiency (of techniques, costs, turned to the outside) that would be opposed in any way to a logic of sentiments off the people being controlled (turned to the interior of the organization and social satisfaction). 'These two strategies are not defined only by the particular interests or stakes or specific rationality of either party. They include one common goal (the perpetuation of the organization), they admit of outside validation which can separate them and temporarily arbitrate their conflict' [JD Reynaud, 1993].

How can the local situation be interpreted using this analysis grid ?

Out of the three industrial groups from the arms sector implanted in the region (DCN, SNPE, Thomson-CSF), Thomson-CSF alone appears to present the conditions of the emergence of a 'local strategy'[10].

Despite having been implanted in the region for such a long time, the SNPE and DCN's local plants have little strategic autonomy.

Both firms have hierarchical structures, almost modelled on the military organization. The level of workers' qualifications is rather low (6 percent engineers at the DCN, less than 8 percent at the SNPE). Research and Development is practically non-existent on these sites which are essentially production units.

The particulary protective nature of the the statute[11] given to the majority of the DCN work force is probably a further hindrance to the development of diversification strategies. Indeed, the statute of State worker is quite justifiable in that the main task, and often the only one, assigned to military shipyards has always been to supply the State with military materials. To that extent, the shift towards civilian markets which weakens the privileged relationship with the State is felt to be*** dangerous by many of the employees.

Conversely, the Brest-based plants of the Thomson-CSF group seems to enjoy other assets:

— the workforce includes a high number of engineers. The global coherence of the group is to be found in its technological culture. Emphasis is placed on Research and Development which is decentralized at unit level.

— affilited companies and plants are specialized in one or more technological competences (submarine acoustics at Thomson Sintra, Radar at Thomson RCM).

— the plants are not merely manufacturing sites, they are also design and marketing centres

Thomson-CSF is a group of 42000 persons and such meets the same problems as any company of this size, problems of coordination, of information transmission. The procedure for the supervision of activity in local units leaves a certain degree of freedom to allow for the development of local strategy projects.

It is interesting to note the importance of the CGT trade union within the Thomson sintra plant in Brest. It aims to promote diversification towards civilian activities. There are two reasons for this: the first being ideological (defending the civilian sector to the detriment of the military), and the second social (defending local jobs). Thus, if the the strategy of the Thomson-CSF group aims at making it the 'Electronics Champion' on the world arms market, local counter-proposals aim instead at exploiting generic technological which make up he firm's partrimony of competences on various civilian markets (Fishing, transport....).

The union is a practical powerful strategic.group. Its power lies in the control of internal and external information and in a network of relationships within local firms liable to become corporate clients or technological partners.

These local counter-proposal cannot be considered as organizational dysfunctions.

First it should be underlined that there exists no real dissociation between top management and the plants, since senior executives have all had to spend some time in operational units. This means that there is a common repertoire of reference available to ease communication between actors.

Local plants, real authors of strategy, thus have their role to play in reflection and experimentation with the added advantage of reduced risks (reduced resources and markets). A local success could lead to the institutionalisation of the local project that is its appropriation by the parent company would mean increased resources which in turn would permit it to pass from an experimental phase to an industrialisation phase.

Thus the integration of local projects could contibuters to the emergence of a new cohernce at the level of the industrial group. Indeed, the project developed on a local scale base their coherence vis a vis a particular trade, ensuring the cohesion of the industrial group, but also vis a vis local constraints and opportunities.

The local environment as strategic regulator

It is possible to detect a real community of interest between the local plant and local communities. Indeed, in a harshly competitive context, the latter are trying to increase the attraction of their territory so as to lure national and foreign investment and, in so doing, find solution to macro-economic preoccupations (job creation for example).

For this to succeed, there needs to be a supply of production factors in favour of quality rather then quantity (skilled manpower, infrastructures, communications systems...) which guarantee the 'efficiency of the firms' environment. However, this supply of factors alone will not suffice. Returning to one of the main characteristics of the Marshallian district [Marshall, 1919, 1920], it is necessary to have a very structuring dominant activity [Gaffard, Roiman, 1990]. Following this analysis, one might be tempted to say that the arms industry is, in the region of Brest, the structuring dominant activity. However, it is our opinion that what could be called structuring is not the activity but the technological competences inherent to the arms industry. The existence of an industrial district which 'results from a process of collective construction of a productive system whose efficiency depends on the variety and complementarity of the human activities and competences united at a given place' [Lecoq, p. 212, 1993] would make it possible to preserve and develop technological competences. In fact it would have a dual function. The first would be to attract but also to fix production factors (labor, capital) which are, in today's climate of economic uncertainty, increasingly mobile. Recognition, as much on an individual level as on a collective level, on an industrial would limit loss of competences. Competences would be tapped, directed towards other entities belonging to the district, towards structuring activities classed as ancilliary (maintenance, sub-contracting, logistics, training).

The second function attributed to the district would consist in creating loops of positive retrospective effects between entities, interactive trading relations (commercial transactions) and non-trading relations (cooperation agreements concerning competences, production and marketing) leading to an evolutive process in terms of technological competences. The development of cooperative realations represents, we believe, a means to exploit potential interdependence and thus increase the 'fundamental competences of the firms[12]. On this subject B. Lecoq highlights the endogenous nature of technology in the sense that it is the product of a specific context [Lecoq, p. 201, 1993]. The exploitation of interdependence would thus create externalities to the advantage of the different actors in the industrial district. In order words, they would make conglomerate external economies which in fact are the 'free services which closely related firms render to each other through their actions on their environment: fight against transaction costs, economies of scale, manpower training, circulation of innovation...' [Courlet, Pecqueur, Soulage, p. 9, 1993]. Consequently local plants would strengthen their competitive 'linked to the network of interdependence which would go beyond the normal limits of the firm' [Prter, 1993]. Furthermore, the plant as a part of the industrial district implies active involvement in the definition of new organizations, manufacturing procedures and products (military and civilian). Both structuring and structured by the district, the plant would enjoy a certain degree of freedom vis a vis top management. The latter would have to respect the socio-economic specificities of the plant in the definition of its own strategy.

At a local level, the industrial district is organizing around the metallurgical and mechanical engineering industries and naval ship building and repair. This productive system whose inside cooperation networks are largely due to the perpetuation of socio-economic tradition [Torre, 1990] contains a large number of small and medium size firms with really more than 200 employees.

The existence of this network of firms certainly a major condition for the efficiency of an industrial district, but even so it is necessary that inter-firm relations do not become too lop-sided and focussed on the determining actor (the local DCN establishment). In this respect, it is vital, in Brest, to break with the dominant/dominated relation which has become inherent to contracting in local markets for enginering and supplies, between the DCN and a number of local firms. The shift to industrial and technological cooperation, i.e. to a realtionship with grater balance and active participation, is necessary. Economic agents would thus become partners in the development of, for example, new civilian producers and products. Thanks to positive interaction, a source of conglomerate external economies, the industrial district would be 'visible,' identifiable for the actors [Johanisson N, p. 133, 1994]. Consequently, they, and more notably the DCN, would more systematically apply their technological competences to the service of large-scale civil engineering projects initiated by local project managers.

In fact the success of a conversion programme involves the joint intervention of the various actors (local communities, State, firms, unions...) but also th definition of new modes of coordination of activities and competences to create and exploit interdependence between sectors, technologies and firms.

Apart from a cross fertilisation between industries and technologies, this line of action would make it possible to attain a higher level of socio-economic coherence between headquarters, the plant and the district.

In Brest, such an approach to the problem of conversion for arms industries would make it possible to create an industrial district and thus to guarantee, for the common good, reconciliation of two imperatives which are today all often opposed: industrial profitability and regional development.

Notes

1. Diversification increases market heterogeneousness (in terms of geography and/or products). It does not necessarily decrease the military aspect of the firm activity. In this way it can be differentiated from conversion which involves a progressive substitution civilian activities for military activities (FONTANEL, 1994).

2. The history and the growth of these firms ar largely based on privileged links with the State (protected markets, subsidies....) (CHESNAIS, SERFATI, 1992).

3. The total of direct (Naval forces, industrial jobs) and indirect jobs (in intermediate consumption and household services) reached more than 40,000 out of the 127000 jobs registered in Brest's employment area in, 1990 (ECONOMIC, BRESTOISE, 1994).

4. The three groups are controlled by the State:

    — the D.C.N. is a state-run plant without pecific legal personality of its own. It is run according to the administrative rules of the Ministry of Defence. Its industrial work force has a specific statute, very similar to that of civil servants.

    — the S.N.P.E. is a public corporation, the State possessing more than 95 percent of the capital shares.

    — Since its main shareholder - Thomson - SA - was nationalized in 1982, the Thomson-CSF group had also under State control.

5. Internal council made up about 150 experts elected by their peers and organised by speciality. These experts have an advisory and explicative role. Nevertheless they remain in the hierarchy and contibute to the construction and maintenance of scientific and technolgical coherence within th group.

6. Aeronautical equipment, communication commands, detection system, missiles systems, services and computing, other activities.

7. The Paris region and the Toulon region for the D.C.N.

8. Research and Development expenditure represents 7 percent of the production value (BOUCHERON, 1985) compared with 23,7 percent of turnover for Thomson-CSF.

9. This is a shock absorber which tightens the belt before the body moves.

10. Strategy set up by local actors, within affiliated companies or plants.

11. 75 percent of the DCN workforce in 1993 was registered as a State worker which, apart from other advantages, ensured them of job security comparable with that enjoyed by civil servants. The last workers at the S.N.P.E. Pont-de-Buis to have acquired a similar statute before the 'Service des Poudres' became a national company in 1971 were regarded at the DCN-Brest in autumn 93.

12. "Fundamental competences comprise the various technological competences, complementary assets and routines which make up the basis of the competitive capacities of a firm in particular activity" [DOCI, TECE, WINTR, P246, 1990]

References

Batsch. L (1993) 'Influence des structures productives et percentage sur le métir,' *Economices et Sociétés*, 19.

Boucheron, JM (1985) 'Rapport due les établissements industrelds du ministe de la Défense,' *Rapport 2755*, Assemblée Nationale, France.

Chesnais, F and C. Serfati (1992) *L'armement en France: genése, ampleur et coût d'une industrie* (Paris: Nathan)

Courlet C, B. Pecqueur and B. Soulage (1993) 'Industire et dynamiques de territories,' *Revue d' Economic Industrielle*, 64.

Crozier, M. and E. Friedberg (1977) *L' Acteue et le Systéme*.

Dozi, G, D. Teece, and S. Winter (1990) 'Les Frontiérs des enterprises: vers une théorie de la cohéerence de la Grande Enterprise,' *Revue d' Economic Industrielle*, 51.

Economic Brestoise (1993) (France: Centre d' Economic Sociale, Université de Bretagne Occidentale).

Fontanel, J. (1994) 'La conversion civile de l'Armement,' *Chaiers de 1' Espace Europe*, 5.

Gaffard, JL and PM Romani (1990) 'A Propos de la localisation des activitiés industrieelles: le district marshallien,' *Revue Francaise d' Economic*, V, 3.

Johanison, B. (1994) 'Building a Global Strategy, Internationalising Small Firms through Local Net Working,' *Small Business and Its Contribution to Regional and International Development*, Proceedings of the 39th ICSD conference.

Lecocq, B. (1993) 'Dynamique Industrielle, Histoire et Localisation: Alfred Marshail Revisité,' *Revue Francaise d' Economic*.

Marshall, A. (1919) *Industry and Trade* (London : Micmillian).

Marshall, A. (1920) *Principles of Economics* Chapter 5 (London: Macmillan).

Morvan, Y. (1991) 'Foundaments d' Economic Industrille,' *Economica*.

Porter, ME (1993) 'L'avatage concurrntiel des nations,' *Inter Editions*.

Reynaud, JD (1993) *Les Régles du Jeu* (Paris: Armand Colin).

Torre, A (1990) 'Interactions techniques et interdépendendances hors-marché: Quelques Réflexions,' *Revue Francaise d' Economie*, 3.

Chapter 16

Crisis in the Armaments Industry and Conversion Problems in the European Union*

Keith Hartley

1. Introduction

Defence industries in Europe, the USA and in the former Warsaw Pact nations are facing a major crisis. Disarmament means reduced spending on defence equipment, as nations seek a Peace Dividend, thereby forcing their defence industries to adjust to change (UN, 1993).

This paper considers some of the theoretical, empirical and policy aspects of the crisis in Europe's defence industries. It presents a microeconomic framework for analysing the adjustment behaviour of defence firms ; this framework is then tested for a sample of UK defence contractors and some of the general implications for conversion are identified. The adjustment process caused by disarmament has major impacts on labour markets raising questions about market failure and the opportunities for state intervention. Prospects for the future of Europe's defence industries also need to be addressed and consideration is given to the possible extension of the Single European Market to defence equipment. The conclusion presents an agenda for further research.

2.   The Analytical Framework : The Economics of Defence Markets

The defence market has a number of characteristics which make it a special market. It is dominated by a single customer, the defence ministry, which determines the domestic market and controls the export market. Typically, decision about defence procurement are taken at an early stage in the development of equipment when the full technical and economic implications are uncertain. Many defence projects are on a large scale, and few companies would be able and willing to assume all the risks entailed in such projects on a private venture commercial basis. Any assessment of the response of companies to defence cuts has to take into account the special features of the defence market.

As a single buyer of monopsonist, **the defence ministry's procurement choices can have a major impact on the domestic defence industry.** Its purchasing decisions can determine :—

*   **technical progress** both directly through its operational requirements for new equipment any, indirectly, through any technical 'spin-off' which results from expenditure on defence research and development undertaken in the domestic economy

*   **the size of an industry**

*   **industry structure** (the number and size of companies)

*   **entry and exit** (for example whether any competition is restricted to domestic companies 'or open to companies from the rest of the world)

*   **location and ownership** (privatisation or nationalisation)

*   **industry performance** as reflected in prices, efficiency, exports and profitability, with profits determined by competitive purchasing or via regulation.

On this basis, the defence ministry can determine the size, structure, conduct and performance of defence industries. Future prospects for the 1990s suggest a reduced threat leading to lower NATO defence spending which is likely to be associated with further restructuring in the European defence markets as well as exits from defence industries. **Such a market environment will force defence contractors to re-assess their likely future income prospects.** They are faced with a much more uncertain future which will require difficult commercial judgements about the potential profitability of defence and civil markets.

2.1 The response of companies

Although the future is unknown and unknowable, it is likely to be characterised by continuing cuts in defence budgets, leading to cancellations of new projects and orders, fewer new projects, shorter production runs and reduced requirements for spares. Nonetheless, there are likely to be some new growth markets resulting from a shift towards defensive rather than offensive forces and equipment. Examples include a greater emphasis on surveillance and early warning systems, rapid reinforcement capabilities, helicopters, fighter aircraft and defensive missiles. A shift towards contractorisation of services traditionally undertaken 'in-house' by the Armed Forces will provide new market opportunities for defence industries. And the arms control process itself will create new markets associated with inspection, verification and the disposal of surplus equipment (for example disposal of nuclear weapons).

The likely cuts in defence spending will force profit-conscious defence contractors to pursue a variety of responses. Defence industries will be characterised by :—

* **Job losses**
* **Plant closures**
* **A search for new military business at home and overseas**. However, export markets for defence equipment are likely to become even more competitive as European and US companies respond to defence cuts and the prospects of excess capacity by seeking overseas markets (for example with various forms of state support). Moreover, overseas markets might be restricted through two new policy initiatives. First, there are prospects for a UN international agreement limiting arms exports and the international arms trade. Second, some donor countries and the international aid agencies (for example Germany, Japan, the World Bank, and IMF) are proposing that aid to developing countries be "tied" to reductions in thier military spending (McNamra, 1991).
* **Diversification** into civil activities for home and overseas markets

* **Mergers** and take-overs both nationally and internationally, particularly within the EU
* **International collaboration**
* **Some companies shifting from prime defence contractor to sub-contractor status**
* **Exits** from the defence sector.

2.2 Short and long-term effects

In theory, the response of companies to defence cuts will differ between the short and the long term. Here, much depends on the costs and expected benefits of different adjustment paths, the constraints of contractual commitments and whether a company's resources are highly specialised and specific to defence production. In both the short and long run, companies can be viewed as entrepreneurs searching for profits in a world of uncertainty about the future. At all times, the entrepreneur has the task of identifying profitable market opportunities before their rivals. But these choices have to be made in a world of uncertainty where the future is unknown and unknowable : today's sunrise industries might be tomorrow's dinosaurs. In this context, defence companies face essentially the same problem identifying potentially profitable markets for their current and likely future plants and workforce. They have to make a judgement about future market and profit prospects in the defence and non-defence sectors. Defence, though, has a distinctive feature in that the market signals are under the control of a single ministry. A failure to signal future equipment plans can add to uncertainty amongst suppliers, so affecting their willingness to remain in the market.

In considering their response to defence cuts, companies will be motivated by profitability. In the short run, their adjustment will be constrained by fixities and contractual commitments. They have to operate with existing plant, labour and their locations, and with existing markets and distributional systems. It takes time to close down a plant and to declare large numbers of workers redundant. Similarly, it takes time to re-think a company's strategy and to identify new profitable markets which might utilise the company's competitive advantages. In the long run, everything can be changed and the most efficient adjustments can be made. Thus, labour, especially variable labour without long-term contracts is most likely to bear the immediate costs of adjustment that is, via reduced hours of work, followed by job losses. Plant closures will take longer to organise since companies will need time to reorganise the geographical distribution of their business, selecting for sale those sites which are marketable for other uses. Another short-run response will be to seek additional sales in the company's existing markets, to withdraw work from sub-contractors and to obtain sub-contract business. In the longer term, a company can invest in the costs needed to enter completely new military or civil markets and it can decide whether to enter such markets by internal expansion or by merger or take-over. Of course, economic theory offers only general explanations of short and long-run adjustment by companies. It does not recommend specific solutions: these are choices which have to be made by profit-seeking companies and their managers.

2.3 Public policy : overview

Government has to decide how to respond to these changes in the size and structure of the defence industrial base. Should it intervene ? If so, how and what would be the objectives of such intervention ? Or, should it leave the industry's size and structure to be determined by market forces ? The answers are important for two reasons. **First**, the defence industrial base and the efficiency with which it provides equipment are major 'inputs' into the continued protection of the nation. **Second**, if the defence ministry decides to allow the industry to be determined by market forces, it has to accept that in the future (say 10-20 years ahead), there might be no domestic capability in certain areas. This is not to argue for maintaining a complete and comprehensive across-the-board technical capability, regardless of cost. The argument is simply that **current** market signals from the defence ministry will determine **future** domestic supply capabilities, so that policy-makers cannot ignore the long term. In which case, consideration might be given to assessing the benefits and costs of two-way **conversion**-that is the prospects of shifting resources quickly from the civil sector back into defence activities (Kirby and Hooper, 1991).

3. Empirical Results : A Case Study

There are few economic studies of defence contractors adjusting to change and the problems of undertaking such a transition successfully (Willett, 1991 ; Hartley and Hooper, 1990). This is an area which raises more questions than answers : there is much economists and policy makers do not known. This section reports the results of an interview survey of UK defence companies, based on 21 companies, comprising 13 suppliers and sub-contractors and 8 prime contractors undertaken in 1991-92. It was very much a small scale pilot study aimed at assessing how UK companies were adjusting to defence cuts the approach was based on the analytical framework outlined above (section 2). The results are presented around three themes, namely, the response to cuts, the adjustment period and problems of adjustment, including conversion (for details see Hooper and Hartley, 1993).

3.1 The response of firms to cuts

(*a*) Many of the responses outlined in the Analytical Framework were evident. In the **short run**, the typical response was to seek new business based on existing products, or to obtain sub-contract work (fill-in business) to retain and utilise its plant, equipment and workforce. **Faced with the need to adjust factor inputs, labour usually bore most of the short-run adjustment costs, typically through short-time and job losses.** In adjusting labour some companies mentioned the need to retain skilled workers and for others, manual production workers appear to have borne most of the immediate labour adjustment. The focus on labour adjustment also reflected a feature of the sample, namely, that a number of suppliers were single plant enterprises. For multi-plant companies, plant closure was a further option for adjustment. Changing capital inputs via plant closures takes more time than adjusting labour inputs, and this response was more typical amongst prime contractors.

(*b*) The UK's new defence policy, known as Options for Change, has forced companies to re-think their traditional defence business (a shock effect). Even companies with assured MoD orders are starting to ask serious questions about their long-term future and the company's strategy. It was also apparent that MoD's competition policy introduced in the 1980s was a further factor which had caused companies to re-think their attitudes to the UK defence market.

(*c*) In the **long run**, companies will respond by seeking **completely new markets**. This response embraced a search for new military markets either in the UK or overseas, and new civil markets in the UK and abroad. Many companies expected that over time they would be able to shift the balance of their business more towards the civil sector and/or towards overseas defence markets. Some companies had decided to specialise largely in the defence business: they have a comparative advantages in defence (for example they still regarded the UK as a large market and they aimed to obtain a larger share of the expected smaller UK defence budget). Some mention was also made of the need to change the culture of an organisation with such a change being achievable over the long run.

(d) The mechanisms for achieving the planned long-run adjustment varied between companies. Small companies which were part of a larger diversified group expected the group to provide support in entering new civil markets. Elsewhere, specialist defence divisions which were part of a large diversified group focused on their defence business, leaving the group to diversify into completely new civil markets. Here, there was a difference between the suppliers and the large prime contractors. Some primes were using **acquisitions** to enter new markets. Elsewhere, at least three companies in the sample mentioned international alliances and joint ventures as a further response to UK defence cuts. Of course, international alliances and joint ventures involving defence companies create the possibility of cartels emerging, with adverse effects on competition for defence contracts.

(e) In selecting new markets, companies usually build on their **existing strengths**, reflected in their expertise and reputation. For some companies, this means expanding their defence interests, either by developing new products or by entering related markets using existing resources and skills. This is a likely response for prime contractors almost wholly dependent on defence sales. A well-publicised example is VSEL which has specialised in nuclear-powered submarines but decided to try to re-enter the warship building and refitting business, as well as entering the land armaments market (AS 90 Howitzer). At the same time, faced with cancellations and reduced orders, it has reduced it labour force and closed its Cammell Laird yard. Other defence companies, especially the prime contractors, have announced plans to diversify around their specialisms such as precision engineering, power systems and acting as systems integrators.

3.2 Time period for adjustment

(a) Contrary to the assumptions of some simple economic models, adjusting to changes takes time: it is neither instantaneous nor costless. Much depends on civil market opportunities but

typically, **an adjustment period of up to 5 years might be
needed.** Adjustment costs will also be incurred in the form of
redundancy payments, rationalisation and restructuring, the
costs of acquisitions and the costs needed to enter new
markets (UN, 1993).

(b) The distribution of responses to the survey for various
adjustment time periods is shown in Table 16.1. For suppliers,
the **median** adjustment period is up to 3 years, varying within
a range of 4 weeks to a maximum of up to 10 years. For prime
contractors, the adjustment period was longer, with a median
of 5 years, varying within a range of 2 years to 10 years.

Table 16.1 : *Adjustment Periods for Conversion*

	Number of responses			
	Up to 1 year	1-3 years	4-5 years	5-10 years
Suppliers	3	3	2	1
Primes	0	2	1	3

3.3 Problems of adjustment (conversion)

(a) For suppliers and sub-contractors, it seems that conversion
was not usually a problem. **Typically, these were companies
where defence might be only part of their total business
and/or the company's resources could be used flexibly and
inter-changeably between military and civil work.**
Examples include foundries which can produce products for
either defence or civil business: track for armoured vehicles
can be used for earth-moving equipment.

(b) For suppliers, it was noticeable that most companies (over 80
per cent of the supplier sample) stressed that their production
facilities were inter-changeable between defence and civil
work; suggesting that the supply side was not a constraint on
adjustment. Instead, a number of companies referred to their
major problem being the difficulties of finding new markets
and the time needed for this search process. **This point was
reinforced by some companies stressing that the general
recession in the UK economy in the early 1990s was
restricting civil market opportunities.**

(c) There was, however, a major difference between the supplier
and prime contractor samples. For 55 per cent of the supplier

sample, defence accounted for under 30 per cent of their total business ; and under 20 per cent of the sample were wholly dependent on defence sales. In contrast, almost 40 per cent of the prime contractor sample were wholly dependent on defence business ; and for 50 per cent of the prime contractor sample, defence accounted for between 55 per cent and 100 per cent of their sales. It was also apparent that some prime contractors were characterised by assets and skills which were highly specific to defence markets with little or no alternative uses (for example sea systems, missile plants).

(d) Amongst prime contractors, there are possibilities for direct conversion that is using the company's defence resources to produce civil goods. Aerospace is probably the best example, where a company's plant and labour force are transferable and can be used to manufacture either military or civil aircraft, helicopters and aero-engines. Other examples of conversion have been less successful. For instance, Vickers efforts at converting from tanks to tractors was a financial failure simply because Vickers were unable to compete with the existing specialist tractor companies (Foss and McKenzie, 1988). There are good reasons for such failure. Vickers are a defence specialist able to compete and survive in their specialist market. Similarly, tractor companies have survived by establishing a competitive advantage in their specialism. If there are profitable opportunities in civil tractor markets, there is every reason to expect the established tractor companies to have identified and exploited such opportunities. Defence companies seeking direct conversion have to identify **profitable civil markets** which are appropriate for their resources. In some cases, the resources and skills of defence companies are highly specific and non-transferable.

(e) There is a related cultural adjustment problem for specialist defence companies, particularly those wholly dependent on defence work. Mention was made of the difficulties of changing the culture of an enterprise from demanding defence requirements to the different demands of civil markets. For example, defence products are often of high specification and such products are difficult to sell in civil markets where specifications are not as high. In specialist defence companies, the MoD dominates and determines the company's culture and that culture tends to be one of dependence rather than enterprise. However, both Options for Change and MoD's competition policy might have 'shocked' companies into changing their culture. But one small supplier wholly dependent on defence estimated that it would take at least 3 years to change its culture (for example to improve marketing and to change attitudes for all the company's staff). If a 3 year adjustment is needed to change the culture of a small defence company it is not surprising that some of the large specialist prime contractors are likely to encounter major adjustment problems.

3.4 General conclusions from the survey

Much of the popular debate has focused on the major prime contractors and issues of conversion. However, this survey showed that there are considerable numbers of suppliers and sub-contractors who are coping quite well with defence cuts. For many of these firms, their plant and labour can be quickly and easily switched from defence to civil business. For such firms, conversion is both technically and economically feasible: their resources are transferable between defence and civil production and they are not wholly dependent on defence business.

There are, however, specialist defence firms wholly dependent on defence business where direct conversion is technically difficult, costly and probably not worthwhile. For such enterprises, their plant, equipment, managers and workforce are highly specific to defence and non-transferable—at least at reasonable cost. In such circumstances, it is probably most efficient to close the specialist defence plant and, if there are willing buyers, redevelop the site (cf redevelopment of town shopping centres and the alternative uses for surplus defence facilities—for example RAF bases).

Advocates of direct conversion for specialist defence plants often claim that there are lots of civil market opportunities available to such firms. Rarely do they address :—

* the costs of converting defence plants and retraining the workforce ;
* the costs of entering civil markets ; and
* whether the civil markets are expected to be profitable.

And if these advocates of conversion are right and there are lots of opportunities not already being exploited by existing specialist civil firms, then there is the ultimate capital market test, namely, defence firms will be taken over and/or their defence plants will be marketable. In other words, when defence plants are offered for sale, they will be bought by firms which believe they can find a profitable use for the assets ; but, typically, what happens is that the original defence plant and its site are redeveloped for more appropriate alternative uses (for example housing, light industry, and so on).

This survey should be regarded as no more than a pilot study which might be the basis for an in-depth research project monitoring defence contractors as they adjust to the changes which will develop over the next few years. Certainly there is an absence of good case studies of conversion in defence firms (both successes and failures). A comparative study of experience in the civil sector might be informative (for example coal, steel, shipbuilding, textiles). There is also scope for further analysis of the internal organisation of firms, their culture and the interest groups which are likely to oppose change.

The results have at least two implications for public policy. **First,** if labour markets fail to work properly there might be a case for considering policy measures to improve the operation of such markets (for example information, mobility, retraining). Second, policy-makers need to recognise that current defence policy decisions will affect the future size, structure and performance of the domestic defence industrial base. Policy-makers need to predict the likely extent of the domestic defence industrial base in, say, 2005 and then decide whether the outcome will contribute to national defence. Some of these policy issues will now be addressed, starting with assumptions about the labour market.

4. Policy Issues : The Operation of The Labour Market

Unemployment is likely to be one of the immediate costs of disarmament and this raises questions about how well and how quickly labour markets adjust to change (that is their efficiency as clearing mechanisms : UN, 1993) ; whether there is a role for state intervention and what such a role might be. Inevitably, there are differences between political parties and economists on how the economy operates and the role of the state. Indeed, a more general question arises as to whether defence industries are "different" and whether the resources released from these industries qualify for special favourable assistance.

The extent to which labour markets remove unemployment will depend upon how well and how quickly they work. Keynesians believe that markets, especially labour markets, take a long time to adjust and to clear and that full employment is not automatic and might take years to achieve. Monetarists believe that markets work well, clear quite quickly and tend to operate around full employment (Begg, Fischer and Dornbusch, 1993). Differences also arise between political parties on the role and extent of state intervention in industry and in the regions. Interventionists advocate an industrial strategy and an active regional policy, whilst free market advocates focus on correcting major failures of the market. Where disarmament results in unemployment, should governments consider the Exchequer impacts of such unemployment in their decision-making (shadow pricing) ?

The British Government's decision on the 1993 pit closures provides insights into its views on the operation of the UK economy, particularly on the operation of labour markets, and the relevance of Exchequer costs. In its Report on the pit closures, the Parliamentary Employment Committee expressed concern over the uncertainty about the total costs of unemployment associated with the proposed pit closures ; and it called"... on the Government to take into account the full financial cost of the pit closures and to publish estimates of the amount involved" (HCP 597, 1993, p. iv).

The Department of Employment's response to the Employment Committee's Report is relevant to the economic evaluation of the arguments about unemployment and financial costs associated with the cancellation of projects and disarmament. In its reply, the Department made the following points :—

(*i*) "Calculations of employment effects outside the enterprise concerned, and of supposed overall Exchequer costs, are not a good guide for the individual employment, purchasing and investment decisions of nationalised industries, any more than they would be in the case of private sector companies. In any case, the wider economic effects of changes in employment patterns are difficult to quantify with any accuracy, and estimates of the financial costs of unemployment are likely to be erroneous and partial" (HCP 597, 1993, p. iv).

(*ii*) "The Government considers that the way in which tax and national insurance (NI) payments are frequently included as an element of the financial cost of closures and of the resulting unemployment is also inappropriate. Such calculations are often quoted in support of mistaken arguments that the Government could maintain employment by means of a subsidy equivalent to the amount the business paid in tax and NI, without any adverse effects elsewhere in the economy. There are at least two ways in which this argument is fallacious. First, if the subsidy were paid in such a way as to enable the subsidised business artificially to reduce its prices below those of its competitors, this would be likely to have an adverse impact on other businesses. Secondly, applying public finances to safeguard employment in one business prevents the finances being applied elsewhere in the economy where they might create equivalent, or even more, employment in the public or private sector" (HCP 597, 1993, p. iv).

(*iii*) "Arguments using the supposed financial costs of unemployment often appear to be based on a static view of the economy, where industries and methods of work remain unchanged in the longer term ; but this does not accord with reality where both technological advance and changing markets demand that the economy be dynamic and responsive. In such a dynamic economy employment will be maximised when the transfer of resources from declining to expanding industries and from less efficient to more efficient methods of production occurs via a competitive market and commercially based decisions. An economic approach which attempts to take account of wider costs outside the business, especially if these exclude wider positive outcomes, is likely in the long run to result in a static economy, in which uncompetitive industries and methods of production are maintained at tax payers' expense, to the detriment of more efficient sectors and ultimately to the detriment of consumers and of employment" (HCP 597, 1993, p. v).

The Department of Employment's analysis is not without its critics, particularly in relation to its assumptions about the operation of local labour markets and whether there is a role for high value, high technology industries, such as aerospace, in maintaining living standards and the UK's future international competitiveness. Much depends on how well and how quickly local labour markets adjust to change. Adjustment problems are likely where change is sudden, unexpected and large-scale, such as the elosure of a major employer in a high unemployment area when the UK economy is in recession (UN, 1993). In such circumstances, local labour markets might take years to adjust to plant closures and job losses, so that state intervention is required to protect workers from the income consequences of job loss and to assist the adjustment process within a socially-acceptable timescale.

Large-scale disarmament will have both macro and micro-economic impacts (UN, 1993 ; Fontanel, 1994). Table 16.2 outlines a range of possible public policy measures which could be used to minimise the adjustment period and costs resulting from reductions in defence spending.

Table 16.2 : *Public Policies to Minimize Adjustment Period.*

	Types of Policies	Examples
1.	Manpower Policy	Training ; retraining; information ; mobility
2.	Capital Policy	Retooling old plant-equipment; new investment
3.	Technology Policy	New civil R & D programmes
4.	Regional Policy	Location of industry policy : infrastructure
5.	A State Conversion Agency	Assisting the conversion of plant from defence to civil markets
6.	Aggregate Demand Policy	Using government expenditure to avoid recessions ; tax cuts
7.	Income Deficiency Payments	Compensating the losers from disarmament (for example unemployment and redundancy pay)

4.1 Is defence different ?

There is, though, the more general question of whether defence industries are different and should qualify for special favourable assistance from governments. Consider some of the arguments which might be used. One view is that for public policy purposes, defence industries should be treated like any other declining sector in the civil economy (for example coal, steel, textiles). After all, its workers will have received wages and salaries reflecting the risks and uncertainties of employment in defence industries ; and a sudden decrease in the demand for defence equipment is, in principle, no different from the adjustment problems faced by workers in other civil industries faced with declining demand (for example fashion houses ; car firms ; tobacco companies). However, it might be the case that defence industries use a relatively high proportion of specific assets and skills which are non-transferable. But here, the appropriate solution requires a public policy towards assisting the re-allocation of highly-specific and non-transferable resources throughout the economy (for example retraining), rather than focusing on the particular resource-specificity problems of defence industries.

Public choice analysis provides a further possible case for special favourable assistance for defence industries adjusting to the changes caused by disarmament. Without adequate compensation (also reflecting the welfare economics compensation criterion for a sociably-desirable change), major interest groups in the military industrial complex will oppose disarmament : hence generous compensation is required to minimise the costs imposed on such groups and to compensate them for their losses of rents. And here, a distinctive feature of defence industries could be relevant, namely that disarmament not only releases resources for alternative civil activities but it can also contribute to peace and stability in the world (Sandler and Hartley, 1995). Nonetheless, caution is required in introducing specially favourable assistance to defence industries. There is the inevitable danger of special pleading with arguments dominated by myths and emotion. Economists need to specify the economic logic of any proposals for favourable assistance towards defence industries, distinguishing between the **efficiency** and **distributional** basis and implications of such policy proposals.

In addition, to the adjustment problems resulting from disarmament, European governments need to address the efficiency of the current market arrangements for providing defence equipment. The prospects of extending the Single European Market to defence equipment could add to the adjustment problems facing Europe's defence industries.

5. The Future : A Single European Market For Defence Equipment ?

Within the EU, independence through supporting a domestic defence industrial base (DIB) is costly. Each Member State's support for its national defence industry has resulted in :—

(i) The duplication of costly R & D programmes (for example the development of three advanced combat aircraft in Europe—Eurofighter 2000 ; Gripen and Rafale, with EF 2000 R & D costs estimated at £ 10.4 billion, 1993-94 prices).

(ii) Relatively short production runs reflecting small national orders and hence a failure to obtain economies of scale and learning (for example production orders are EF 2000 = 602 units ; Gripen = 340-350 units ; and Rafale = 328 units).

(*iii*) A number of domestic markets are characterised by :—

* Domestic monopolies (for example aerospace).
* Government protection (barriers to entry and exit).
* Non-competitive markets and cost-based contracts.
* State ownership, subsidies and government regulation of profits (for example firms pursuing non-profit objectives).

Falling defence budgets and rising equipment costs will compel European nations (especially France, Germany, Italy and UK with major defence industries) to reappraise their policies towards their DIB. Economic pressures will require more serious consideration of alternative procurement policies, namely :—

(*i*) Competitive procurement

(*ii*) Importing with or without offsets

(*iii*) International collaboration

(*iv*) The extension of the Single European Market to defence equipment.

Extending the Single Market to defence equipment is expected to result in major economic benefits reflecting :—

(*i*) Increased competition both within and between nations (the competition effect).

(*ii*) Savings in R & D costs by reducing duplication in R & D.

(*iii*) Economies of scale and learning from longer production runs (the scale effect).

The scenarios under which these benefits will be achieved are summarised in Table 16.3. Three broad scenarios are outlined and for each it is assumed that there will be a liberalised competitive market either restricted to Member States of the EU or open to the world.

Table 16.3 : *Single Market Scenarios for Defence Equipment*

Scenario	Liberalised Competitive Market	
	EU	World-wide
I National Procurement by national defence ministries.		
II A centralised purchasing agency buying standardised equipment and replacing national defence ministries.		
III A twin track approach combining competition and collaboration		

Each scenario offers potential efficiency improvements. To illustrate the orders of magnitude, UK experience suggests that a typical competition effect leads to cost savings of at least 10 per cent (with greater costs savings if European markets were opened-up to the rest of the world), with typical scale and learning effects offering unit cost savings of around 10 per cent for a doubling of output. On this basis, with both competition and scale effects, scenario II is likely to lead to the greatest cost savings—it is also likely to be the most difficult policy both politically and economically to implement (this analysis does not consider how best to reach each of the three scenarios). Moreover, creating a Single European Market for defence equipment is not without its problems and costs :—

(*i*) There are problems of creating a "level playing field".

(*ii*) There is potential for the creation of monopolies, cartels and collusive tendering.

(*iii*) Some firms, industries, towns and regions will be loser.

(*iv*) The maintenance of competition in the Single Market is likely to require the Market to be opened-up to the world. Here, US firms would be a major threat to the European DIB which will create European pressures for "fair competition" and "balanced" two-way trade ; or, alternatively, the creation of Fortress Europe characterised by protectionism, cartels and inefficiency !

6. Conclusion : A Research Agenda

The economics of disarmament remains an under-researched area. Current cuts in defence spending throughout Europe raise a variety of research questions :—

(*i*) What are the effects of disarmament on the employment and unemployment of both defence industry labour and former military personnel ?

(*ii*) What are the regional impacts of defence cuts and how successful is public policy in minimising the adverse employment effects of disarmament ?

(*iii*) How marketable are the labour and capital resources released ? For example, some of the labour skills and capital resources might be highly specific and non-transferable.

(*iv*) What are the characteristics needed for the successful conversion of defence plants and facilities from military to civilian use ?

(*v*) What are the lessons from previous disarmament and from the experience of the civil sector adjusting to change ?

There is no shortage of theoretical, empirical and policy questions. The challenge is for the economics profession to allocate more of its scarce resources to the study of defence, disarmament, conversion and peace.

* This paper is the result of a research project on the Economics of Disarmament in the UK, funded by the Leverhalme Trust (F/224/S). Thanks are due to Nick Hooper and Barbara Butler for comments on the paper: the usual disclaimers apply.

References

Begg, D., S. Fischer, and R. Dornsbuch (1993) *Economics* (London : McGraw-Hill).

Fontanel, J (1994) 'The Economics of Disarmament : A Survey' *Defence and Peace Economics*, 5, 2.

HCP 197 (1993) *Employment Consequences of British Coal's Proposed Pit Closures* (London : Employment Committee, House of Commons, March).

Hartley, K and N Hopper (199) *The Economics of Defence, Disarmament and Peace* (Elgar, Aldershot).

Hooper, N. and K, Hartley (1993) *UK Defence Contractors : Adjusting to change* (York : University of York).

Kirby, S. and N. Hooper (eds) (1991) *The Cost of Peace* (Chur : Harwood).

McNamara, RS (1991) 'The Post Cold War : Implications for Military Expenditure in the Developing Countries,' Annual Conference on Development Economics (Washington DC, World Bank).

Sandler, T and K. Hartley (1995) *The Economics of Defense* (Cambridge : Cambridge University Press).

UN (1993) *The Economics Aspects of Disarmament : Disarmament as an Investment Process* (New York : United Nations).

Willet, S (1991) 'Defence Employment and the Local Labour market of Greater London' in Paukert L and Richards P (eds), Defence Expenditure, *Industrial Conversion and Local Employment*, (Geneva : ILO).

Chapter 17

The Effects of Disarmament on the UK Economy

Ron Smith

Introduction

Like many other economies the UK is having to adjust to lower military spending, though not, as yet, to full disarmament. This paper provides a review of the effects of these cuts in military spending and defence employment. After a discussion of some previous work on the effects of military spending, the paper examines the evolution of the UK defences budget, the macro-economic environment in which the adjustment is taking place ; the transformation in the defence industry ; and policies for adjustment.

Reductions (or increases) in military spending have a three stage impact. There is the immediate short-run effects of the change in the structure of final demand ; there is a transition process as resources are either transferred to different uses or left-idle ; and there are longer term supply side effects that the new allocation of resources has on the economy as a whole. These impacts have to be evaluated within the context of a particular theory of the operation of the economy. This raises the problem, which has nothing to do with the military nature of the cuts, that there are disagreements about how the economy operates in the short, medium and long-run, particularly about what determines longer run productivity growth. An excellent review of the economic aspects of the disarmament process is provided in UNIDIR (1993).

Studies of the immediate demand side impact of cuts in military spending have tended to use large econometric models. There has been a large number of studies of the effects of cutting military expenditure on the economy, since the classic study by Leontief (1965). Most of the main UK macro models do not include military expenditure as an explicit variable, aggregating it within total public spending. One exception is the highly disaggregated Cambridge Econometrics (Cambridge Growth Project) Model, which contains a detailed Input-Output structure. Dunne & Smith (1984) and Barker, Dunne & Smith (1991) used this to investigate the effect of reduced military spending. Like most other studies that have involved simulating large macro models, these found that given appropriate fiscal and monetary adjustment the macro-economic effects of cuts in military expenditure are rather small, though probably positive, (for example lower unemployment because military spending tends to generate less jobs than other forms of public spending which could replace it). This conclusion fits with a wide range of historical evidence, discussed in Dunne & Smith (1984) which covers much larger and much faster reductions in military spending than are now contemplated. Individual industries and regions would be adversely affected, but the impact of cuts of the size now envisaged would be small relative to other recent structural adjustments in the UK such as in the coal and steel industry.

An alternative to simulating existing models is direct econometric investigation. Dunne & Smith (1990) conducted an exhaustive econometric study of the effects of the share of military expenditure in output on the unemployment rate. The study used a dynamic model to examine 100 years of data for the US and UK ; and post World War II data for 11 OECD countries, in cross-section, time-series and pooled. None of these data suggested that there was any significant persistent effect of military expenditure on unemployment (or conversely that there was any counter-cyclical pattern of increasing military expenditure in response to high unemployment) beyond the obvious fact that unemployment is low during major wars. This is not to say that there were not time when cuts in military spending were followed by high unemployment. The classic case is the end of World War I in the UK which was followed by mass unemployment. However, the cause was not the reduction in military demand (unemployment did not increase in the US despite large cuts in military spending) but the policy of deflation to allow sterling to return to gold at its pre-war parity.

The UK experience after World War I indicates how the adjustment process following cuts in military spending will depend on the macroeconomic policy that is implemented at the time. The speed of adjustment will also depend on the inherent flexibility of the economy concerned ; the degree to which new demands for resources match the characteristics (location, skill and so on) of the old resources freed by the reduction in military spending ; and any regional or industrial policies which will aid adjustment. For instance, in the UK, the peculiarities of the housing market tend to inhibit regional movement of labour, compared to the US, and this may slow adjustment. In addition, the Conservative Government has been reluctant to intervene in market adjustment with industrial or regional policies.

Whereas the short-run effects of military spending seem to be rather small, its long-run effects on growth may be more significant. What has become known as the 'burden' hypothesis, that high shares of military expenditure tend to reduce the rate of growth of an economy, has been the subject of a large amount of dispute and empirical investigation. In cross-section for the OECD countries, there appears to be a systematic pattern : countries with high military spending as a share of GDP (like and US and UK) have had lower savings and investment ratios and lower growth rates than countries (like Japan and Germany) which had low shares of military spending in GDP. Other countries seemed to fit the same relationship albeit with a large dispersion. However, the relationship is much less obvious in time-series for individual countries, and there are some clear counter-examples, South Korea and Taiwan, fast growing countries with high shares of military spending.

One version of the argument suggests that over the medium run, the share of consumption, public plus private, is independently determined and relatively stable and thus increases in military spending are at the expense of investment (again public plus private). The stability of the sum reflects the fact that normally, military spending like investment is regarded as provision for the future. However, in time of major war and national survival, it becomes a provision for an immediate need, therefore consumption is sacrificed. Lower rates of investment tend to be associated with lower rates of growth, though the relationship is not that close, depending on the cost, productivity and composition of the capital acquired by the country concerned.

There have been a variety of studies of the direct relationship between the shares of military spending and investment, both across and within countries. These include Smith (1977, 1980), Rasler & Thompson (1988) and Cappelen et al (1984). There have been fewer studies of the Mechanisms by which military spending may influence investment. An exception is Findlay and Parker (1992). They find, on quarterly post-war US time-series, that military expenditures have a significant positive effect on interest rates of 5 maturities and that this effect is significantly larger than for non-military government expenditures. They interpret this as evidence in favour of the hypothesis that military expenditures have a higher crowding out effect on investment than other government expenditures.

Although the 'burden' seems to be statistically significant, it is not the case that the effect of military spending is quantitatively large. This would explain why it is more difficult to detect in time-series where the within country variation in military spending and growth is smaller. The estimate in Smith (1978) from an OECD cross-country simultaneous equation system was that a reduction in the share of military expenditure by 1 per cent of GDP would increase the growth rate by 0.13 per cent. So had the UK spent 1 per cent of GDP rather than 5 per cent of GDP on the military its average growth rate would have been about 3.2 per cent rather than 2.5 per cent. Differences in military spending can only account for a small part of the observed differences in growth rates.

One possible positive effect of military spending is technological spin-off from military R & D. While this may have been significant in the 19th Century and immediately after World War II, most of the evidence suggests that spin-off has been minimal for the last three decades, rather military R & D may have displaced commercial R & D in the US and UK. Lerner (1992) provides references to the literature. That the effect of military spending on growth is disputed is not suprising, since we do not have good theories of growth within which the marginal effect of military spending can be examined. Thus while the 'burden' hypothesis, of a negative effect of military spending, has been extensively investigated by many researchers and is consistent with a large amount of disparate evidence, including the detailed historical study of Kennedy (1988) as with most economic hypotheses, the evidence is not always clear-cut and there are a number of facts that appear inconsistent with the hypothesis. Even if the burden hypothesis is correct, the signal-noise ratio would make it difficult to detect the effects of cuts in the share of military spending of the size we now observe. The share of military expenditure was cut from 5 to 4 percent between the mid 1980s and 1990. This might, when the adjustments have worked through, raise the growth rate by about 0.1 per cent. Since the standard deviation of the UK growth rate is over 1 per cent, such an impact would be difficult to detect.

Budgetary Response

Despite devoting a larger share of GDP to defence than most of its European partners, British defence policy has always faced a delicate balancing act between inadequate cash and excessive commitments. Figure 17.1 plots the historical share of defence in GDP together with the Government's current projections to 1995-96. From a post-war low in 1978-79 of 4.4 per cent, it rose under Mrs. Thatcher's government to a peak of almost 5.3 per cent, before falling sharply, steadying around 4 per cent at the turn of the decade, then falling to a projected 3.2 per cent in the mid 1990s. In 1990 the UK Ministry of Defence (MOD) announced its planned restructuring in response to the end of the Cold War. This plan, known as "Options for Change", projected that force numbers would fall from 63,000 to 55,000 in the Navy ; 156,000 to 116,000 in the Army ; and 89,000 to 75,000 in the Airforce, by the mid 1990s.

The first point to note is that the UK has already had much of the Peace Dividend : military spending has been falling in real terms since 1984-5. Table 17.1 provides some figures. The figure for the defence budget are actual figures plus the plans announced in the 1992 Treasury "Autumn Statement" on public expenditure, all expressed in constant 1991-92 prices. The growth and inflation figures are those given by the Chancellor of the Exchequer in the Autumn Statement. The 1992-93 defence budget is likely to be about 12 per cent lower in real terms than the peak of 1984-85, and equipment has fared worse being 27 per cent lower. The total budget planned for 1995-96 is 10 per cent less than 1992-93. Although the MOD does not publish its plans for the equipment budget it is possible to make estimates. The estimates for the equipment budget it Table 17.1 are obtained by calculating what is left over after personnel and other costs are met. It is assumed : that Options for Change targets will be met by 1994-95 and the number in the armed forces will fall from 300 thousand to 234 thousand ; that the number of civil servants will fall from 170 to 134 thousand ; that wages will grow in line with the economy as forecast by the Treasury ; that the new pensions arrangements follow the pattern in the Autumn

Statement ; and that other costs will take their usual share of the budget.

Table 17.1 : *Defence Statistics: Actual and Projected.*

	84-85	90-91	91-92	92-93	93-94	94-95	95-96
GDP Growth %	2.0	—0.5	—2.0	—0.8	1.9	3.7	3.5
Inflation %	5.1	8.0	6.9	4.3	2.8	3.2	2.7
Share of Defence %	5.25	3.93	3.96	3.96	3.74	3.52	3.24
Defence Budget in 1991-92 prices							
Total	25950	23302	23015	22830	21957	21474	20433
Personnel Costs	9068	9408	9948	9450	9000	8550	8459
Armed Forces	4905	5131	5480	5228	5046	4751	4696
Pensions	1255	1503	1615	1433	1370	1298	1284
Civilians	2908	2773	2853	2789	2584	2502	2479
Other Costs	5005	4970	4839	4714	4611	4509	4291
Equipment	11879	9448	9753	8667	8346	8414	7683
Less Gulf Payments		524	1525				
EMPLOYMENT							
Armed Force	336	313	298	285	270	245	234
Civil Servants	207	173	169	165	150	140	134
Industry	670	545	507	452	420	399	359
TOTAL	1213	1031	973	902	840	784	727

Sources : Defence Statistics, 1992; Autumn Statement 1992, and authors calculations.

On the basis of the scenario set out in Table 17.1, the equipment budget in 1995-96 will be 15 per cent below the 1992-93 figure and only 65 per cent of its 1984-85 peak. Equipment is not all that will suffer ; on previous experience, training, spares, readiness and sustainability will all be cut to the bone ; damaging military effectiveness. There are also a range of costs that are directly associated with cutting military spending. The Ministry of Defence estimates that redundancy payments will amount to £1.29 billion and capital expenditures to restructure and reorganise facilities to cope with the new smaller deployments will amount to £1.14 billion over the period 1992-95. These will be met out of the planned defence budget.

The balance between cash and commitments looks likely to become yet more delicate in the coming years. Although in Spring 1993 there are signs that the UK's longest post-war of recession has

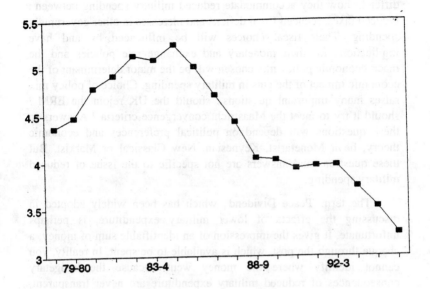

Fig. 17.1 : Share of Defence Percentage of GDP

ended, the Treasury is facing an escalating Public Deficit. Even with
the tax increases promised in the 1993 budget, it will balloon to almost
£60 billion, putting further pressure on the planned defence budget. At
the same time, disorder in Europe and elsewhere is likely to provoke
yet more demands for the use of British Armed Forces, which will slow
the pace of the manpower cuts planned in Options for change. The
MOD has already had to find an extra £80 million to pay for cuts in
3,000 army posts that will not now be made. Almost inevitably,
equipment spending will continue to be squeezed between the declining
budget and the essential personnel costs.

Macroeconomic Response

Since countries face different priorities and constraints, they will differ in how they accommodate reduced military spending between : reduced taxes, reduced fiscal deficits and increases in other government spending. Their fiscal choices will be influenced by and have implications for their monetary and exchange rate policies and the macroeconomic policy mix chosen will be the major determinant of the economic impact of the cuts in military spending. Choice of policy mix raises many important questions : should the UK rejoin the ERM ? should it try to meet the Maastricht convergence criteria ? Answers to these questions will depend on political preferences and economic theory, be it Monetarist, Keynesian, New Classical or Marxist. But these questions and answers are not specific to the issue of reduced military spending.

The term 'Peace Dividend', which has been widely adopted in discussing the effects of lower military expenditure, is perhaps unfortunate. It gives the impression of an identifiable sum of money, a cheque through the post, which is available to be spent. In reality, you cannot identify where the money went because the budgetary consequences of reduced military expenditure are never transparent, either *ex ante* or *ex post*. An ex post analysis or cuts in military spending involves a comparison of the observed trajectory, in which UK military expenditure fell by about 12 per cent in real terms over 1985-1991 with a speculative counter-factual alternative in which it did not fall and there was some alternative mix of other government expenditures, taxes and deficits. This different policy mix would have had effects on output, employment etc. and these would have fed back onto government revenues and expenditures further complicating the comparison. Similarly, in an *ex qnte* analysis some hypothetical alternative has to be specified. This has implications for assessing the macro-economic effects of the change: to what extent are the observed consequences that follow cuts in military spending a result of the cut in military spending itself or a result of the different macroeconomic policy that happened to be followed at the same time? In the simulations described above we held public expenditure constant while cutting military spending. This was done not because it was necessarily the correct macroeconomic response, but because it allowed us to identify the effects of changing military spending with the minimum change in macro-policy. If the UK now uses the lower defence demands as an occasion to cut the public sector deficit, as seems likely, this will be associated with higher unemployment than would otherwise have occurred. But this is a consequence of cutting the deficit, not cutting military spending.

The first stage of the reduction in military spending from 1985 to 1990 was very easy. The effects of the cuts were hardly noticed because the economy was booming, particularly in the South where defence jobs tend to be concentrated. The second stage, since 1990, has been very difficult because the UK was in its longest post-war recession, which hit the South East particularly hard, and because the defence industry was shedding excess capacity at the same time.

Industrial Response

To understand the industrial response, a historical perspective is useful. Ten years ago in 1983, the world was starting to recover from a major recession; real defence spending in the US and UK was still expanding quite rapidly as was the World Arms market. The LDC debt crisis cut credit to many buyers and arms exports peaked in the mid 1980s as did US and UK real military spending. Arms exports and defence spending then turned down, though there were some profitable opportunities, such as the UK sale to Saudi Arabia of Tornados and other equipment under the Al Yamamah contract, which was possibly worth about £20 billion over 1985-1993. The UK government also introduced a new procurement policy emphasising fixed-price competitive tenders which increased risk and reduced return in the defence market. Smith (1990) examines the effects of these changes on defence industry structure. The most common corporate response to the change in the trend of the market was inertia, do nothing. This was not, at the time, self-evidently wrong. Defence is a volatile market, both at the aggregate level and at the firm level, where occasional large contracts tend to be the rule. Thus it is often sensible to sit out down-turns in demand, like the earlier period of detente, and maintain the capacity to respond to the next upturn. However, in the 1980s this inertia left the industry with a large amount of excess capacity, even before the fall of Berlin Wall. The Gulf War provided a brief surge in demand, and the Asian market for weapons continued to grow in the 1990s, but these were not enough to offset the need for substantial restructuring. Worse the adjustment had to be made at a time when related commercial markets, like civil aerospace, electronics, automobiles and shipbuilding were also suffering from the effects of world recession and increased competition. Acquisitions, Divestments, Redundancies and closures have changed the shape of the industry. The adjustment process is far from finished, SIPRI estimates that the global defence workforce of about 15 million will decline by about a third over the next five years. These estimates are given in Wulf (1993) which provides a detailed review of the global arms industry.

Defence dependent companies have responded to the decline in demand by restructuring the industry in many different ways. Adjustment was aided by the fact that although the medium term outlook for defence is not bright, in general it is still profitable. Corporate strategies have ranged from contract and consolidate to expand and diversify.

The most notable advocate of the contract and consolidate strategy is a US company, General Dynamics (GD). In 1992 GD Chairman, William Anders, dismissed the benefits of diversification as illusory, a waste of management time and shareholder funds because the failure rate was unacceptably high. The usual estimate is that around 80 percent of the attempts by defence manufacturers to convert to commercial production fail. In 'The Myth of Conversion' (Wall Street Journal 29 March 1993) Jerrold T Lunquist of McKinsey, cites Norman Augustine of Martin Marietta as saying that the defence industries record of unrelated diversification is 'unblemished by success'. Instead of diversifying, GD developed a plan of contraction, selling businesses where it did not have a sufficiently strong competive position to dominate the market and returning the money to shareholders. Among the businesses it shed were the Cessna civil aircraft company and its missile and tactical aircraft divisions. The military divisions were acquired by other producers, Hughes and Lockheed, who hoped to be able to consolidate their positions in a market with fewer competitors. Divestment is not always possible. In the late 1980s, the UK companies Thorn-EMI and Racal wanted to divest their defence subsidiaries but neither were then able to find buyers at acceptable prices.

Few UK companies took the divest and consolidate route so explicitly as General Dynamics, most chose to attempt to diversify. Some of the defence producers were already diversified conglomerates who could expand existing commercial businesses through organic growth. Others could only diversify by entering new lines of business either through acquisition or organic entry. With few exceptions, these diversifications tended to confirm Ander's analysis. In general they proved a waste of management time and shareholder funds with an unacceptably high failure rate. BAe and Dowty provide well publicised examples of the dangers of diversification. These corporate strategies are discussed in more detail in Smith & Smith (1992). As Lundquist concluded 'In the end, the task is not converting the defense industry from military to commercial markets but changing the industry : from one made up of diversified defense contractors in too many declining segments to one made of highly focussed companies in sustainable segments. Only then can defense-related industries maintain the quality of the defense industrial base while providing for their stakeholders.' (Wall Street Journal 29 March 93).

A distinction is usually drawn between diversification and conversion. Diversification reduces a company's dependence on defence by acquisition or organic growth of non-defence operations. Diversification in the face of declining demand is difficult enough; those civilian 'Merchants of Death', the tobacco firms, failed repeatedly despite their huge cash flows. Conversion, actually getting defence plants and establishments to produce profitable non-military products, is yet more difficult. Proponents of policies to convert defence plants have tended to emphasise the technological capacity to change—the ability to create alternative products. But given the nature of military production the major constraint is managerial capacity to change—the ability to create an alternative culture.

There are certain common, though not universal or unchanging, patterns in the corporate cultures of defence companies. These operate both at the formal level of explicit goals, procedures and systems and at the informal level of the norms, attitudes, values, styles and skills that predominate. The goals of the defence companies are heavily product oriented and technology driven. The aircraft and weapons have an intrinsic value, almost irrespective of demand. Time horizons are long, projects can last a decade, so rapid response to fast moving markets is not a priority. Innovations are radical rather than continuous, great leaps between generations and, unlike civil aircraft, little commonality between generations. The priority given to maximising performance and the prevalence of cost-plus contracts in the past means that cost minimisation is rarely a primary goal of the culture. A certain amount of slack is potentially valuable, since it allows the firm to respond quickly to crises like the Falklands and Gulf Wars, when cost is no object. The procedures and systems of the defence companies are shaped by dealing with a single customer, the government. They have little need to understand more diffuse markets, tend to be tolerant of beuraucracy and delay, and are heavily rule driven. They also have great lobbying and negotiating skills—the best are very good at getting money out of government.

These companies were well aware that they had cultivated a culture which was optimised to the military environment and which would inhibit adjustment to normal civil work. It would also inhibit adjustment when the military environment changed. The example of British Aerospace, the largest European arms producer, is interesting. BAe consolidated particularly through the acquisition of Royal Ordnance Factories in 1987 and strengthened its military profitability through the large Al Yamamah sale to Saudi Arabia of Tornados and other equipment. Its civil aerospace business, though large, was only marginally profitable and it adopted an explicit strategy of diversification through acquisition of businesses in new areas. The main fields were construction, Ballast Nedam; cars, Rover; and property, Arlington. There were certainly tactical, if not strategic, justifications for these choices. The Saudi export contract involved a large amount of specialist military construction and BAe was well placed to win the orders if it had the construction capability provided by Ballast Nedam. BAe had a large property portfolio, aircraft and ordnance production are very land intensive, which was being freed by redundancy and restructuring, therefore it seemed sensible to acquire the expertise to develop it in house. However, it bought Arlington at the peak of the property market, and acquired not merely expertise but a series of expensive developments vulnerable to the subsequent collapse. BAe acquired Rover from the Government at what appeared at the time to be a relatively low price, since the alternative would have been foreign ownership. Belief in the synergies between cars and aircraft was fashionable, it was advertised by Saab and demonstrated in the US by General Motors acquisition of Hughes Aircraft in 1985 and in Germany by Daimler's acquisition of MBB in 1987.

Rover and Arlington had been bought on the expectation that they would be cash positive but the recession in the car market and the collapse in the property market meant that this was not the case. In the Autumn of 1991, unexpectedly low profits, a failed rights issue and the enforced departure of its chairman Roland Smith left BAe appearing directionless. BAe suffered severely from acquiring businesses that the management did not know how to run, and allowing diversification to divert their attention from cost and financial control in their core areas, particularly civil aerospace; as a result it lost £1.2 billion in 1992 and cut back its workforce substantially. The new Chairman, John Cahill, was optimistic that BAe's revised strategy—to concentrate on core businesses of military aircraft, Airbus, cars and properties, divesting peripheral areas such as company and regional aircraft—would restore the company to profitability. But he has now been replaced and more of the "core" divested.

Adjustment Processes

Table 17.1 gives figures for employment in industry on defence contracts, from MOD and exports. The official figures for industrial employment only go up to 1990-91, and the main contraction was in 1991-93. The estimates given in the table are based on the projections for MOD spending, the assumption that real export demand will decline slightly in the face of increased competition in the market and the assumption that productivity in the defence industry grows at 5 percent per annum. This productivity growth is about twice the historical rate and reflects my assessment of the impact of redundancy and industrial restructuring. These assumptions suggest that employment in the defence industry declined by about 100,000 between 1990-92 and 1992-93. Trade Union Estimates of the number of redundancies are rather higher than this. Using these estimates and the projections for the number in the armed forces and civil service discussed earlier, Figure 2 plots total employment in defence. It is clear that employment is continuing the trend decline observed through the 1980s. But whereas the cuts in the 1980s were hardly noticed because the economy was booming, those in the 1990s provoked pain because the economy was in recession.

Given these cuts in employment, there has been pressure for Government intervention to help defence firms to convert to civilian production. In the UK both the Labour and Liberal Democrat parties have advocated a Defence Diversification Agency which would use public money to assist defence companies to develop new markets, Voss (1992). Conversion policies are also under active discussion elsewhere., The US position is discussed in Markusen & Yudkin (1992) and Weidenbaum (1992) and Renner (1992) provides a global survey.

The evidence, on the physical conversion of defence plants is by its nature fragmented and anecdotal. The information that gets into the public domain is the small proportion of attempts, successes or disasters, which are sufficiently eye catching to attract media attention. The Hungarian use of two jet fighter engines mounted on a tank chassis to blow out Kuwaiti oil fires and the Grumman buses which fell apart on New York streets are atypical, the bulk of attempts go unreported. There are a few examples of successful UK conversion—GEC-Marconi converted an existing defence plant to produce satellite dishes for Amstrad, who provided the market expertise and access to outlets—but they are notable for their rarity and this seems to be the international pattern. McKinsey concluded that US success rates in conversion have ranged from low to terrible (Business Week April 20, 1992, p. 65). A Rand Corporation study concluded for the US that attempts to shift into civilian products 'will generally be unsuccessful because of mismatches between defense and non-defense experience, skills and technology, but primarily because of their defense oriented company culture and management style which military militate against civilian success.' Alexander (1990, p. 8).

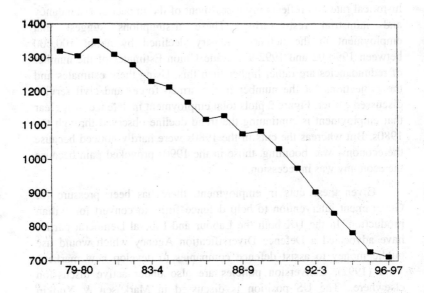

Fig. 17.2 : Total Employment in Defence Thousands

Conversion becomes fashionable when military demand turns down. But if the companies could produce profitable civilian products, why should they choose to wait till times of crisis to develop them? If defence product and process technology gives them an innovative advantage they should be able to 'spin-off' this innovation whether the defence market is growing or declining. But most studies have concluded that cultural obstacles make spin-off from defence technology rather small. Companies make similar judgements, and defence plants complain that new civil ideas which emerge from them get transferred to non-defence subsidiaries which are seen as having the relevant skills to exploit them. A major plan for conversion was developed in the 1970s by a shop-stewards combine at Lucas Aerospace. Although this Plan has become famous in the conversion literature, (for example Voss 1992), Lucas chose to reduce its defence dependence through the safer route of growing its existing commercial businesses.

To be successful in civil operations, defence establishments need markets, organisational flexibility and a cost advantage. But they have little experience at responding to commercial markets, which is why so many conversion plans emphasise the need for government demand for the new products. Their rule based relations with MOD have inhibited the necessary organisational flexibility and their skill lies in performance maximisation not cost minimisation. When you have put a lot of effort into establishing how to do one thing well, it is very difficult to switch to doing something completely different. Such a switch is hard enough for individuals, it is an order of magnitude more difficult for organisations where procedures, hierarchies and social interactions all reinforce the old attitudes, expectations and behaviour patterns. Individuals are much more likely to adjust rapidly in a new commercial culture than in their old teams. Follow-up studies in the US indicate that displaced defence workers do somewhat better than their civilian counterparts in getting new jobs, primarily because of their higher average skill levels and succeed in them. When the UK economy starts to grow again, defence workers could supply the skills needed in other high technology industries.

Thus policies to convert arms factories to alternative production may be no more sensible than a policy of converting coal mines. In the case of coal-mines the obstacles are geological, in the case of arms factories the cultural obstacles may be equally constraining. It is true these plants contain highly skilled workers and expensive equipment capable or producing output. But while this human and physical capital is not physically obsolescent it is economically obsolescence if there is no profitable market for the output it is capable of producing.

The market solution, closure of the defence plants and redundancy for the defence employees may appear extremely wasteful, particularly when, as in the recent UK experience, the bulk of the redundancies have occurred at a time of rising unemployment. But in the longer-run the market solution may be more productive than attempting to convert defence plants. Redundancy and re-employment allow individuals to refocus their technical skills within new teams outside the cultural constraints of the defence companies. There is certainly a role for policy but it is at the locality not at the company or plant level. In the US the Office of Economic Adjustment has been effective at job creation following the closure of military bases. A DOD study of civilian reuse of 97 former military bases 1961-1990, concluded that closure had caused the loss of 93,000 civilian jobs but the creation of 158,000 new ones. Local enterprise initiative such as those used in the UK at former steel towns like Corby and Consett would also play a role. UNIDIR (1993) Table 12, lists a range of other policy options which could aid adjustment. Because of its non-interventionist stance, the UK government has used relatively few of them.

Conclusion

There were quite large cuts in defence expenditure between 1985 and 1990, the share of defence was reduced by 1 percent of GDP, and total defence employment fell from about 1.2 million to 1 million. These cuts attracted little economic or political notice because of the buoyant macro-economic climate. Between 1990 and 1993 defence employment probably fell by about 150,000 but this was at a time when total unemployment was rising from 1.7 million to 3 million, so defence redundancies could not be so easily absorbed. There will probably be another 150,000 job losses over the next three years, as the share of defence falls to about 3 percent, but over this period the economy should be growing and unemployment falling, easing the adjustment. As UNIDIR (1993) describes it, disarmament is an 'investment process', there are short term costs of adjustment but potential long term benefits if the resources can be relocated. The benefits will be both direct, the demands met by those resources in their alternative use, whether consumption, investment or public spending; and indirect, the higher growth allowed by higher investment. However, the indirect effect on the growth rate, while welcome, is unlikely to be very large.

References

Alexander, Arthur, J (1990) 'National Experiences in the field of Conversion: A Comparative Analysis,' Rand Corporation working paper presented to the UN Conference on Economic Adjustments in an Era or Arms Reduction, Moscow.

Barker, Terry, Paul Dunne and Ron Smith (1991) "measuring the Peace Dividend in the UK', *Journal of Peace Research*, 28, pp. 345-358.

Cappelen, Adne, Nils Petter Fleditsch and Olav Bjerholt (1984) 'Military Spending and Economic Growth in OECD Countries' *Journal of Peace Research*, 21, pp. 361-373.

Dunne, J Paul (1990) 'The Political Economy of Military Expenditure : An Introduction,' *Cambridge Journal of Economics*, 14, pp. 395-404.

Dunne J Paul and Ron Smith (1984) 'The Economic Consequences of Reduced UK Military Expenditure,' *Cambridge Journal of Economics*, 8, pp. 297-310.

Dunne, J Paul and Ron Smith (1990) 'Military Expenditure and Unemployment in the OECD,' *Defence Economics*, 1, pp. 57-73.

Findlay, David W and Darrel Parker (1992) 'Military Spending and Interest Rates,' *Defence Economics*, 3 pp. 195-210.

Kennedy, Paul (1988) *The Rise and Fall of the Great Powers* (London: Unwin Hyman).

Leontief, W (1965) 'The Economic Impact—Industrial and Regional— of an Arms Cut,' *Review of Economics and Statistics*, 48.

Lerner, Joshua (1992) 'The Mobility of Scientists and Engineers between Civil and Defence Activities,' *Defence Economics*, 3, pp. 229-242.

Markusen, Ann and Joel Yudkin (1992) *Dismantling the Cold War Economy* (New York: Basic Books).

Rasler, Karen and William R. Thompson (1988) 'Defence Burdens, Capital Formation and Economic Growth,' *Journal of Conflict Resolution*, 32, pp. 61-86.

Renner, Michael (1992) *Economic Adjustment after the Cold War : Strategies for Conversion* (New York: Dartmouth, Aldershot for UIDIR).

Smith, Ron (1977) 'Military Expenditure and Capitalism,' *Cambridge Journal of Economics*, 1, pp. 61-76.

Smith, Ron (1978) 'Military Expenditure and Capitalism: A Reply,' *Cambridge Journal of Economics*, 2, pp. 299-304.

Smith, Ron (1990) 'Defence Procurement and Industrial Structure in the UK', *International Journal of Industrial Organisation*, 8, pp. 185-205.

Smith, Ron and Dan Smith (1992) 'Corporate Strategy, Corporate Culture and Conversion: Adjustment in the Defence Industry,' *Business Strategy Review*, Summer 1992, pp. 45-58.

UNIDIR (1993) *Economic Aspects of Disarmament: Disarmament as an Investment Process* (New York: United Nations).

Voss, Tony (1992) *Converting the Defence Industry* (Oxford, Oxford Research Group).

Weidenbaum. Murray (1992) *Small Wars, Big Defence* (Oxford: Oxford University Press).

Wulf, Hubert (ed) (1993) *Arms Industry Limited* (Oxford : Oxford University Press).

Chapter 18

Geo-Spatial Developments in the UK Armaments Industry and Problems of Adjustment in a Post-Modernist Era

Susan Willet

This paper sets out to examine the forces which have defined the evolution of the UK defence industrial base with particular reference to the Greater London defence industry. Ultimately the UK defence industry has emerged within the context of a complex relationship between state policies, the development of the productive forces and the form of the Britain's insertion into the international economic and security system. These factors have determined not only its size, structure and characteristics, but also its geo-spatial distribution. These issues are dealt with in the first part of the paper. The second piece examines the structural nature of the profound changes taking place in the defense sector as present. The final section of the paper examines appropriate responses to defence industrial adjustment, with particular emphasis on regional adjustment strategies.

The Geo-spatial Evolution of the UK's Defence Industry

In common with other manufacturing sectors, the development of the defence industry has not proceeded along a continuous and regular upward path but rather its evolution has followed a series of cyclical fluctuations. The frequency of imbalances between the total volume of production and consumption, dictated by war and international tension, accounts for the fluidity of these cycles. Because the demand function for weapon systems is subject to fundamentally different influences than the demand for civilian goods, defence production cycles tend to follow a course independent from the rest of industry. As a general rule the military production cycle is governed by political factors, with peaks of production taking place during wartime and troughs being experienced during peace. The booms in military production invariably run counter to the booms in civilian production, creating a tendency for firms with potential military-technology capabilities to expand arms production activities when there is a downturn in demand for civilian goods and to reduce defence production when the demand for weapon systems collapses.

Over time, a pattern has emerged in the UK's defence sector in which the prime contractors of one era which tend to be related to a specific set of technologies are superseded by a new set of prime contractors based on a different and emerging cluster of technologies. This pattern finds its counterpart in Edgerton's historical taxonomy of British militarism, namely, navalism, airforceism, nuclearism and the fourth emergent phase as an active but minor part of a new American-led "Pax Technologica" (Edgerton 1991). The characteristics of these phases are :

Navalism—in the 19th century the Royal Navy was the central component of Britain's war fighting machine. In 1914, the Royal Navy was the largest in the world with the biggest submarine force and the largest naval air service. In this period, warships and guns reflected the dominance of heavy industries such as steel, engineering and shipbuilding. These industrial sectors were mainly based in the North and Scotland where companies like Vickers and Armstrongs led the world in innovation and trade. But these regions began their long decline as the military procurement shifted its emphasis towards the use of combat aircraft.

Airforceism—by the end of the First World War the airforce was to emerge as the strategically important force due to its important role in the strategic bombing of German cities. The RAF was in fact created in 1918. At that time Britain was unique in having a separately organised RAF and Air Ministry. By the end of the 1930's the airforce had become the principal force used against Germany. The bomber became known as the "dreadnought of the air". By 1939 the airforce was the largest spender of the three services, whereas in the 1920's it had been the lowest. In 1940 the UK was the largest manufacturer of aircraft in the world. The period of airforceism coincided with the application of scientific knowledge to industry and technology. The emergence of a new set of defence contractors associated with the aircraft and aero-engine sectors which were located in the urban centres of the South of England, particularly in and around London. The period from the 1930s to the 1960s saw the Greater London defence industry reach its zenith with the growth of aircraft companies such as Handley Page and Sopwith in Kingston and the Short Brothers in Battersea and electronic companies such as GEC, Plessey and Marconi. These companies all required highly skilled labour in engineering and sciences, which proved another factor in their choice of location, as highly skilled staff tended to be located in the South rather than the industrial North. By the early 1970s most of the London defence production sites were engaged in long-term military projects. The nationalisation of the aerospace and shipbuilding industries in 1977 had few significant short-term effects. Little restructuring in fact took place with the predominant pattern being amalgamated companies remaining at individual sites.

Nuclearism—from the 1950s onward Britains defence policy involved, for the first time during peacetime, a commitment to conscript troops. At the same time, it had begun to develop an independent nunclear programme and in 1952 the Global Strategy Paper committed the UK to building up a nuclear strategic air offensive capacity, this involved the MoD in funding the hugely expensive V-bomber programme, But with rearmament, during the Korean War, the costs of this broad-based approach become too high. In 1957, the Defence White paper announced that conscription was to be abolished, as was much of the airforce, the V-bomber programme and a significant part of the Navy. Instead, Britain was to rely on an independent nuclear force. The possession of nuclear weapons had become a symbol of potency in the post-war security environment, and helped a declining UK maintain its influence in global affairs. In the late 1950s, Britain's strategic force consisted of nearly 200 long-range nuclear bombers, giving it a nuclear capability that was comparable with that of the US and USSR. With the adoption of Flexible Response doctrines in the late 1960s the defence electronics industry grew in importance. New developments in communications and guidance systems as well as the advent of the micro-circuit revolutionised many weapon systems. The age of nuclearism was also associated with the growing R&D intensity of weapon system and the establishment of research centres like Aldermaston located in greenfield sites away from centres of population and concern.

Pax technologica—the ending of the Cold War reflects a turning point. While the nuclear element remains, the key new instrument of military strategy is the multinational Rapid Deployment Force equipped with ever more sophisticated and high-tech weaponry, particularly command, control and communication systems. Such forces are rapidly being deployed to police the "new world order" (or disorder). The new "pax technologica" articulates the material interests of certain sections of the military industrial complex, ideologically justified by a new form of universalistic idealism. In this phase British defence policy is subordinate to the USA, but the characteristics of militarism remain very much British.

This phase is based on technologies such as computers, new materials, fibre optics and software. The high technology companies producing such military goods tend to be located in high technology "enclaves" in greenfield sites along motorways such as the M4 and M11. They have contributed to the phenomenon of high tech "agglomeration" economies based around Newbury and Cambridge. The growing importance of electronic based systems generated a new generation of defence contractors in the electronics sector, such as GEC, Plessey, Ferranti and Racal which became predominant in the 1980s and more recently software houses and computer susyems suppliers such as CAP Scientific Ltd, Systems designers plc, Scion Ltd and Logica. A predominant feature of this period is the blurring of the industry's national characteristics by the tendency towards internationalisation of production and ownership patterns. The inducement to globalise operations through a complex web of arrangements such as licensed production, offsets, sub-contracting, full-scale collaboration including R&D co-operation and even mergers and takeovers, is particularly strong among those companies operating in advanced technology markets with high R&D costs and short production runs. These new patterns of international corporate linkages have resulted in of form of technology networking known as 'techno-globalism'.

During the boom period in the early 1980s the defence sector was largely able to shield itself from the impact of post-Fordist patterns of production and organisation due to the highly protected nature of domestic defence markets. The UK increased its military expenditure by an average of 3 percent per annum in real terms, when many other areas of public expenditure were under attack. The rapid increase in military expenditure was largely targeted at procurement programmes creating buoyant conditions for the defence industry. The high level of military expenditure which was sustained during this period was only achieved by imposing large cuts on other government spending programmes. This led to a structural weakening of the economy due to the defence industry's consumption of scarce investment resources. The effect was to create what Smith and Smith have called a 'dual sector economy', in which the military sector boomed while the civil sector stagnated (Smith and Smith 1983 p. 38). High military expenditure artificially expanded industries that would otherwise have contracted. It absorbed resources that might otherwise have been used for investment and innovation in newer and more dynamic industries. It distorted ideas of what constituted technical advance, emphasising complex custom built products instead of mass market process innovations. It contributed to the slow down in capital investment and productivity growth and to the long term decline of the UK economy.

By the mid-1980s the Thatcher government was no longer able to sustain high military expenditure and defence spending began to decline. As a result of the state's in ability to maintain its protective mantle over the defence sector, defence companies became increasingly subjected to the influence of normal market forces. Of this process Lovering noted that the continuity which had characterised the industrial, corporate and geographical structures of the Cold War defence industrial base came to an end. A smaller post-Cold War defence industry began to emerge based on different political and economic relationships, with new social and geographical implications (Lovering 1993).

The Defence Industry in a Post-Modernist Era

A number of recent authoritative studies exist on the changing nature of the defence industrial base (Hayward and Taylor 1989, Brzoska and Lock 1992, Wulfe 1993). While providing a detailed insight into the contemporary restructuring and re-orientation of national defence industries they provide little perspective on the evolution of regionally based defence industries. *Defence Expenditure, Industrial Conversion and Local Employment*, Paukert and Richards (1991) produced the first major international study to address the regional impact of disarmament. Since then the interest in regional defence industrial adjustment has grown. However, a general tendency of these studies is to remain at the level of description, examining the defence sector in isolation from the broader trends and influences which determine regional, national and international economy and security trends.

This can partly be explained by the fact the defence industry has been marked out from the rest of the manufacturing sector because of its strategic importance to the state during the Cold War. Despite the high degree of protection afforded by the state, the defence sector has never been wholly insulated from the broader economic and social upheavals that are characteristic of the capitalist system. Supply and demand dynamics, innovation, economies of scale, globalisation of production and profitability are all factors which have determined the sectors structure and performance. The fact that defence production more often than not takes place within companies that have large civil operations means that they are never wholly insulated from commercial market influences. The Recent trends disarmament has intensified defence contactors exposure to market forces.

Moreover in the final decade of the twentieth century profound changes have been taking place in the global economic and security order. The collapse of the bi-polar world has turned much conventional thinking on its head and caught many academics and policy-makers off guard. Neo-liberals were quick to hail the end of the Cold War as the triumph of capitalism over communism. Francis Fukuyama went so far as to declare it an Hegelian 'End of History' (Fukuyama 1992)-discontinuation of the 'history of thought about first principles, including those governing political and social organisation'. This triumphalism was short lived, however. The capitalist system did not emerge from the end of the Cold War in good shape. The depth of the world recession, characterised by a contraction of global markets, the slowing down of economic growth and widespread structural unemployment, exacerbated the transitional problems of Eastern Europe, giving rise to the ugly face of right wing nationalism. The 'New World Order' fast disintegrated into regional and inter-ethnic conflicts.

The confidence which characterised the Western security alliance during the Cold War rapidly dissipated into uncertainty, confusion, complexity and ambiguity. As Heisbourg observed, the postwar security structure was defence orientated, clear cut, comprehensive and rigid, but in the new system, confusion and complexity prevail. Defence has been losing its centrality and the nature of the threat has become ambiguous (Heisbourg 1991).

What theoretical underpinnings can inform an analysis of such profound and fundamental shifts in the global order. Post-modernism and Post-Fordism are different ways of trying to characterise and explain the dramatic, even brutal break with the past. The sense of being at an end of an era finds expression in the post-modernist discourse in which the relationship between modernity, progress and rationality has been brought into question. Hebdige argues that at the core of this debate lies the 'legitimisation crisis' 'If modernity is a condition—' in which all that's solid melts into air '—then the old institutions and centres of authority—from religion to royalty—which guaranteed stability and continuity in earlier epochs and more traditional societies are prone to crisis and contestation.' (Hebdige 1989 p. 78). According to Lyotard the French philosopher the legitimisation crisis of the past had been solved through the invention of 'meta-narratives' of the modern period. These are the overarching belief systems originating in the Enlightenment—from the belief in rationality, science and causality to the faith in human emancipation progress and class struggle. These great stories have been used to legitimise everything from war, revolution, nuclear weapons, concentration camps and social engineering, Taylorism, Fordist production models to the gulag. The present collapse of faith in these meta-narratives heralds what Lyotard terms the 'post-modern condition'. None of these centres of authority legitimated by the meta-narratives has survived the transition into the new era. Just as in the realm of art and architecture where the post-modern debate first found expression so too in the world of defence and security a 'crisis of legitimisation' is taking place.

While the post-modernist discourse provides some powerful insights into the disjuncture of the Cold War, it is against a background of economic transition characterised by the move from Fordism to post-Fordism that the changes taking place within the defence industry can best be understood. Post-Fordism is a term suggesting a whole new epoch distinct from the era of mass production, with its standardised products, concentrations of capital, and its 'Taylorist' forms of work organisation and discipline (Murray 1989). Hall provides a list of characteristics which define Post-Fordism (Hall 1989 p. 117-119) ;

First it is characterised by a technological paradigm shift, the beginning of a long wave of capital accumulation based on the emergence of 'information technologies'. This trend is distinguished by a shift from the old arenas of accumulation in the automotive, chemical and electrical industries which propelled the second industrial revolution from the turn of the century onwards—heralding the advance of the American, German and Japanese economies and the relative decline of the British economy.

Secondly, it represents a shift towards a more flexible specialised and decentralised form of labour process and work organisation and as a consequence, a decline of the old manufacturing base (and regions and cultures associated with it) and the growth of sunrise computer-based, hi-tech industries and their regions.

Thirdly, one of its central features is the hiving-off or the contracting-out of functions and services hitherto provided 'in-house' on a corporate basis.

Fourthly, consumption plays a leading role, reflected in such things as greater emphasis on choice and product differentiation, on marketing, packaging and design, on targeting of consumers lifestyle, taste and culture rather then by categories of social class.

Fifthly, there is a decline in the proportion of the skilled male manual working class and the corresponding rise of the service and white collar classes, In the domain of paid work itself, there is more flexi-time and part-time working, coupled with the 'feminisation' and 'ethnicisation' of the workforce.

Sixthly, a dominant feature is the globalisation of production. National and international economies are increasingly dominated by multinationals, which impose a new international division of labour and operate with relative autonomy from nation state controls.

Seventhly, the transnationalisation of production has been accompanied by an internationalisation of financial markets and the emergence of new financial centres.

Finally, there are new patterns of social division—especially-between public and private sectors, and between the two-thirds who have rising expectations and the 'new poor' and underclasses of the one-third that is left behind on every significant dimension of social opportunity.

The Greater London Defence Industry

The reduction in UK defence expenditure from the mid-1980s onwards has a disastrous effect on the Greater London economy, and has been a major factor contributing to a deep and prolonged recession in the nation's capital. For a detailed description of the Greater London defence industry in the 1980s see Willett in Paukert and Richards (1991).

The response of defence companies to declining profitability was to cut production costs by reorganising its manufacturing activities. Typically this has been achieved by mergers and takeovers followed by the closure of older more labour intensive plants and concentrating production into larger more capital intensive units employing a smaller workforce.

A significant number of defence workers have lost their jobs as a result of restructuring. In the mid-1980s the Greater London Council estimated that about 95000 jobs were supported by defence spending within the Greater London region. Apart from military personnel, this included an estimated 32000 defence workers employed in about 70 companies, both prime and subcontractors. As a proportion of total manufacturing jobs in the Greater London area, defence employment accounted for roughly 7 percent of manufacturing employment. About two-thirds of all defence industrial jobs, roughly 22000, were located in nine major companies, GEC-Marconi, British Aerospace, Plessey, Thorn-EMI, Muirhead Vatric, Racal, the Royal Small Arms Factory, Lucas Aerospace, and Dowty plc. By 1992 two of these large defence contractors had disappeared from the Greater London industrial landscape. The Royal Ordnance sites at Enfield and Waltham Abbey and the BAe Kingston plant were closed down with a loss of 5500 engineering jobs. These large defence contractors were major employers in their localities maintaining a relatively advantage core of workers with above average incomes, which had a significant multiplier effect in their local economies. Thus the negative impact of plant closures to these localities far outweighs the number of job losses.

The major characteristics of the defence labour force in the 1980s was scarce skills, high mobility and high wages. The defence sector provided a protective haven for this skilled labour shielding it from competitive pressure and providing a refuge for organised labour, which was undergoing a concerted attack by a hostile Conservative government. Collective bargaining procedures and formalised internal labour markets governed employment status and rewards, generally ensuring good pay conditions and chances of advancement. However, the reorganisation of defence production which began in the late 1980s involved large scale job losses as the contractors addressed overmanning on the shop floor and within middle management grades. Long standing employment practices were undermined as firms attempted to modify promotion ladders, collective bargaining agreements, demarcation patterns and recruitment practices.

Amongst the numerous sub-contractors in London that existed in the early 1980s there is some limited evidence that a number relocated, diversified out of the defence market or have simply gone into receivership. (Willett 1990) Relocation of production facilities has been a central feature of London's industrial decline since the 1960s. The factors which have given rise to this trend include high land prices and rents, poor road access, archaic production facilities and highly paid organised labour. An added incentive for relocation during the 1980s property boom was the ability to finance new investments from the sales of assets due to the high premium on land for redevelopment.

The ability of displaced defence workers to find alternative employment has been severely undermined by the depth of recession in the engineering industry and the profound restructuring of the regional labour market over the last decade or so. There has been a profound shift in demand for labour reflecting the decline of the manufacturing sector and the growth of the service sector.

Engineering output in the UK has been affected by the fall in demand for capital goods and the increased market share taken by imports. The industry's investment and market share have fallen dramatically and competitiveness has declined. A major component of this trend has been the introduction of new technology based production systems, such as CAD/CAM, which have reduced the need for skilled operatives and drastically cut labour inputs into the firms' production costs.

The overall contraction of the engineering industry means that there is a significant reduction in employment opportunities between different occupational groups within the industry. In particular there has been a massive reduction in the numbers of production operatives who are employed which has affected the employment opportunities of women and ethnic workers particularly adversley. There has also been a direct impact on the apprentice training opportunities due to the dramatic cuts in intake of craft and technician apprentices. The government's closure of existing skill centres as part of a reduced public sector involvement in skill training has further exacerbated this problem.

The general decline of manufacturing employment in the Greater London area and the country at large has meant that many defence workers have joined the long-term unemployed or have been forced to take occupations in the service sector in jobs unrelated to their skills.

Reactions to the Defence Industrial Decline

On the whole, the UK government has reacted to defence industrial restructuring by stressing the role of market forces, shying away from more interventionist strategies. But the scale of the problems being experienced in certain industrial sectors and the way in which the down-sizing of the defence sector has compounded recessionary trends has raised questions about the wisdom of this approach. Pressure has existed for some time to use the 'peace dividend' to help defence industries adjust to the new market environment by converting or diversifying into civilian production. Conversion (as understood as the transformation of military resources into civil activities and production) is by no means a new phenomenon. During World War Two conversion from the civil to the military and then vice versa was a widespread occurrence. However, the conditions which favoured conversion in the past no longer exist. Dumas (1989) has pointed out that military and civil production have widely diverged since World War II, and while in the past military production generated a number of significant spin-offs into the civil economy, the present norm is the reverse with '8pin-in' dominating the patterns of technology transfer between military and civilian industries. In addition, there are now a whole generation of scientists and engineers who have worked solely on military products, isolated from the competitive discipline of market forces.

In the 1970s and 1980s the advocates of conversion, including the Greater London Conversion Council, were largely concerned with winning the moral and political arguments for disarmament. Conversion was a means to an end. It was a way of gaining political support for disarmament by proposing a solution to the negative economic consequences of defence cuts, particularly job losses and skill displacement. However, such proposals were characterised by a *priori* assumptions and advocacy and limited by the specific concerns of pacifism, labourism or companies and regions seeking to alleviated the effects of defence cuts (See Lovering 1991a).

With the end of the Cold War conversion has become an all encompassing concept open to many interpretations. Willett (1990), Southwood (1991) and Cronberg (1992) have attempted to classify the different approaches to conversion and their implications. A number of strands can be identified which work at different levels, starting with macro-economic adjustment, then corporate policies, plant-based approaches, community responses and finally economic regeneration strategies which emphasise Industrial structural adjustment. In general, the macroeconomic studies of conversion concern themselves with the demand side of the problem, analyzing what the impact of the cuts in defence spending will be, either without considering the problems of adjustment, or simply assuming that the market will deal with them. In contrast, the other approaches focus on the supply-side adjustments required at the level of the company, plant and community. Industrial structural adjustment policies attempt to deal with all levels in a coherent and consistent manner and have the explicit goal of demilitarisation.

The **macro-economic approach** to conversion focuses on the evidence of the negative relation between military spending and economic development. Pioneering examples of this approach include Melman (1974) and Smith (1976). While there is still some debate, see Dunne (1990), it is generally accepted that military spending represents an economic burden and that reducing it can lead to improved economic performance. This school of thought argues that equivalent levels of investment in the civil sector creates more employment than in the military sector. Thus the transfer of defence savings to investment in the civil economy would not only create jobs for those displaced by military expenditure cuts, but would also generate more employment opportunities than previously. At the core of this approach is the focus on macro-economic adjustment policies. The government can use the defence savings to reduce the budget deficit, to increase other government expenditures (current and capital), or to cut taxes. All of these will have different effects and may create different problems of adjustment. See Oden (1990) for a comparable review of this debate in the USA.

Barker, Dunne and Smith (1991) provide an example of the macro-economic approach applied to the UK situation. The Cambridge Growth Project inter-industry model of the UK economy was used to investigate the impact of cutting military spending by one half by the end of the century. Chalmers (1990) provides a detailed discussion of the force structure consistent with such a reduction. This implies an 8-9 percent per annum cut in military expenditures, considerably more than the 2 percent per annum being considered at present, though from a lower level. The result of such a cut without reallocation the saving, is a reduction of about 200000 defence jobs and almost a 4 percent reduction in GDP. This is close to what would be expected to happen if the defence expenditure savings were used for deficit financing. When this expenditure is reallocated to the other categories of expenditure 600000 jobs are generated, giving a net increase of 400000 jobs. Simultaneously a net increase of almost 2 percent of GDP is generated. This evidence suggests that at a macro economic level there should be no problems in adjusting to lower levels of military expenditure and that with sensible policies of adjustment there can be improved economic performance. The study also discusses the context of these changes and some of the likely problems of adjustment. In particular it present estimates of the likely industrial impact, and suggests where there may be problems.

Underlying the simulations, however, is an implicit assumption that the defence contractors are able to adjust without government assistance and that the defence workers, with their particular skills, will to able to gain employment in the jobs created by the increase in expenditure in other parts of the public sector, and those created in the private sector by multiplier effects. This may in practice be problematic and may lead to structural unemployment. In addition, the costs of troop relocation, retraining, rehousing, and the restructuring and re-equipping of remaining troops could be costly. This was not taken into consideration in the simulations.

Unless the macroeconomic policy of resource allocation is part of a well-orchestrated economic plan for restructuring, the UK an economy could risk losing scare high technology capabilities, particularly technical skills and know-how which could result in a long-term economic cost in the form of worsening balance of trade and payment deficits, technological dependency and thus loss of control over the future direction of economic development. In addition, reallocating resources to other items of public expenditure does not address the considerable social and regional economic disruption caused by plant closures. It is therefore important to consider other approaches to conversion, within the context of the general macroeconomic, demand side, picture.

The company-based approach refers to corporate adjustment strategies to defence market decline. At the corporate level, the terms conversion and diversification are often used interchangeably but are actually quite different. Conversion means the transformation of military production facilities into the production of civilian goods and implies that a defence firm stops making military products altogether. Diversification, on the other hand, involves a company shifting into other areas of production without converting defence industrial facilities and does not necessarily imply the substitution of military work. This has the advantage of making a defence contractor less vulnerable to a downturn in the defence market, but has the disadvantage that it may just shift into other defence areas.

At the level of the firm, diversification is the norm. It can either occur in the form of mergers and acquisitions or through organic growth. There is only limited evidence of UK defence firms successfully diversifying. British Aerospace attempted to diversify into care manufacture through the acquisition of the Rover Group, which had the effect of reducing its defence market dependency from 60 percent of total turnover to 30 percent. But in 1994 BAe abandoned this strategy when it sold Rover to the German car manufacture BMW. Corporate diversification has proved marginal in terms of compensating for production and employment losses taking place in their defence divisions, see Schofield, Dando and Ridge, (1992).

The successful conversion of existing military plants to civil production is even less in evidence. The Russian and Ukrainian governments have both introduced conversion policies which emphasizes plant-based conversion, but so far there is very little evidence of success. Cronberg (1992) argues that the contemporary problem inherent in military conversion, irrespective of economic system, relates to the military-industrial product concept-producing for one customer based on a 'performance at any cost' principle—and the specific product development culture created by the closed nature of military secrecy. The general failure of defence companies to adjust to the changing market environment arises from the considerable barriers to exit, which are essentially due to the orientation of defence contractors' culture and management style. (See Alexander 1990). These include : inexperience in commercial marketing, emphasis on product rather than process innovations, a hierarchy of skills and a technological orientation structured by the weapon systems acquisition process, a management organisation isolated from commercial practices, a production structure defined by large scale systems integration, protected markets through government procurement, and risk aversion through government subsidies in R&D and capital investment.

This combination of barriers to exit from the defence market explains the fact that despite attempts at conversion, since the 1960s no country in the world has succeeded in converting military industries to civil production. This is in spite of the fact the military and civilian production more usually than not coexist within the same firms.

Lower down the defence production hierarchy those contractors producing generic technologies which have dual-use applications, mainly subcontractors, are unlikely to need to convert their production processes or products. For this group of contractors the crucial issue is to identify alternative markets for their outputs, which is dependent on healthy demand from civil industries. In this context a demand stimulus is far more important than supply side intervention.

Plant-based conversion is concerned with the transformation of resources which have already been accumulated by the military-namely defence production facilities, skills and scientific and technical knowledge and in some instances military products and components. It is the traditional 'swords into ploughshares' approach which aims to replace defence production capacity with civil activities. It seeks technical solutions to existing production facilities and skills. It tends to rely upon government or local government markets to guarantee survival. The most well known of such proposals was the Lucas Corporate Plan which was produced by defence workers in 1975-76. Central to the Lucas Plan was the notion of reshaping technological priorities through the development of 'socially useful products' which could be made utilising the existing resources and skills of defence plants. Although never implemented, the Lucas Plan provided a model for many such workers' plans and workplace conversion committees. (See Wainwright and Elliott 1982).

At the centre of the plant-based approach to conversion is the alternative use committee. Ideally such a committee would include management, labour, and community representatives who would join forces to draw up an alternative business plan. While the interest in alternative use committees has gained widespread acceptance amongst trade unions, local authorities, and peace groups so far there are few examples of successful alternative use planning to be found. Alternative use committees have found a lack of cooperation from management, but there are other significant barriers to plant-based conversion. The social groups with make up alternative use committees do not always seek the same objective. Defence workers' and the trade unions' priority is to retain jobs and this often means continuing to lobby for defence contracts. Peace activists, in their moral commitment to achieving disarmament, are often insensitive to defence workers plight and vulnerabilities. Local authorities may actively promote alternative production at local defence plants with the resources to back up workers' initiatives.

In general alternative use committees operate under unrealistic assumptions about how business and markets operate in the commercial world. The lack of technical expertise and knowledge of the innovatory process, may lead alternative use committees to make technical proposals with little realistic consideration of costs of production, market opportunities, or required investment capital. Plans are often drawn up with little acknowledgment of the problems of raising investment funds, as few of the participants have entrepreneurial expertise. The management of the defence companies also tends to share these characteristics and problems.

Unless new markets are identified, entry into existing markets is highly risky because competing firms as well up the learning curve and already have the advantage of economies of scale. In this context the provision of government markets would represent an opportunity cost as they would not be purchasing from the least cost source of supply. Furthermore, new EC legislation rules out the possibility of reliance on government markets.

In the final analysis the capital/labour endowments of most defence facilities are rarely appropriate for volume commercial production. Prime contractors tend to be produce custom build products to a very high standard of specification. The technological characteristics of the product are the major selling point rather than costs. This is because the market is guaranteed and the customer, not the producer, bears the brunt of costs for product enhancement. Production runs are short and the production process is skill, labour and capital intensive. In contrast, the production of all but the most exotic of luxury commercial goods is characterised by economies of scale, production techniques which have a low labour to capital ratio, and is above all driven by unit cost consideration. Because governments guaranteed both the markets and profits for defence output, the production and labour processes within defence plants are often archaic in comparison with those employed in commercial enterprises. In addition commercial companies have widespread experience in marketing and selling in highly competitive markets and dealing with a multiplicity of customers. Defence companies or divisions, in contrast, are used to dealing with one major customer and have built up the networks and experience to deal with what is essentially a highly bureaucratised market.

Alternative use committees rarely have experience in commercial production or marketing, they tend to lay emphasis on product rather than process technologies, they seek to preserve the existing hierarchy of skills and military-technology culture which is isolated from commercial disciplines, reflecting an industrial and technological culture which has been nurtured by the structure and nature of defence production and markets.

The civil counterparts to the particular combination of capital, skills and technology culture are to be found in civil aerospace, satellites, nuclear power, power generation, space technologies and off-shore oil rig platforms. Most of the major defence contractors are already involved in these types of civil programmes, and if anything, overcapacity exists in these markets. In addition, such programmes are often publicly funded and likely to prove very expensive as well as involving considerable opportunity costs by diverting resources from other areas of public expenditure or the economy. Moreover, given the notable absence of venture capital in the UK, the opportunity for firms to self finance such programmes is severely constrained.

The emphasis in conversion programmes on product solutions to employment problems reflects a poor understanding of the present restructuring of the global economy, the new forms of innovation, technological change and market penetration as described is the section a Post-Fordism. Attempts to produce civil technological systems within the context of production structures which have evolved out of the military production and procurement system are likely reproduce the same technological characteristics of weapon systems, namely capital intensive, complex, over specified and costly (see Kaldor 1983). As such they reinforce the existing and outmode of technologies industrial and technological patterns of production rather than transform them, representing a missed opportunity in terms of the potential for economic restructuring. Conversion programmes tend to ignore completely the innovatory implications of the diffusion of existing process technologies. The great technological successes of the Japanese other Asian Pacific economies result not so much from the evolution of radical new products, but rather from they innovative application of existing techniques and technologies to a wide range of existing products, ie, digitalizing televisions, miniaturising products through the application of microchips, improving the productivity of traditional industries through the use of flexible manufacturing systems, and the application of information technologies which have given rise to 'just in time production' all contributing to the highly competitive nature of Japanese products. See Best (1990).

Local community conversion accentuates the role of local authorities and activist groups like trade unions and peace groups working together to develop strategies to reduce the vulnerability of local economies on defence expenditure. The emphasis is on local communities and local authorities taking the initiative in supporting workers at local plants or bases to convert their facilities or to diversify their local economies away from defence dependency. This approach has largely found its inspiration from the USA where local state initiative supported by the Department of Defence's Office of Economic Adjustment were successfully implemented following the downturn in military expenditures at the end of the Vietnam war.

In the UK and Germany a number of local authorities, including the Greater London Council in the 1980s, set up conversion committees to actively supported local conversion initiatives. But the majority of these initiatives were are engaged in promoting plant-based conversion programmes, which are subject to all the limitations outlined above. So far there have been no successful conversions of defence facilities to civil production, although, in a couple of cases, the effect of local campaigns have prevented plant closures such as the Plessey military ratio plant at Ilford in London and the navel stores depot in Llangenerch, Wales.

The failure to convert defence plants to civil production has led local governments to re-examine their initiatives. In the last three years there has been a noticeable shift of emphasis away from the defence plant to the broader context of regenerating the local economy. The transformation of local economies through small innovative firm-led disarmaments, modelled on the experience of the north and central Italy, the Basque country and Southern France, offers and alternative strategy for local economic regeneration. (See Best 1990). But for such local initiatives to work there needs to be either central government support for such initiatives, as in the United States, or a high level of local autonomy in politics, finance and legislative power.

In the United States a number of State initiatives have adopted the regional and community approaches. In California for example the Career/Pro programme is training people laid off from base closures through a community college programme, in association with both commercially orientated and defence dependent contractors, to clean up the closed down based using experimental technologies. The programme also helps to build companies in a growing market—environmental remediation—which is predicted to be strong well into the next century. Already the consortium has won a number of federal contracts to clean up toxic waste sites. Another example of a community which is effectively taking the lead in community based conversion is in St Louis where McDonnell Douglas has laid off tens of thousands of employees. An Economic Conversion Project has been set up to assist displace employees. At least $250000 have been allocated to the programme. See Lall and Tepper Marlin (1992) for details of successful local programs in the US.

Industrial Structural Adjustment—At the national level strategies for adjustment need to addresses simultaneously the problems created by demilitarization, demobilization, reduced defence expenditure and reductions in weapons production. In this context conversion presupposes a general shift in national priorities, away from the military imperatives which have dominated national technological and industrial objectives of the high military spenders into a new trajectory of economic development (See Sandholtz et al 1992). This approach is part of a broader social and economic agenda of reconstruction. (See Willett, 1990, Renner, 1992, Yudkin and Black 1990, Cronberg 1992).

Central to this approach is the recognition of 'market failure' in advanced capitalist societies in meeting basic human and environmental needs, and the urgent requirement of shifting industrial, technological and scientific resources away from militarily defined objectives and instead targeting 'national needs' like industrial renewal, environmental restoration, sustainable agriculture and renewable energies. Such a national needs policy, although initiated by the government should operate in partnership with industry, finance and local and regional authorities, workers and consumers.

The mechanisms available to government to encourage such developments include direct, subsidies as well as incentives, for industries to conduct basic research, improve technology transfer, support job training, education and environmental regulation. If the government were to use the peace dividend to meet national needs the economic returns would offset the short-term dislocations caused by defence spending cuts. However, bringing a country's industrial and technology policies in line with its national needs requires more than just drawing up a shopping list of industrial and social needs and then shifting funds from defence towards other public expenditure programmes. More fundamentally an economy requires a new framework for guiding R&D policies, scientific and technological priorities and relationship between the private and public sectors. Such a framework would provide a filter for selecting critical industries and technologies that would receive R&D and other forms of government assistance.

R&D programmes targeted at high-tech performance in computing, advanced materials and optoelectronics, for example, could be designed to complement technology initiatives in metal processing, public transport, renewable energy system, environmental monitoring and pollution control. At the same time superconductors and ceramics which are at present being developed largely under military programmes have potential application in fuel-efficient engines, cutting tools, power turbines and many consumer items.

The ability to generate technological innovations and to incorporate them quickly into industrial processes and commercial products—in other words the process of technological transfer—is the key factor in improving productivity and competitiveness. But, a high rate of technological innovation and diffusion requires a strong civil science and technology base. In the high military spenders such as the US, the UK and the former Soviet Union science and technology have been driven by 'mission orientated' programmes largely funded by military or space programmes. In contrast, in Japan and German science and technology have been has been orientated towards civilian concerns and has been driven by diffusion orientated R&D strategies, which are largely responsible for the innovative successes of both countries in the last decade.

Because of the structural barriers to conversion within the defence sector, mechanisms such as retraining, start-up incentives, tax breaks, civil R&D programmes and investment programmes should be designed to restructure capital and labour away from the defence industrial base and into more dynamic and innovative forms of production and labour processes, by encouraging spin-out companies. This would help create a radical break with the pervasive culture of the defence industrial base. The use of such mechanisms need not be prohibitively expensive and offers more positive benefits to the whole of society in the long run. Critical to this strategy is support for community-based technology projects, technology extension services and the establishment of technology advisory centres which can assist local businesses with the application of new process technologies.

Another institutional step would be to increase the R&D advisory process within the government system. Regional and local interests should have far greater representation on such advisory groups and panels. In addition, local and regional planning authorities would need to be empowered to aid regional development programmes which address the totality of a local economy and not just the declining defence sector. This would benefit from research into the experience of localities which have gone through major factory closures, through the contraction of steel and coal.

In the European Parliament there is strong support for the formation of a structural fund for regional economic regeneration in defence dependent communities, comparable with that of previous EC programmes for regenerating regions previously dependent on the shipbuilding and steel industries. However the EC Council and the European Commission are less enthusiastic for such proposals. A major obstacle to EC involvement in a structural adjustment programme for the defence industry is Article 233 of the Treaty of Rome which recognises national sovereignty in defence and security issues thus excluding national defence industries from the open market policies of the European Community.

The evidence of the economic successes of 'the Third Italy', and the Basque region of Spain where groups of small companies using flexible technology compete successfully in highly competitive world market, give much credence of strategies which argue for regional regeneration through the application of process innovations to mature industries. (See Cooke, 1990, Best 1990). At the same time that mature industries are being revived, several 'strategic' high-tech industries—microelectronics, computers, telecommunications, advanced materials, robotics and numerical controls—should also be targeted as they are crucial to the performance of other industrial sectors. These industries would no longer be dominated by military objectives, R&D policies directed towards these sectors would instead reflect commercial as well as social objectives.

Conclusions

The combined impact of the end of the Cold War and Post-Fordism has hastened the decline of the Greater London defence industry. No longer is it a major location for combat aircraft production, small arms of explosives. A handful of the larger sub-components suppliers remain, but even their future remains uncertain as the pace of restructuring redefines the arms industry on 'an international stage.

Laissez-faire policies adopted towards the defence industrial base have been clearly seen to have failed as have the traditional plant-based conversion programmes advocated by the Greater London Conversion Council in the 1980s. The alternative of restructuring production and economic priorities towards national and environmental needs implies the creation of entirely different industrial and technological structures. In this context conversion represents much more than a strategy to save defence workers' jobs or industrial capacity ; it provides an important lever for the establishment of different goals in which social and environmental values have a central place.

References

Alexander, A. (1990) 'National Experiences in the Field of Conversion. A comparative Analysis', Paper presented at the UN Conference on Conversion, Moscow, 13-17, August.

Anthony, I. Courades Allebeck, A. and Wulf, H. (1990) West European Arms Production (SIPRI.)

Barker, T. Dunne, P. and Smith, R. (1991) 'Measuring the Peace Dividend in the United Kingdom' *Journal of Peace Research, 28, 4,* November, pp. 345-358.

Baylis, J. (1989) *British Defence Policy : Striking the Right Balance* (Macmillan)

Best, M.H. (1990) *The New Competition : Institutions of Industrial Restructuring* (Polity Press.)

Brzoska M, and Lock, P. (1992) *Restructuring of Arms Production in Western Europe.* (SIPRI, Oxford University of Press.)

Campbell, M. (ed) (1990) *Local Economic Policy* (Cassell Educational Ltd.)

Cooke, P. (1990) *Manufacturing Miracles : The Changing Nature of the Local Economy* in Campbell (1990).

Cronberg, T. (1992) The Social Reconstruction of Military Industries and Technologies : Seen from a Danish Perspective, Paper presented at the International Institute of Peace Conference on Conversion, Vienna, February.

Driver, C. and Dunne, P. (1992) *Structural Change in the UK Economy* (Cambridge University Press.) forthcoming.

Dumas L. and Thee, M. (1989) *Making Peace Possible* (Pergamon Press, Oxford.)

Dunne, P. (1990) 'The Political Economy of Military Expenditure : An Introduction, '*Cambridge Journal of Economics*, Vol 14, No 4, Dec, pp. 395-404.

Dunne, P. and Smith, R. (1992) 'Thatcherism and the UK Defence Industry' Chapter 5 in Michie, J. (1992) *1979-92 The Economic Legacy*, Academic Press.

Edgerton, D. (1991) *Liberal Militarism and the British State* New Left Review No. 185, Jan/Feb.

Fukuyama F. (1992) *The End Of History and the Last Man* London Penguin Books.

Hall S. (1989) "The Meaning of New Times "in Hall S. and Jacques M. (eds) *New Times : The Changing Face of Polities in the 1990s* London Lawrence and Wishart.

Heisbourg F. (1991) Conference, Royal United of International Affaires, London, Chatham House, Council of Foreign Relations, December.

Hebdige D. (1989) 'After the Masses' in (Hall and Jacques (eds) *New Times : The Changing Face of Polities in the 1990s* London Lawrence and Wishart.

Hilditch P. (1990) 'Defence Procurement and Employment : The Case of UK Shipbuilding' *Cambridge Journal Economics*, Vol 14, No 4, Dec. pp. 453-468.

Hayward K. and Taylor T. (1989) *The UK Defence Industrial Base* (London, Brassey's.)

IPMS, MSF, TGWU (1990) *The New Industrial Challenge- The Need of Defence Diversification.*

Kaldor, M. (1982) *The Baroque Arsenal* (London, Andre Deutsch.)

Lall, B. Tepper Marlin, J. (1992), *Building a Peace Economy* (Boulder Westview Press.)

Lovering, J. (1990) 'Military Expenditure and the Restructuring of Capitalism : The Military Industry in Britain' *Cambridge Journal of Economics*, Vol 14, No 4, Dec. pp. 453-468.

Lovering J. (1991a) *The Defence Industry After the Cold War* CND.

Lovering J. (1991) 'The British Defence Industry in the 1990s : a Labour Market Persepective', *Industrial Relations Journal 22, 2* 103-116.

Lovering, J. (1993) 'Restructuring the British Defence Industrial Base after the Cold War : Institutional and Geographical Perspectives.' *Defence Economics 4,* 123-139.

Melman, S. (1987), *Profits Without Production*, (University of Pennsylvania Press.)

Melman, S. (1985), *The Permanent War Economy* (Simon and Schuster.)

Melman, S. (1981), 'From Military to Civilian Economy : Issues and Options', Occasional Paper Series No. 8 (Centre for the Study of Armament and Disarmament, California State University.)

Miller, D. (1990), 'The Future of Local Economic Policy : A Public and Private Sector Function' in Campbell (1990).

Murray R. (1989) 'Fordism and post-Fordism' in *New Times : The Changing Face of Politics in the 1990s.* (eds) Stuart Hall and Matirn Jacques, (Lawrence & Wishart, London.)

Oden, M.D. (1988), *A Military Dollar Really Is Different : The Economic Impacts of Military Spending Reconsidered.* Lansing, Michigan, Employment Research Associates.)

Paukert and Richards (1991) *Defence Expenditure, industrial conversion and local employment* (ILO Geneva.)

Renner, M. (1992), *Economic Adjustment After the Cold War : Strategies for Conversion*, UNIDIR, (Dartmouth Publishing Co.)

Sandholtz, W et al (1992) The Highest Stakes : *The Economic Foundations of the Next Security System.* A BRIE Project, (Oxford University Press.)

Schofield, S., Dando, M. and Ridge, M. (1992), 'Conversion of the British Defence Industries" *Peace Research Report No 30* (Department of Peace Studies, University of Bradford.)

Schofield, S. (1991), 'Arms Conversion in the 1990s', unpublished mimeograph, Dept. of Science and Technology Policy, University of Manchester.

Smith, R.P. (1976) 'Military Expenditure and Capitalism' *Cambridge Journal of Economics.*

Smith, D & Smith, R. (1983) *The Economics of Militarism.* Pluto Press.

Smith, R.P. (1990), 'Defence Procurement and Industrial Structure in the UK' *International Journal of Industrial Organisation 8* 1990 pp. 185-205.

Southwood, P. (1992) 'The UK Defence Industry : A Case of Severe Neglect...With Worse to Come ?' *Peace Research Reports.* Department of Peace Studies, (University of Bradford.)

Southwood, P. (1991) *Disarming Military Industries* Macmillan.

Taylor, T. and Hayward, T. (1989) *The UK Defence Industrial Base : Development and Future Policy Options* Brasseys Defence Publishers.

Tepper Marlin J. (1993) 'The Seven Laws of Conversion' paper presented at the Oxford Research Group seminar, Oxford, June.

UNIDR (1993) *Economic Aspects of Disarmament : Disarmament as an Investment Process* United Nations Institute for Disarmament Research, (New York, United Nations.)

Voss, T. (1992) 'Converting the Defence Industry' *Current Decisions Report No 9,* Oxford Research Group.

Wainwright, H. and Elliot, D. (1982) *The Lucas Plan* (Allison and Busby.)

Walker, W. and Gummett, P. (1989) 'Britain and the European Armaments Market' *International Affairs.*

Willett, S. (1990) 'Conversion Policy in the UK' *Cambridge Journal of Economics.* Vol 14, No 4, Dec. pp. 460-482.

Willett (1990) *Defence Contracting in the Borough of Ealing.* Consultancy Report for the London Borough of Ealing.

Wulf, H. (ed). (1993) *Arms Industry Limited* (SIPRI Oxford University Press.)

Yudkin, J. and Black, M. (1990) 'Targeting National Needs : A New Policy for Science and Technology' *World Policy Journal.* Spring.

Part III

Security and Development

Chapter 19

Does Military Spending Have An Impact on the Economy ?

Carlos Seiglie

I. Introduction

Benoit's (1972, 1973) finding of a positive association between defence spending and economic growth· generated a considerable literature investigating this relationship. Most of this literature has failed to provide the theoretical underpinnings for a possible relationship, but for the most part has been empirical in scope (see for example Chatterji, 1992 and Adams, Behrman and Boldin, 1992 for surveys). These models focus on the supply side of the economy and assume either government expenditures on defence reduce the resources available to other sectors of the economy, and therefore adversely affect the rate of economic growth, or that there are some positive externalities from these expenditures which spillover to increase the output from other sectors of the economy. These are what I will call the dynamic implications from defence spending.

Concurrent with this literature there is the belief amongst many that increases in defence spending have a positive impact on the economy from the demand side of the economy as aggregate demand increases from government spending. This literature is static in nature in that the increases in aggregate demand are assumed to increase output once and for all. Yet more comprehensive studies of the dynamics underlying the relationship between defence spending and economic growth also require understanding the theoretical underpinning between defence spending and aggregate demand. For to understand the relationship between growth and defence spending requires a modeling of the economy which is structural in nature. Since the rate of growth of an economy is dependent on the initial conditions, as well as on the current state of the economy, then defence spending which has an impact on the real variables of the economy, that is real income, interest rates and exchange rates, will impact on the growth rate of the economy. Another way to state this is that is that if an economy has a series of potential growth paths each of with is dependent on the current state of the economy, then defence spending which alters this state will alter the economy's growth path.

It is the purpose of this paper to establish the conditions under which defence spending can have an impact on the economy. This is done by presenting a theoretical framework grounded in rational altruistic agents living in a world where generations overlap. Contrary to the 'military Keynesianism' literature, it is shown that it is quite possible for defence spending to have no impact on the aggregate demand of the economy outside of altering the composition of output within the economy. To do so, we assume the altruism exists on the part of parents towards their children, that parents' bequest motives are operative, and that capital markets are perfect in order to highlight that Ricardian equivalence can lead to debt-financed defence expenditures having no impact on the real variables of the economy. The discussion is initially motivated by using a standard overlapping generations framework, which allows us to be able to characterise the preferences of individuals by dynastic utility functions since the bequest motive is assumed operative.

The key to whether defence spending increases aggregate demand hinges on whether intended bequests to one's offspring are fully realised since there always exists the possibility that international or domestic conflict can occur and that a portion of the amount bequeath may never reach one's heirs as these are confiscated by an adversary or lost in conflict. If defence spending is perceived by individuals as a form of protecting against these attacks or increasing the likelihood that bequests to future generations are realised, then increases in public debt to finance defence expenditures will not be fully offset by increases in savings. Consequently, in this case aggregate demand increases. Alternatively, if intended bequests are not perceived as being in jeopardy then Ricardian equivalence will hold and aggregate demand is unaltered.

II. An Intergenerational Model

We begin by reviewing the overlapping generations model of finitely-lived individuals developed by Samuelson (1958) and extended further by Diamond (1965). Each individual is assumed to live two periods, working during the first and retiring during the second. Generation t is young (denoted by 1) in period t and old (denoted by 2) in period t + 1. At any given moment in time the old of generation t overlap with the young of generation t + 1. Each individual is endowed with a unit of labor which they supply inelastically and are paid a real wage of w. The real rate of interest is denoted by r and is paid at the beginning of each period on both public debt and private debt used to finance capital accumulation as both as assumed to be perfect substitutes in an individual's portfolio. We assume that population is stationary and that all individuals have identical preferences, that is they possess the same time-invariant utility function.

Output at time t, Y_t is produced by means of a constant returns to scale production function

$$Y_t = F(K_t, H_t L), \qquad (1)$$

where K_t denotes the aggregate stock of capital, L the total labor force (population of the young) and H_t is the rate of labour augmenting technological progress.

The real rate of interest and wage rate are determined by the marginal productivity of capital and labour, respectively.

All individuals of generation t, have their preferences represented by the utility function

$$W_t\left(c_t^1, c_t^2, W_{t+1}^*(c_{t+1}^1, c_{t+1}^2)\right) = U\left(c_t^1, c_t^2\right)$$
$$+ \beta\, W_{t+1}^*\left(c_{t+1}^1, c_{t+1}^2\right) \qquad (2)$$

where c_t^1 and c_t^2 are the consumption of an individual of generation t while young and old, respectively ; c_{t+1}^1 and c_{t+1}^2 are that of generation t + 1, and $W_{t+1}^*\left(C_{t+1}^1, c_{t+1}^2\right)$ is the maximum attainable utility of generation t's offspring, generation t + 1. $U(\cdot)$ is assumed to be continuous, twice differentiable and strictly concave with marginal utility positive but decreasing in consumption.

Since the bequest motive is operative we can state these preferences by the dynastic utility function for a representative extended family

$$W_t = U\left(c_t^1, c_t^2\right) + \beta\, U\left(c_{t+1}^1, c_{t+1}^2\right)$$

$$+ \beta^2\, U\left(c_{t+2}^1, c_{t+2}^2\right) + \dots \qquad (3)$$

$$. = \sum_{i=0}^{\infty} \beta^i\, U\left(c_{t+i}\right)$$

We assume that the role of government is to act as an intermediary in the provision of national defence. Government expenditures in period t are financed by a lump-sum tax on each member of the young at the rate, τ_t, and by the issuance of bonds, b_t which pay a rate of r_t. More specifically, individuals of generation t are taxed at the rate of τ_t to provide for per capita national defence spending at time t of m_t. If the government derives all revenues from taxation then the individual's (dynastic) budget constraint requires that the present value of their consumption stream be equal to the present value of net income plus the present value at time t of the bequest received by the individual from the previous generation net of the present value at time t of government debt outstanding in period t. More formally,

$$\sum_{i=0}^{\infty} \frac{c_{t+i}}{\prod_{j=0}^{i}\left(1 + r_{t+j}\right)} = \sum_{i=0}^{\infty} \frac{w_{t+i} - \tau_{t+i}}{\prod_{j=0}^{i}\left(1 + r_{t+j}\right)} + B_t - b_t, \qquad (4)$$

where r_t is the interest rate at time t, B_t is the amount of bequest received by generation t from the previous generations and b_t is the value at time t of the government debt outstanding for which they are liable.

The government also has the budget constraint that the present value of total expenditures per capita on defence spending, m_t must equal receipts, that is,

$$\lim_{T \to \infty} \sum_{t=1}^{T} \frac{m_t - \tau_t}{\prod_{i=1}^{t}\left(1 + r_{i-1}\right)} = 0. \qquad (5)$$

Both these constraints hold at any time and for each generation, that is, for members of generation $t, t+1, t+2,\ldots$, and so on. We can therefore state the constraint faced by a member of generation t given the constraint faced by government on their military expenditures as :

$$\sum_{i=0}^{\infty} \frac{c_{t+i}}{\prod_{j=0}^{i}\left(1 + r_{t+j}\right)} = \sum_{i=0}^{\infty} \frac{w_{t+i} - m_{t+i}}{\prod_{j=0}^{i}\left(1 + r_{i+j}\right)} + B_t - b_r. \qquad (6)$$

Therefore, the young of generation t solve the maximisation problem :

$$\underset{\left\{c_t, c_{t+1}, \ldots, c_{t+j}, \ldots, \lambda^t\right\}}{\text{MAX}} \sum_{i=0}^{\infty} \beta^t U\left(c_{t+i}\right) - \lambda^t$$

$$\left[\sum_{i=0}^{\infty} \frac{c_{t+i}}{\prod_{j=0}^{i}\left(1 + r_{t+j}\right)} - \sum_{i=0}^{\infty} \frac{w_{t+i} - \tau_{t+i}}{\prod_{j=0}^{i}\left(1 + r_{t+j}\right)} - B_t + b_t \right], \qquad (7)$$

or equivalently,

$$\underset{\left\{c_t, c_{t+1}, \ldots, c_{t+j}, \ldots, \lambda^t\right\}}{\text{MAX}} \sum_{i=0}^{\infty} \beta^t U\left(c_{t+i}\right) - \lambda^t$$

$$\left[\sum_{i=0}^{\infty} \frac{c_{t+i}}{\prod_{j=0}^{i}\left(1 + r_{t+j}\right)} - \sum_{i=0}^{\infty} \frac{w_{t+i} - m_{t+i}}{\prod_{j=0}^{i}\left(1 + r_{t+j}\right)} - B_t + b_t \right]. \qquad (7')$$

The above optimisation problem assumes that the level of military spending is viewed as exogenous by the individual, but we could equally assume that the political equilibrium level of defence spending is either determined by the preferences of the median voter or by those of some other decisive individual depending upon the political structure of the country. In this case the individual's maximisation problem given by equation (7') must include choosing the optimal sequence of defence spending as well (see Seiglie, 1993b).

We would now like to show that an increase in defence spending financed by the issuing of bonds, that is, debt financing of expenditures can be neutral. In other words, using a differential incidence approach the substitution of debt for tax financing of defence spending can have no affect on the real economy, outside of altering the composition of assets in individuals' portfolios and the allocation of resources towards a greater share in defence production. To show this, note that the maximisation of equation (7) or (7') leads to some optimum consumption path $\left(c_t^*, c_{t+1}^*, \ldots\right)$ for a representative individual of generation t. If at time $t+i$ the government decided to finance m_{t+i} by reducing taxes and increasing debt (incur a fiscal deficit) generation t's taxes would be reduced by this amount which I denote by b_{t+i}. Yet this debt has to repaid at some future time period $t+J$, and therefore, taxes at this time have to be increased to pay off this debt. The amount of the tax increase at $t+J$ required to retire the debt is $b_{t+i}\left(1+r_{t+i}\right)\left(1+r_{t+i+1}\right)$ $\ldots\left(1+r_{t+j+-1}\right)$. But as is obvious the present value of this tax increase at time $t+i$ is in fact the reduction in taxes, b_{t+i}. Therefore, since the individual's intertemporal budget constraint is unaltered by switching from tax financing to debt financing of military expenditures the initial path of consumption and therefore savings is unaltered. This implies in the aggregate that since aggregate savings is unchanged real interest rates and output for the economy are unchanged. In this sense is debt financing of military spending neutral and therefore, if the path of economic growth is dependent on the initial levels of real output then the growth rate is invariant to the method of financing defence spending.

The main point here is that much of the literature which may be termed 'military Keynesianism' assumes that defence spending can stimulate an economy via the standard multiplier analysis. Yet as shown above that once we model the aggregate economy more formally, this argument does not necessarily hold. In order for military spending to potentially have real effects requires either the creation of a unique function for military expenditures or more specific assumptions about how resources are employed by the military. The following are some assumptions which may serve useful in establishing the non-neutrality of military spending.

(1) Resources devoted to military purposes cannot be converted over time to civilian purposes, or that the rates of depreciation of military hardware differs substantially from that of other forms of capital. It is not obvious that for labor the former assumption would hold, but for resources devoted to weapons production one may believe that the transfer of resources from military to civilian use is less than `smooth.' It should be pointed out that this argument is not unique to resources employed by the military, since a similar one could be made when one compares many other civilian uses of resources. Addressing this area requires the introduction of distributional considerations into macroeconomics, something which is rarely done.

(2) Military spending serves the purpose of insuring or protecting desired intergenerational bequests for example, as in Seiglie (1993b). The argument hinges on parents partly offsetting desired bequests since they are assumed to view defence spending as a substitute for assets which underlie physical capital.

(3) One can assume that individuals are bequest constrained, that is, they desire to bequest zero or negative amounts (liabilities) to their children. It has been shown that this is more likely to be the case for low income individuals by Cukierman and Meltzer (1989) and if we extrapolate, the bequest motive would be less likely to be operative in developing then in the developed countries. In this scenario, defence spending is more likely to be non-neutral in the developing world.

(4) We could assume that weapons are accumulated for offensive purposes and therefore may be viewed as an investment by a country towards that end of confiscating another country's resources. In other words, this scenario requires the assumption that certain nations systematically employ war as a means to `redistribute' or confiscate, as well as the assumption that the probability of their success in doing so is high relative to the success of potential victims in defending against such attacks. This leads one into exploring arms races between nation, including the research and development of both offensive and defensive technologies to achieve or prevent these conflicts (see for example, Intriligator, 1975 and Seiglie, 1993a).

(5) Some of the above are necessary but not sufficient conditions. For example, even if defence spending is viewed by current generations as a form of insuring bequests, if some future generation is not constrained in their ability to prevent an attack by an adversary that is, they do not require the `gift' of military resources from past generations then it is not obvious that defence spending is not neutral.

(6) Finally, we could introduce uncertainty in the timing of conflicts or we could assume that individuals have imperfect information as to when wars will occur. This assumption introduces uncertainty about an individual's income stream which can lead to real effect (see Feldstein, 1988).

III. Summary

This paper has analysed the effects of military expenditures on the aggregate demand of the economy. It has been shown that it is quite possible for defence spending to have no real effects on the economy outside of altering the composition of output. It has outlined more sophisticated mechanisms which may cause defence expenditures to be non-neutral. Even taking into account these stronger assumptions at best only provide necessary, and not sufficient conditions. These conclusions are important in studying the linkage between defence spending and economic growth in so far as paths of economic growth may be dependent on the state of the economy. This paper has shown that this state may in fact be invariant to the defence expenditure policy of a country.

References

Adams, F. Gerard, J. Behrman, and M. Boldin (1992) 'Defence Expenditure and Economic Growth in the Less-Developed Countries : Reconciling Theory and Empirical Results,' in *Disarmament, Economic Conversion and Management of Peace* edited by Manas Chatterji and Linda Forcey (New York : Praeger).

Barro, Robert J (1974) 'Are Government Bonds New Wealth ?.' *Journal of Political Economy*, 82, pp. 1095-1117.

Chatterji, Manas (1992) 'Regional Conflict and Military Spending in the Developing Countries.' in *Economics of Arms Reduction and the Peace Process*, edited by Walter Isard and C. Anderton (Amsterdam : North Holland).

Cukierman, Alex and A. Meltzer (1989) 'A Political Theory of Government Debt and Deficits in a Neo-Ricardian Framework,' *American Economic Review*, September, 79, pp. 713-732.

Deger, S. (1986) 'Economic Development and Defence Expenditures,' *Economic Development and Cultural Change*, 35 pp. 179-195.

Diamond, Peter A. (1965) 'National Debt in a Neoclassical Growth Model,' *American Economic Review*, 55, pp. 1126-1150.

Drazen, Allan (1978) 'Government Debt, Human Capital, and Bequests in a Life-Cycle Model,' *Journal of Political Economy*, 86, pp. 505-516.

Feldstein, Martin S. (1988) 'The Effects of Fiscal Policies When Incomes are Uncertain : A Contradiction to Ricardian Equivalence,' *American Economic Review*, 78, pp. 14-23.

Intriligator, Michael D. (1975) 'Strategic Considerations in the Richardson Model of Arms Races,' *Journal of Political Economy*, 83, pp. 339-353.

Isard, W. (1988) *Arms Races, Arms Control, and Conflict Analysis* (New York : Cambridge University Press).

Samuelson, Paul A. (1958) `An Exact Consumption Loan Model of Interest with or without the Social Contrivance of Money,' *Journal of Political Economy*, 66, pp. 467-482.

Seiglie, Carlos (1993a) `Technological Progress, Alliance Spillover and Economic Growth in a Disaggregated Arms Race Model,' *Defence Economics*.

Seiglie, Carlos (1993b) `Defence Spending in a Neo-Ricardian World,' Rutgers University, Department of Economics, Working Paper # 93-01.

Seiglie, Carlos (1992) `Determinants of Military Expenditures,' in *Economics of Arms Reduction and the Peace Process*, edited by Walter Isard and C. Anderton (Amsterdam : North-Holland).

Chapter 20

Deferred Costs of Military Defence : An Underestimated Economic Burden

Petra Opitz and Peter Lock

Flawed Perceptions

Ending the Cold War would alleviate the major industrial nations from paying high opportunity costs for the prevailing doctrine of military security based on deterrence. The proportion of military outlay in government spending would considerably shrink. As long as the end of the Cold War remained but a theoretical proposition, a broad consensus unfolded about the welfare to be gained should the bi-polar confrontation be phased out.

In parallel the United Nations was the forum of a political debate on the intricate relationship between armaments and development. The manifest incapacity of the prevailing world order to cope with the imperative need for development was often conceptually linked to the on-going arms race. The rhetoric of this debate on development yielded the catching concept 'peace dividend.' It its Human Development Report of 1992, the United Nations Development Programme saw the world already moving towards a 'new global compact,' based on US $1.2 trillion 'peace dividend' during the last decade of the century.[1] Other UN agencies were more prudent about the effects of reduced military expenditures and emphasised that "part of these reductions in defence spending will be needed to fund new investment in labour and capital to facilitate adjustment and reallocation of resources.[2] But the logic of a peace dividend became a pervasive notion not least because the arms build-up under Reagan had further inflated the opportunity costs of the prevailing security doctrine.

Indeed, the public good 'external security' was narrowly defined in a military sense ever since the Cold War had started. The ensuing armament dynamics instilled the danger of arbitrary nuclear annihilation at global scales. The last two decades, however, were marked by an increasing perception of global threats of non-military nature. Security developed into a much broader concept. Global survival is perceived as being structurally at stake. A consensus is building that survival presupposes at minimum a concentration of available resources for combating underdevelopment and further environmental degradation and for supporting the exploration of strategies towards sustainable development. Systemic exhaustion of the Soviet Union and a broadened concept of security combined and reinforced the view that military expenditures constitute a reserve to be tapped in order to meet the new challenges.

Statistically a segment of the supposed reserve became already available and global military expenditures keep shrinking considerably. However, the heralded peace dividends haven not materialised so far, because the structural dimensions of the burden of military expenditures have been systematically underestimated in the past. Particularly the structural deformations of productive forces were being overlooked. Only as disarmament becomes operational, costs to restructure the productive forces are accounted for.

Also, the considerable costs of disposing of the huge surplus of weapons are just beginning to enter the balance sheet of disarmament. In some fields like chemical weapons for example not even approximate cost parameters are known as viable technologies of large scale disposal are not yet operational.[3] Sums amounting to thousands of million dollars of deferred costs of the armament dynamics were omitted from the original defence budgets. Just as environmental destruction caused by industrial and agricultural production is insufficiently being accounted for through the pervasive operation of the market, the costs to repair environmental hazards caused by military activities as well as the consumption of apparently cost free resources have, if ever, only partially been covered by the defence budgets.[4]

Since arms reduction and disarmament began to appear on the agenda, these deferred bills of the arms race commence to present themselves and will, for the time being, absorb whatever peace dividend might have been calculated on paper. Looking back we begin to learn that the price for the public good 'external security' as defined at the time was actually much higher than the figures presented in voluminous defense budgets and voted for, often unanimously. With hindsight it is obvious that the political decisions on security policy which were taken in the eighties and earlier were not based on a realistic accounting of the true costs. They were, as matter of fact, deferred upon future generations.

Environmental damages

The damage for the future generations is cumulative because the leading industrial nations were focusing their industrial efforts to such an extent on the arms race that the already available knowledge about minimum 'imperatives' to support globally sustainable development was deliberately neglected. The widespread ecological disaster in and around many of the closed cities in the former Soviet Union where defence production was the dominant economic activity documents the destructive mode of production prevailing during the Cold War.

But the Cold War damage is not restricted to the former Soviet Union. The General Accounting Office of United States reports: 'As of February 1993, federal agencies reported owning or operating over 1900 potentially contaminated facilities, including military installations, research laboratories, maintenance facilities, landfills, and nuclear weapons plants. Each facility includes one or more hazardous waste sties that may require cleanup. Some of these facilities now rank among the largest and most heavily contaminated in the nation. Cleaning up this hazardous waste legacy will take decades and will cost hundreds of billions of dollars.'[5] The same report estimates total cleanup costs for the Departments of Defense and Energy (nuclear warhead program) to come close to US $ 200 billion. In 1988 823 sites, 58 per cent were the result of direct military activity, were on the docket to be inspected with respect to the National Priority List (NPL), by July 1992 the Environmental Protection Agency had evaluated only about 500 sites of the original list, while the list grown in the meantime to 1930 sites suspected of critical environmental contamination. While many investigations of the damage had not been terminated, 45 per cent were removed from further NPL consideration, but of the 100 put on the national priority list 94 were military sites.[6] If one takes these findings as an indication of the levels of contamination to be disclosed in Russia as it slowly develops into an open society as well, the conclusion is inescapable.

There is an enormous heritage of severe environmental damages accumulated during the Cold War which clearly transcends national borders.[7] At the same time the life-cycle costs of all military systems after decommissioning was never being accounted for.

Seemingly less conspicuous than the nuclear waste often carelessly disposed of behind the top secret screen, the nuclear arms race afforded, the case of Germany also provides ample evidence of the mortgages the military sector took up while engaged in the arms race and leaves future generations to pay. Germany was the geographic focus of the East-West military confrontation. One in twenty m[2] or more than five percent of the territory in the two former German states was set aside for the exclusive use of the military. In addition the imprint of military priorities penetrated the entire infrastructure[8] and in the GDR the industrial design[9] as well. Finally the military forces made extensive use of the civilian infrastructure and consumed a significant proportion of all other economic activities.

By the end of 1995 close to one million military personnel will have disappeared from the German territory, about 300000 family dependents as well. The on-going large-scale process of reverting military bases and production sites to civilian use all over Germany[10] brings to the attention of the public the unrestrained use of natural resources practiced by the military while they were protected from public scrutiny behind the curtains of military secrecy and the exceptional prerogatives 'national security' still convey to the military institution.

As a large proportion of these military areas becomes available for civil use the environmental rehabilitation must be addressed. But so far not even the dimension of the ecological damage to be repaired is known. The assessment itself is a time consuming and expensive process, it was started only in 1991. Preliminary estimates are available, thus the liquidation of ecological damages left by the Russian troops alone will require DM 25 billion. These rehabilitation costs definitely exceed the financial capacity of the new "Bundeslander" as well as the local authorities.[11]

In Brandenburg, one of the new five federal states (Lander), the on going investigation of former National People's Army and Soviet troop's installations so far revealed 96 different types of contamination that need rehabilitation before full civilian use will be possible.

Determination of costs and distributing them will remain a controversial issue on the German political agenda. The "Lander" have requested the Federal Government to take over the costs for the decontamination. The Federal Government refused and instead the "Lander" receive a slightly increased share of the value added tax. However, this will not fully cover the costs which have not even been determined in full. The costs of decontamination may increase considerably[12] if it is postponed for lack of finance. In some cases decontamination costs may decrease as time passes and natural rehabilitation takes over part of the job. In these cases, however, the local authorities claim the foregone utility. For would it be possible to use the territory or real estate immediately, income could be generated.[13]

Other dimensions

The absence of a peace dividend permitting the distribution of additional welfare in the terms traditionally envisioned by a wide spectrum of social groups requires a renewed analytical endeavour in order to comprehend fully the accumulated structural impact of the military sector on the development potential of a less militarized future.

Firstly, with arms reduction on the agenda the major industrial countries began to assess the strategic value of their military high-tech industries which was taken for granted as long as the arms race provided the military sector with an unquestioned legitimacy as well as an easy access to government contracts. The Office of Technology Assessment[14] in the United States and several expert groups in Great Britain[15] were commissioned to assess the contribution of military R & D to enhancing national industrial competitiveness.

The results were sobering. In spite of a few industrial highlights related to military R & D, the underlying equation that military R & D supports to the development of strategic innovations which would provide the respective industries with leading edge in the permanent struggle for shares in the world market had to be abandoned. The OTA study observed that military-bureaucratic procedures and their secretiveness inhibit a technological spin-off, even within individual companies the military and civilian branches are hermetically separated.[16] The industrial paradigm has reversed, civil technology, generated within transnational corporate networks, is spearheading global innovation.

The former Soviet Union

In the former Soviet Union it was assumed that the military sector would have the best chances to take the lead in the transformation process because it commanded the best equipment and monopolized qualified labour. The evidence available by now indicates that the privileged military high-tech manufacturers in the former Soviet Union performed badly and were not able to launch viable civil production lines beyond the ones that were already under their command before.

The absence of multi-domestic corporations[17] forced to prove themselves continuously in a violent world-wide oligopolistic competition over global markets for civil products has permitted a virtual dominance of the Russian military-industrial complex (VPK). It has implanted catastrophic inefficiencies, uncontested duplications and an insufficient division of labour into the Russian economy.

The intricacies of central planning provoked centrifugal strategies, particularly illicit hoarding of resources at every stage of production making for substantial cumulative diseconomies. Since the VPK could not draw on innovation generated by the civilian sector as is increasingly the case in the military production in Western countries, the VPK was desperately trying to circumvent CoCom in order to selectively acquire western production technology. Raw material exports, particularly the increased oil price levels in the late seventies, provided the necessary hard currency. But the insufficient division of labour, combined with the secretiveness of military production, provided for a systematic underutilization of the imported production technology. The development of civil high tech structures was almost impossible because virtually all sophisticated civil production was under the command of the VPK.

An exact determination of the size of the military-industrial sector is rather difficult as it does not constitute a separate branch in national statistics. The methodological discussion is advancing[18] but methods still vary greatly and available data still shows large differences. It appears save, however, to assume a share of total Soviet expenditures for armaments in the range of 19 percent to 30 percent before the sizeable reductions in 1991.[19] Approximately 80 percent of the scientific potential and 11 per cent of the total labour force employed in industry were part of the military-industrial sector.[20]

90 percent of the arms production were concentrated in Russia and the Ukraine; this compares with a share of 77 percent of the two states in total Soviet output. The old structures proved neither capable of the necessary innovation nor allowed for sweeping reforms within the system. The announced plans for conversion never materialized and turned out to be a strategy of the planning bureaucracy to fend off the transformation toward a market economy. Though there is no viable alternative to new economic structures geared to provide the institutional frame for a market economy.

However, the mere political proclamation and the dissolution of the old economic command bureaucracy as the only steering mechanism can not automatically be replaced in its functions by market interactions and provide immediately a market-based regulation of production. Instead the old system is followed by a vaccum. All the indispensable subsystems of a market economy are first to evolve in a slow process before the market can really perform its regulating functions satisfactorily. Presently there exists an incomplete and inflexible monopolistic market, close to what existed in some developing countries after long periods of high levels of protection. Barter trade is still expanding and being accepted as an important mode of transactions; cumulative arrears cripple the economy and efficient entrepreneurship is still the exception.

First indications of a spontaneous evolution of structures, which resemble very much a person-to-person network, are already visible. It is marked by pragmatic adaptation in order to survive. These networks help to avoid the total breakdown of the industrial production; they tend to reinforce, however, for the time being monopolisation and a continuation of other forms of non-market regulation. More importantly, some sectors of the VPK are not considered to be of strategic importance and thus entitled to form joint ventures and other forms of cooperation with foreign companies which will certainly speed the development of market behaviour.

The tasks ahead

What are the tasks ahead in order to reach an unbiased assessment of the total sum of deferred costs? How can the economic and social potential of disarmament be activated in a situation were the immediate expectations are being bitterly frustrated?

A first priority will be to develop a realistic understanding of the economic implications of reduced military expenditures. Putting it succinctly the conversion towards peace requires additional investment which presupposes additional savings. As a consequence no increase of consumption is likely to materialise in the near future. Simply stopping or rather reducing the reckless 'non-productive' use of resources of all kind for military purposes only means that the damages and structural deficiencies left for future generations to repair will not build up further. The amount of oil contaminating the soil will not continue to accumulate at the same rate. Nevertheless, the simple stopping of wasting valuable resources will be a significant result of disarmament, though short of bringing about a peace dividend in the expected form.

At the beginning of the decade UNDP held a somewhat optimistic view of the peace dividend:

'For the industrial countries, it looks as though military spending could be reduced by 2 percent to 4 per cent a year during the 1990s, if the present understandings between the superpowers come to fruition and if a lasting peace in the Golf comes soon. This would translate into savings of $ 200-300 billion a year by the year 2000, and savings during the decade of as much as $ 2 trillion.'[21]

These savings, however, will not release significant funds for other areas of social priority. Deep cuts in armed forces and in hardware spending require significant retraining costs for labour before it can be diverted to other sectors, as well as significant payments for unemployment benefits. The defence industries themselves must write off considerable capital investment—and face reduced production along with falling sales. But all these cost are essentially short-term—not different from those incurred in other market imposed forms of adjustment and restructuring.

Because the idea of a peace dividend is to divert the savings gathered by demilitarisation to more productive development, the costs of retraining and alternative investment in the military sector should not really be deducted from the savings. The costs were simply not expected to occur and the expenditures should be regarded as fulfilling the purposes of the peace dividend.

Two prime candidates for the peace dividend are the urgent social problems in many industrial nations, from homelessness to drug addiction, and the wide range of development needs in the Third World. The immediate prospect, however, is that the statistical peace dividend will be soaked up in the national accounts as a budget balancing item which reduces, or prevents, deficit spending. This should not raise too much concern, for the alternative might have been cuts in either domestic social programmes or in foreign assistance budgets.

Continued role of the state

The consequences of former mis-allocation must be remedied, this calls for the allocation of considerable financial resources. Earmarking the necessary means in the budget to cover these reconversion costs does not constitute, as it is frequently claimed,[22] a case of undue state intervention in the economy. It is incorrect to assume that the rehabilitation of deformed structures constitutes an offence against the principle of maximizing market forces wherever feasible. It is quite obvious that market forces care little about the hazardous heritage of the military, they rather prefer to move elsewhere in the world economy.

Thus, if the state originally providing the economic base for all military activities now attempts to remedy the damages inflicted, by definition, this does not constitute a counterproductive attempt to expand the state sector. It should rather be considered as a necessary investment to rehabilitate an asset, so that it can be reincorporated into the market economy. The transfer of budget positions from defence to the reconstruction of the devastated military heritage can not possibly be labelled as a case of undue state intervention. It is more like finally paying unpaid bills or making the polluter pay to restore the habitat as it may be the case.

The budgetary price of the public good 'external security'[23] is definitely sinking. Nevertheless, an unexpected consequence of finally achieving security arrangements at lower levels of arming throughout Europe may well be that increases in government expenditures will be required temporarily in order to bring human and material resources back into the 'civilian' economy while aiming at environmentally sound and sustainable conditions.

Given the dimension of problems governments begin to realise that the disarmament process does not bring about automatically a diminished role of the state. To the contrary, even the American government moves towards a declared industrial policy, almost an anathema only a few years ago.

The economic effects of the civilian use of resources of military technology may on the whole be expected to remain within fairly narrow limits. In the case of Russia saved raw materials, aluminium in particular, flood the world markets.[24] But the scrap gained from the delaboration of weapon systems does not pay for the separation and recycling of the materials.[25] The possibilities of commercializing the recycling of ammunition are generally overestimated. Though it is technically feasible to delaborate tanks, ammunition, so that the final product consists of basic substances. But the marketing of the substances is economically hardly viable. In order to meet the conditions of purity required by potential customers, the costs soar considerably beyond competitive price levels. At the same time the quantities produced are likely to be insufficient to enter the market of raw materials. Also, the output is not steady which poses additional difficulties for marketing.

Though some strategic missiles are being converted for the use in civil orbital missions.[26] But the total balance of inactivation[27] and recycling will be grossly negative. For economic reasons (high running costs) and in most cases also for ecological reasons, a conversion of military technology to civilian use appears to make sense in a very few cases only.

Thus, the state remains inevitably a major actor whose role is to compensate for the appropriation of natural resources by the armaments dynamics by means of supporting measures towards a more sustainable development.

Overcoming inherited structures

The organisational pattern and the decision making procedures prevailing in the military-industrial sector of the economy require profound changes before the enterprises and the research facilities can competitively enter the civilian market. The privileged, but isolated R & D establishments cannot easily be reorganized in order to pursue successfully non-military ends at acceptable cost levels. They used to absorb the most valuable elements of the human capital, the respective societies generated. In addition, military R & D was a major pull factor generating the brain drain from the South. However, personnel engaged in military R & D also proved highly mobile and chose to emigrate if circumstances deteriorated. Proliferation[28] of sophisticated military technology was the consequence. Will the partial dissolution of military research laboratories follow a similar path or are no absorptive niches on offer for migrating weapon specialists ?

The task to release the military R & D personnel from perfecting weapons of mass destruction presents itself as an extremely difficult task. What might be the concrete civil research projects these scientists and engineers are capable to perform and the society is willing to finance? It may well turn out that their productivity is not sufficient to convince the sovereign that is the tax payer to maintain the research laboratories or that the civilian products they propose do not correspond to the priorities the market sets. Whereas in western countries the sanctions of the market are straightforward and dynamic efficiency[29] is the key for successful performance in global markets, it will apparently take the Russian VPK still time to adjust to the dire economic realities it has to face entering the global economy.

Therefore, it may well turn out that an economic structure will evolve in Russia for an intermediate period which does not conform fully to the principles of a market economy because non-market regulation reflecting political power relationships will survive for quite some time. While in the leading industrial nations of the former west the restructuring and downsizing of the defence industrial capacity continues to proceed with an extraordinary momentum.

Common responsibility

The cost of the Cold War and the boost of military expenditures under the Reagan administration, in particular, were internationalized since governments have long since ceased to be the sovereigns of their national economies.[30] The militarily overburdened economy of the United States and the dissolution of the former Soviet Union are just two sides of the same coin.

The NATO-countries are likely to have greater capacities to absorb the economic shock associated with the disarmament process than the former WTO-countries. They are not faced with the task to transform their economic system at the same time. However, the profound changes will bring about new challenges to the established market economies as well because adding Eastern Europe to the international market will provoke a considerable reallocation of production.

The over-militarisation of the Soviet economy accumulated structural deadlocks which will not be overcome except for close international cooperation. The necessary re-allocation of resources takes dimensions that will change the entire economy in its regional division of labours and, of course most importantly, the shift towards the markets economy will common and profound charges of the political culture. Parallel to the economic transformation the entire society and its institutions are in a process of dissolution. The legal and institutional framework and the reconstruction of the economy have to be accomplished simultaneously, a task of historically unprecedented dimensions.

What is required is an international consensus that the disarmament process can only be carried out successfully if the western industrial nations accept the need of international cooperation and of funding to cope with the economic impact of reducing the military sectors in national economies which is truly dramatic in some cases.

Measured with the criteria of a sound ecodevelopment[31] or of a sustainable global environment the present generation in the industrial nations lives on credit, nobody, least the earth with its limited resources, can guarantee. In certain regions of the former Soviet Union in particular there are indications that the reckless contamination of past produces serious health hazards and debilitates the entire population.[32]

But there are recent indications that the Russian military have begun to tackle the dramatic ecolocgical problems of their country. A proposal entitled 'guarantee of ecological security by the armed forces' is tabled at the legislative level. It designates two main areas of activity 1. elimination of damages and 2. prevention of damages. The armed forces envision an active role in monitoring and disposal of damages and ecological training.[33] It was resolved to form environmental units (Ekologitschedije Voiska), but these units do not face exclusively damages triggered by military activities. The exhausted industrial system of the Soviet Union did not pay attention to environmental consequences of its production. It was marked by an unilateral productivity orientation and the illusion of unlimited natural resources.

The ecological initiatives may reflect a strategy of corporate survival of the armed forces, but even so, the international community should actively support such activities as a necessary step in support of transforming the successor states of the former Soviet Union. It is a constructive step towards demilitarising the future.

Notes

1.   UNDP, Human Development Report 1992, New York (Oxford University Press) 1992, pp.86 f.

2.   United Nations GA, United Nations Institute for Disarmament Research, Economic aspect of disarmament: disarmament as an investment process, New York A/47/346, 27 August 1992, p. 97.

3. See: United States General Accounting Office, Chemical Weapons Destruction Advantages and Disadvantages of Alternatives to Incineration, Washington D.C. March 1994. The report confirms that the mandated deadline of 2004 for the destruction of the entire stockpile can not be met as the technologies are not yet developed to support the destruction of large quantities.

4. The virtually permanent and hardly repairable damages and deaths caused by nuclear testing are only too obvious examples of costs in the United States reveal regularly that the military activities leave behind heavy contamination whose costs of rehabilitation the US Government Accounting Office refuses to determine as neither full evaluation of damages has been completed not are reasonable estimates of the true costs of the clean-up feasible for lack of experience in most cases. United States General Accounting Office, Military Bases—Transfer of Peace Air Force Base Slowed by Environmental Concerns, GAO/NSIAD-93-111FS, Washington, D.C. (Feb.1993).

5. United States General Accounting office, Superfund Backlog of Unevaluated Federal Facilities Slows Cleanup Efforts, Washington D.C. July 1993, p. 2.

6. Ibid.p.17.

7. Given the dimension of the military contamination it is astonishing that a recent authoritative textbook on "Ecological Economics" does not refer to the military dimension. Robert Constanz ed., Ecological Economics, New york (Colombia University press) 1991.

8. Investments in roads and railways were co-determined by the perceived utility for the military posture of the two alliances operating on German territory.

9. A conspicuous example is the design of the trucks used in the former GDR. Within the Warsaw Pact division of military industrial production the GDR was assigned to produce a medium sized, high wheeled (all terrain) military truck. The civil variant of this inefficient truck, too small for economic transport, became the industrial standard, because in the context of limited resources the economies of scale in production were turned into a law.

10. In 1990 the US-Headquarter in Heidelberg listed already 358 sites as serious damaged. Rehabilitation costs, including compensations were roughly estimated to amount to US$ 200 million. See: Berliner Zeitung, 22 December 1992, p. 6.

11. For example, the rehabilitation of former tank depot of the Soviet Army (550 000 square meters) was calculated to require DM 450 000 for the decontamination of the soil and DM 1,76 million to rehabilitate the ground water.

12. This is one of the findings of the United States General Accounting Office, Department of Energy Cleaning up Inactive Facilities Will Be Difficult, Washington D.C. June 1993, p. 8 f.; see also: Superfund—Backlog of Unevaluated Federal Facilities Slows Cleanup Efforts, Washington D.C. July 1993, p. 3.

13. Calculations of foregone utility were made in several regions of the Land Brandenburg, where communities are pressing for federal support.

14. US Congress, Office of Technology Assessment, Arming Our Allies: Cooperation and Competition in Defence Technologies, DPO Washington D.C. 1990 as well as: U.S. Congress, office of Technology Assessment, Holding the Edge: Maintaining the Defence Technology Base, GPO Washington D.C. 1989.

15. See in particular: Advisory Council on Science and Technology, Defence R & D: A National Resource, HMSO: London 1989.

16. See: US Congress, Office of Technology Assessment, Holding the Edge: Maintaining the Defence Technology Base, GPO Washington D.C. 1989.

17. For an elaboration of this concept see: OECD-International Futures Programme, Strategic Industries in a Global Economy: Policy Issues for the 1990a, Paris 1991.

18. See in particular: Alan Smith, Russia and the Word Economy—Problems of Integration, London (Routledge) 1993.

19. For a summary of the discussion see: K. Schrader, Die Volkswirtschaftliche Bedeutung der Rustungskonversion in der Sowjetunion, in: Die Weltwirstschaft, 1/1991, p. 166-168; L. Bulton, The switch from guns to butter, in: Financial Times 16.7.1991.

20. Vgl. A.Ruzkoi in Moskowskije Nowsti of 29.9.1991, p. 13; M.Spekler, A. Oshegov, V. Malygin, Konversia oboronnych predpriatii: vybor strategii, in: Voprosy Ekonomiki, 2/1991, S.13.

21. UNDP, Human Development Report 1991, New york (Oxford University Press) 1991, p. 81 f.

22. See: M.Ky, K.Lobe, Verteidigungsausgaben der Bundesrepublik und volkswirtschaftliche Auswirkungen verminderter Rustungsausgaben, RWI-Mitteilungen, vol.41 (1990), nr.4, p. 334.

23. See: Wohin geht die Friedensdividende? Deutsche Bank, Frankfurt am Main 1991.

24. See: Kenneth Gooding, US aluminium industry looks to Clinton, Financial Times, Nov. 5, 1993, p. 30; also: L'Aluminium Europeen accuse la Russie, in:L'Echo, Sept. 30, 1993, p. 10.

25. Statements to the contrary aim at government support for delaboration and are not based on sound calculations. Pruning of valuable components may be highly profitable if separated from recycling. See: Nevskoje Vremia, Dec. 10, 1993.

26. See: Philip Clark, Converting Soviet into Russian Space Launchers, in: Jane's Intelligence Review, Sept. 1993, p. 401 f.

27. For parameters of costs and totally unresolved technical problems see: United States General Accounting Office, Nuclear Submarines, Navy Efforts to Reduce Inactivation Costs, Washington D.C. July 1992.

28. After the two world wars German military engineers and scientists continued to practise their trade abroad. Sweden, Japan, the USSR after World War I and the United States, the USSR, France, Spain, Argentina, Egypt and finally India were among the recipient countries after 1945. When Britain abandoned its TSR 2 fighter aircraft in the early sixties Israel and later South Africa hired members of the development team.

29. Margaret Sharp and Keith Pavitt define this Key variable "dynamic efficiency" as "the speed with which an economy develops and uses new technologies". Even a superficial view of the Russian R & D establishment reveals that the structural preconditions for speed do not exist. See: Sharp M. and Pavitt, K., Technology Policy in the 1990s: Old Trends and New Realities, in: Journal of common Market Studies, Vol.31, No.2, June 1993, p. 139.

30. For an interpretation of the implications for the global order See: Stanley Hofmann, Delusions of World Order, in: New York Review of Books, April 9, 1992, p. 37-42.

31. For a convincing argument to use this term see: Ignacy Sachs, Comment concilier ecologie et prosperite, in: Le Monde Diplomatique, December 1991, p. 18f.

32. For an overview see: Georgii S. Golitsym, Ecological Problems in the CIS during the Transitional Period, in: RFE/RL Research Report, Vol.2, Jan.8, 1993, p. 33-42.

33. See: Voenno ekonomitscheski shurnal, 1/1994, p. 73 f.

Chapter 21

Arms Race and Disarmament in Multilateral Interdependence

Hiroyki Kosaka

1. INTRODUCTION

The beginning of a new era post the Cold War has been changing the world situation. For example in the US, drastic change of the US foreign strategy may be seen in the statement of the Secretary of Defence of the Clinton Administration, the first administration after the end of the Cold War. The Secretary of Defence Aspin proposed comprehensive military strategy to the military commission of the Senate, January 1993:

(1) conventional force is directed to menace appearing everything in the world and not directed to invasion of the former USSR, and

(2) nuclear strategy is changed to prevent threat by nuclear proliferation of the other countries shifting from nuclear attack by the former USSR. Conventional menace threatening the US national interest mainly focused on the former USSR, but it is now dispersed in the world.

In the meanwhile, the political situation in Japan has been chaotic. Recently the Socialist Party of Japan (SPJ) has organised a cabinet jointly with Liberal Democratic Party just before the Napoli Summit Meeting of July 1994. Prime Minister Murayama, chairperson of SPJ, stated admission of the Treaty of Mutual Cooperation and Security between Japan and US and Japan Self-Defence Force abandoning the past policy of SPJ. The Defence budget may be depressed in order that its increase rate may be less than 1 percent in the 1995 budget although once its budget exceeded over 1 percent of nominal GNP in the Nakasone Administration.

Reflecting such a tendency, the present paper tries to evaluate the disarmament effect in the world economy. For that purpose we pose arms race models, and degrees of disarmament are measured by the difference from the state of arms race. This kind of military spending model, namely arms race model, will explain international interrelationship of individual country's military budget in a numerical sense; therefore it is easy to assess how disarmament of some countries may have repercussion effects on the other countries' defence spending. One of the aims of the present paper is, therefore, to investigate the applicability of action-reaction type model in describing the state of the Cold War on the main OECD countries' defence spending. Moreover, it will would be made clear how the disarmament affects these countries' military spending, which is followed by the tendency of the latter half of the Reagan Administration. The model used therein was action-reaction type, which depicts the determination of one country's defence spending in response to that of enemy country. Section 2 is dealing with two kinds of actions-reaction models with some extensions. Section 3 shows the estimated results of US and USSR, and section 4 is devoted to show estimated result of the other OECD countries with effect of the world economy concerning the current world recession. Finally we conclude.

2. TYPICAL ARMS RACE MODELS AND DISARMAMENT

2.1 Richardson's Model

The Richardson's first work came out in 1998 in a small thin monograph: his model attempted to describe arms races of dreadnought race between the Alliance and the Allies during the World War I. [L.F. Richardson 1960] The model designed to interpret a spiral effect of dreadnought race within the framework of action-reaction responses. Dreadnought construction now extended to overall disbursements of related items in government budget. He formulates dreadnought race in the following two derivative equations.

$$dx/dt = ky - ax + g \quad k > 0, \ a > 0 \tag{2.1}$$

$$dy/dt = lx - by + h \quad l > 0, \ b > 0 \tag{2.2}$$

x : defense spending of nation A

y : defense spending of nation B

k : defense coefficient of nation A

l : defense coefficient of nation B

a : nations A's fatigue and expense of keeping up defenses

b : nation B's fatigue and expense of keeping up defenses

g : grievances and ambitions of nation A

h : grievances and ambitions of nation B.

Coefficients (a, b) not only represent the cost of actual defence expenses, but also represent psychological cost that the nation would experience. These expenses work to suppress the defense spending. In economic terminology, the existence of the coefficients a, b represent 'flow adjustment' for the defence spending co-efficient (g.h) represent the autonomous level of defence spending which is independent of the rival nation's defence spending, and of internal economic condition. Richardson clarifies several patterns between armament and disarmament. In system (2.1)-(2.2), if g, h, x, y are all zeros, this ideal condition would be called 'permanent peace by disarmament and satisfaction.' If k, l, a, b are zero, the condition is called 'mutual disarmament without satisfaction.' 'Unilateral disarmament' is the case where k and b are zeros. 'Race in armament' could be said as the case where a, b, g, h are Zeros. However, the real world would be; k, l, a, b, g, h # 0. A number of articles have been contributed to extend the model, to estimate the relation, and to apply the model, to the real world. (For example, see [W.R. Caspary 1967] [K.J.Gantzel 1973] [M.D. Intriligator & D.L. Brito 1975]) [R.P. Strauss 1971] [D.L. Wagner, R.T. Perkin & Taagepera 1975]

One discrete version of the Richardson's model (2.1) (2.2), for the empirical application, would be the following in which the force of decreasing $(x_t - x_{t1})$ is the last defence spending x_{t-1} instead of x_t. The country's judgement on $(x_t - x_{t-1})$ would be based on information of x_{t-1} rather than that x_t.

$$x_t - x_{t-1} = ky_t - ax_{t-1} + g \qquad (2.3)$$

$$y_t - y_{t-1} = lx_t - by_{t-1} + h \qquad (2.4)$$

x_t : defence spending of nation A at period t

y_t : defence spending of nation B at period t

Then the system (2.3)-(2.4) can be rewritten in the following first following order autoregressive form

$$x_t = ky_t + (1-a)x_{t-1} + g \tag{2.5}$$

$$y_t = lx_t + (1-b)y_{t-1} + h \tag{2.6}$$

In view of the current nation's long-term foresight for the rival nation's defence spending which means that x_t is influenced by the past defence spending y_{t-1}, it would be much more plausible to specify the spending equation in a distributed lag form.

$$x_t = \Sigma k_i y_{t-i} + (1-a) x_{t-1} + g \tag{2.7}$$

$$y_t = \Sigma l_i x_{t-i} + (1-b) y_{t-1} + h \tag{2.8}$$

In the last we deduce the Richardson model in an optimisation framework by assuming the welfare function of defence spending explicitly. Then the Richardson's model could be worked out in an alternative way. We assume unimodal welfare of defence spending which is expressed in quadratic form.

$$W_x = -ax^2/2 + b_t x + c \qquad b_t = g + \Sigma k_i y_{t-i} \quad a > 0 \tag{2.9}$$

In the above the, optimal level of defence spending exists, which is affected by the rival nation's defence spending. When the enemy nation's defence spending increases, the optimal level shifts in the right direction, and vice versa. The actual increase of defence spending is determined by 'flow' adjustment scheme

$$x_t - x_{t1} = \lambda (x_t - x_{t-1}) \qquad 0 < \lambda < 1 \tag{2.10}$$

x_t^* : optimal defence spending

using the peak value $x_i^* = b_t/a$ of welfare function. Although government authority knows the optimal amount $x_i^* = b_t/a$, it never realises it. The optimal level may not be achieved, and actual deter-mination is considered to be an searching process for optimum level.

2.2 Catastrophe Model

Catastrophe model, which is derived from the catastrophe utility function, has its origin in the Zeeman's argument. [E.C.Zeeman 1977] Zeeman took two kinds of catastrophes of 'cusps' and 'butterfly' catastrophes as utility function for describing military action between two countries, and investigated a background of the appearance of the dove group and centrist group. This theory intends to describe drastic change of structure by the use of nature of curve. It is now known that there are seven kinds of forms in terms of changing points. This model is an adequate model in describing a drastic change, and is not restricted in the description of military behaviour. The catastrophe is more special than Richard's model in the sense that it has a possibility of drastic change in its nature. Isard and his colleague, in line with Zeeman's argument, set forth cusp catastrophe as a welfare function for defence spending which is a model of determining defence spending. They posed a defence spending model which is a different type of the Richardson's model.[W.Isard & P.Liossatos 1974] Now, use their model in showing theoretical possibility of mutual disarmament. 'Cusp catastrophe' welfare W for the defence spending z by W.Isard et al. is set forth in the 4-th order polynomial with using the same coefficients (a,b) as in the (2.1)-(2.2) for saving notation:

$$W = -\frac{z^4}{4} + az^2/2 + bz \qquad (2.11)$$

a : splitting factor

b : normal factor

As is mentioned in later paragraph, splitting factor prescribes the structure determining whether it causes catastrophe effect or not in virtue of normal factor. On the contrary, normal factor directly causes catastrophe effect by change of the factor. Welfare W changes its shape by the shift of combination (a,b). Then three kinds of different shape could be distinguished.

(α) welfare is unimodal in the right

(β) welfare is bimodal

(γ) welfare is unimodal in the left

To see a pattern of shape, we take the derivative of welfare with respect to z.

$$\frac{dw}{dz} = -z^3 + az + b \qquad (2.12)$$

Now it expresses the slope of welfare. As in $z^3 + az + b = 0$ expresses the slope zero, it is a point of welfare flat. Case (a) has one real root in $-z^3 + az + b = 0$, case (b) three real distinct roots, and case (g) also real root. Hence, $D = 4a^3 - 27b^2 > 0$ guarantees that the welfare W is bimodal, and D 0 unimodal. The combination (a, b) lies in phase 1, phase 2, phase 3 control-plane (a, b) of Figure 21.1, graph of W shows three different patterns in Figure 21.2.

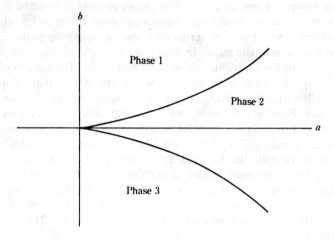

Fig. 21.1 : Control plane (a, b)

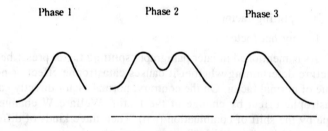

Phase 1   Phase 2   Phase 3

Fig. 21.2 : Phase shift of W_z

Phase 1 shows that the peak lies in the upper region of defence spending, which implies the hawk gaining power. Phase 2 shows the hawk and the above the dove are balanced. Phase 3 tell us the dove is gaining power. Shift from one phase to another sometimes may cause a catastrophic change, and sometimes may not. In a < o, catastrophe never occurs even by a change of 'b' because of D > 0. Catastrophe change may occur in region a > 0. However, smaller 'a' causes smaller catastrophe change. In this sense splitting factor splits the situation. The condition a = 0 is the splitting line. The optimal point maximising the welfare varies by the shift of phase. And, phase shift is brought about by the rival nation's defense spending. As a matter of fact, strategy may possibly change phase shift in W. The following graph depicts the control plane (a,b) in the base vertical direction of the optimal value x of the welfare function. In the inside of triangle there have three distinct real roots, and in the outside one real root. When decreasing of b, namely locus of (a,b) crosses the lower line of the triangle, catastrophic decreasing change occurs.

Inversely, when crosses the upper line, drastic increasing change does. (Fig. 21.3)

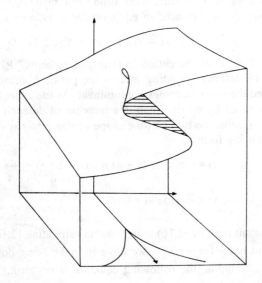

Fig. 21.3 : Catastrophic change

Now the actual change of defence spending is, due to W. Isard et al., set equal to the derivative of W.[W. Isard & P. Liossatos 1974]. If the derivative is positive, the optimal level is located in the right direction of the present level and the nation would increase its defence spending, and vice versa. Perhaps, this adjustment mechanism leads to over-shooting.

$$dz/dt = dW/dz = z^3 + az + b \qquad (2.13)$$

Change in z is the derivative of catastrophe welfare. Again, one discrete version of (2.13), assuming the existence of time lage of response, would be the following :

$$z_t - z_{t-1} = -z_{t-1}^3 + az_{t-1} + b$$

$$\text{or} \quad z_t = -z_{t-1}^3 + (1 + a) z_{t-1} + b \qquad (2.14)$$

Attention has to be paid; catastrophe theory in the above and its modification by W. Isard have to be clearly distinguished. Isard's modified model is to determine current increase of defence spending aiming at the level whereas the optimum level is realized in pure catastrophe model. In Isard's the optimum level is never to be achieved. Assuming spliting factor is constant overtime, only normal factor is supposed to shift over time, and makes an element of catastrophe change. Hence, an equation for estimation is below

$$z_t = z_{t-1}^3 + (1 + a)z_{t-1} + b_t \qquad (2.15)$$

This form could be called 'catastrophe equation.' Rival nation's increase of defence spending affects the public opinion of current nation, and the hawk becomes predominant. As the change of welfare shape is caused by b_t, rival nation's increase of defence spending is assumed to influence b_t. The time shape of this effect is expressed in distribution lag form.

$$z_t = - z_{t-1}^3 + (1 + a)\, z_{t-1} + (\sum_{i=0} k_i y_{t-i} + f) \qquad (2.16)$$

where $$b_t = \sum_{i=0} k_i\, y_{t-i} + f$$

Constant term of (2.16) is in f. As, in estimating (2.16), we have coefficient unity for $- z_{t-1}^3$, we have to make some correction. In practical estimation, the following equation is set forth.

$$x_t = \lambda\, x_{t-1}^3 + (1 + a)\, x_{t-1} + (\sum_{i=0} h_i y_{t-i} + g) \qquad (2.17)$$

Statistical significance of λ justifies the fitting of Isard's catastrophe model. Multiplying $\lambda^{1/2}$ by the left on both side, we get the below :

$$(\lambda^{1/2}\, x_t) = - (\lambda^{1/2}\, x_{t-1})^3 + (1 + a)\, (\lambda^{1/2} x_{t-1}) + (\lambda^{1/2} \sum h_i y_{t-i} + \lambda^{1/2}\, g)$$
$$(2.18)$$

Putting $z_t = \lambda^{1/2}\, x_t$, $k_i = \lambda^{1/2}\, h_i$, $f = \lambda^{1/2}\, g$ into (2.18) makes

$$z_t = - z_{t-1}^3 + (1 + a)\, z_{t-1} + (\sum k_i\, y_{t-1} + f). \qquad (2.19)$$

This equation is again (2.16). The variable that follows catastrophe model is z_t. Naturally we must check catastrophe change in terms of the variable in (2.19).

Validity of 4-th order welfare can be indirectly justified by the validity of behavioural equation of defence spending which is derived by optimising 4-th order welfare function of defence spending [for example (2.16)]; if we accept (2.16) behavioral equation by statistical fitness using empirical data, we automatically accept 4-th order welfare function indirectly. This is true for Richardson's case; if we take (2.7) as the best equation, we have to accept (2.9) at the same time. Both models could be derived from the point of view of optimisation in decision making. Favourably the 4-th order graph has minimum order one of exhibiting both unimodal and bimodal welfare by the shift of parameters. The reason why we investigate to apply the catastrophe model is that 4-th order welfare function expresses a possibility of sudden change for armament or disarmament by the shift of parameters of the function.

Under the depth of drastic change depicted by catastrophe model, we would have negotiating process, which process is unknown for the people other than negotiators and particular governmental officials. If these data of negotiating process are available, other kinds of model analysis such as using qualitative simulation model or qualitative game may be possible. [A. Tanaka 1981] [H. Kosaka & K. Numakami 1993]. Then both models may be possibly linked with each other.

Endogenising Normal Factor

It is possible to formulate the normal factor dependent on other variables. It is considered that an increase of government deficit makes it easier for the government to sit at negotiation table of disarmament. Inversely economic affordance leads big armament. Then, we assume normal factor is dependent on ratio of government deficit with respect to nominal GNP (GNPN).

$$a_t = a_0 + a_1 B_t / GNPN_t \qquad (2.20)$$

$$a_1 > 0 \qquad B_t : \text{government deficit}$$

Putting (2.20) into (2.17) gives us the following :

$$x_t = -\lambda x^3_{t-1} + (1 + a_0) x_{t-1} + a_1 (B_t / GNPN_t) x_{t-1} + (\Sigma h_i y_{t-i} + g) \qquad (2.21)$$

This is the equation for estimation. As in the same way of (2.18), we change (2.21).

$$\lambda^{1/2} x_t = -(\lambda^{1/2} x_{t-1})^3 + (1 + a_0) \lambda^{1/2} x_{t-1} + a_1 (B_t / GNPN_t) \lambda^{1/2} x_{t-1}$$
$$+ (\Sigma \lambda^{1/2} h_i y_{t-i} + \lambda^{1/2} g) \qquad (2.22)$$

Equation (2.22) can be rewritten below :

$$z_t = -z^3_{t-1} + (1 + a_0) z_{t-1} + a_1 (B_t / GNPN_t) z_{t-1} + (\Sigma k_i y_{t-i} + f)$$
$$= -z^3_{t-1} + [(1 + a_0) + a_1 (B_t / GNPN_t)] z_{t-1} + (\Sigma k_i y_{t-i} + f) \qquad (2.23)$$

This is catastrophe equation of (2.21).

Effect of Disarmament on Catastrophe Welfare Function

To incorporate an effect of disarmament into the catastrophe equation, we simply reduce constant term $(\Sigma h_i y_{t-i} + g)$ by the amount of disarmament in (2.17). And that, we have to distinguish the amount of disarmament of the current and of rival nation. Both disarmament is evaluated in (2.17).

$$x_t = -\lambda x_{t-1}^3 + (1+a) x_{t-1} + [\,(\sum_{i=0} h_i y_{t-i}) + (g - d_{x,t})\,] \quad (2.24)$$

$d_{x,t}$: amount of disarmament in current country

Current nation's disarmament is evaluated explicitly in constant term as in (2.17) whereas disarmament of rival nation is implicitly in y_t of $\Sigma h_i y_{t-1}$. Quite in the same way, $z_t = \lambda^{1/2} x_t$, $k_i = \lambda^{1/2} h_i$, $f = \lambda^{1/2} (g - d_{x,t})$, we get (2.25).

$$z_t = -z_{t-1}^3 + (1+a) z_{t-1} + (\Sigma k_i y_{t-1} + f)$$
$$= -z_{t-1}^3 + (1+a) z_{t-1} + b_t \quad (2.25)$$

Criteria of Predominance of a Dove

Consequently, whether disarmament causes phase change on welfare or not depends on the shift of (a, b_t). Now let us calculate the amount of disarmament which brings about predominance of a dove. Determinant equation incorporating disarmament is below.

$$D = 4a^3 - 27b_t^2 = 4a^3 - 27\,(\Sigma k_i y_{t-1} + f)^2$$
$$= 4a^3 - 27\,[\,\Sigma \lambda^{1/2} h_i y_{t-1} + \lambda^{1/2} (g - d_{x,t})\,]^2 \quad (2.26)$$

Determinant equation gives us the critical disarmament point of phase change. Putting $D = 0$ gives us the following.

$$d_{x,t} = \Sigma h_i y_{t-1} + g \pm (2.3^{-3/2})\, a^{3/2} \lambda^{-1/2} \quad (2.27)$$

Above equation shows us two points, which are both points of phase changes. It is possible to expect rival nation's disarmament, which is evaluated in y_t implicitly.

Relationship with the Richardson Model

Treating the Richardson model by introducing optimization of welfare function $W_x = - ax^2/2 + bx + c$ in (2.9) suggests the following relationship.

$$x_t - x_{t-1} = \lambda \left(x_t^* - x_{t-1} \right) \tag{2.28}$$

$$x_t^* = - ax_{t-1} + b = dW/dx \mid x_{t-1}$$

$$W = - ax^2/2 + bx + c$$

In the above, $dW/dx \mid x_{t-1}$ means that the derivative dW_x/dx is evaluated at x_{t-1}. In the same way of (2.28), catastrophe model could be derived.

$$x_t - x_{t-1} = \lambda \left(x_t^{**} - x_{t-1} \right) \tag{2.29}$$

$$x_t^{**} = - x_{t-1}^3 + ax_{t-1}^2 + b + dW_{xx}/dx \mid x_{t-1}$$

$$W_{xx} = - x^4/4 + ax^2/2 + bx$$

From above simple argument it is shown that both models have welfare function (W_x, W_{xx}) and the optimal level of defence spending (x_t^*, x_t^{**}) respectively. It should be noted that the meaning of coefficients (a, b) in (2.28) (2.29) are different.

3. EMPIRICAL ANALYSIS ON USA/USSR DEFENCE SPENDING

In empirical studies in this and the next sections, we first use the Richardson model basically which has some variations. Yet, as this model may fail to explain empirical data, we adopt catastrophe model. This process follows according to usual econometric modeling approach. Therefore the estimated result cited below is only a part of whole estimation work.

3.1 US Model

(a) Estimated equation

Data for US defence spending is retrieved from President Economic Report of the Congress (1990 edition). For USSR defence spending we use Prof Niwa's estimate, which is available from 1958 over 1985. [H. Niwa 1989] Estimated result for Richardson model is below.

$$\text{USAMIL} = -6.4346775 + 0.8362987 * \text{USAMIL}_{-1} + \tag{3.1}$$
$$0.1758835 * \text{USRMIL}_{-1}$$

t values : (–0.9097)   (14.737)   (8.5009)   (3.1)

adj. R^2 = 0.9906 D.W. = 2.0107 SE = 3.046 sample : 1973–1986

USAMIL : US defence spending USRMIL : USSR defence spending

The t-value for constant term is statistically less significant.

(b) utility of US defence spending

From the estimated equation, the US welfare function is easily obtained.

$W_x = - ax^2 / 2 + bx = -(1-0.8362987) *x^2 / 2 +$
$$(- 6.436775 + 0.\,1758835*USRMIL_{-1})$$

$= - 0.08185065*x^2 +(- 6.4346775+0.1758835*USRMIL_{-1})$ (3.2)

where x= USMIL

Welfare is unimodal because of the 2nd order polynomial, and its peak shifts in accordance with that of coefficient 'b'.

(c) US optimum defence spending

The optimum defence spending can be directly derived from the welfare function. It depends upon USSR defence spending.

$x^* = (- 6.4346775 + 0.1758835*USR_{-1}) / (1 - 0.8362987)$

$= - 39.307430667 + 1.0744172465*USRMIL_{-1}$ (3.3)

Figure 21.4 shows the actual and optimum level of military spending.

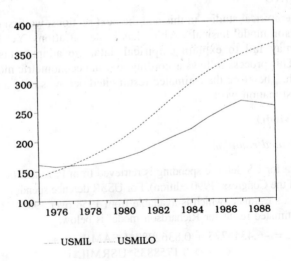

USMIL ----- USMILO

Fig. 21.4 : US Actual & Optimal Defense Spending

The optimal values were over the actual values except those of 1975 and 1976. FIGURE 21.4 shows the difference becomes bigger since 1977. The difference first emerged in 1977, and it stimulated US defence spending. The difference became serious in 1979, and become much more serious in the first Reagan Administration, which accerelated big expansion of defence spending.

3.2 USSR Model

(a) *Inadequacy of Richardson's model*

We tried to fit Richardson model to USSR data, but in vain. Two estimated equations are illustrated for reference.

$$USRMIL = 5.1282845 + 1.0191977*USRMIL_{-1} + 0.0357137*USMIL_{-1}$$

t values : (0.1152267) (9.7799529) (0.1076966)   (3.4)

adj. $R^2 = 0.9458$ D.W. $= 2.3067$ SE $= 15.2666$

sample : 1973-1985

$$USRMIL = 2.7863672 + 1.0327071*USRMIL_{-1} + 0.0276140*USMIL$$

t values : (0.0797031) (9.3950584) (0.0973420)   (3.5)

adj. $R^2 = 0.9531$ D.W. $= 2.2641$ SE $= 14.7771$

sample : 1972-1985

In both equations the effect of one year delayed USSR defence spending is too powerful to get statistically significant effect of US defence spending. It is considered Richardson model is inadequate for describing USSR defence spending.

(b) *Catastrophe model*

Then we tried to fit catastrophe model to USSR data.

$$USRMIL = -0.4019781*USRMIL_{-1}^{3}/100000 + 1.5257242*USRMIL_{-1}$$

    (-1.6901)   (4.9388)

$$-157.90774 + 0.6099218*USMIL$$ (3.6)

(-1.5723)   (1.4083)

adj. $R^2 = 0.9598$ D.W. $= 2.226$ SE $= 13.66$ sample : 1972-1985

Statistical significance of $USRMIL_{-1}^{3}$ and USMIL is a bit weak, but more adequate than Richardsonian. Multiplying

$(0.4019781/100000)^{1/2} = 0.002005$ on both side, we get catastrophe equation for the above welfare.

$$z_t = -z_{t-1}^3 + (1+a) z_{t-1} + b_t \qquad (3.7)$$

$z_t = 0.002005*\text{USRMIL}$

$a = 0.5257424$

$b_t = -0.3165954021 + 0.001222856*\text{USÁMIL}$

(c) *graph of welfare of USSR defence spending*

Corresponding to the catastrophe equation (3.7), we easily obtain the 4-th order polynomial welfare function.

$$W_x = -z^4/4 + az^2/2 + bz$$

$$= -z^4/4 + (0.5257424) z^2/2$$

$$+ (-0.3165954021 + 0.001222856*\text{USAMIL}) z \qquad (3.8)$$

Accordingly its derivative is below.

$$dW_z/dz = -z^3 + az + b$$

$$= -z^3 + (0.5257424) z$$

$$+ (-0.3165954021 + 0.01222856 * \text{USAMIL}) \qquad (3.9)$$

Within the sample period, the 3rd order equation $dW_z / dz = 0$ has three distinct real roots; the largest and the smallest roots show upper and lower peaks respectively, and the third root shows the bottom. FIGURE 21.5 illustrates time series of roots with actual defence spending.

Then the figure suggests the following.

(α) Upper optimum defence spending, which corresponds to the upper real root, is always increasing and is over actual defence spending. This difference draws up actual defence spending. Yet the difference is very small since 1983. Defence spending of USSR is considered to have reached the upper limit since then.

(β) Lower optimum defence spending stays at about –400. The value itself has no significant meaning. It has a roll of a kind of 'black hole' that makes USSR defence spending decrease downwards if actual defence spending lies in the left of bottom.

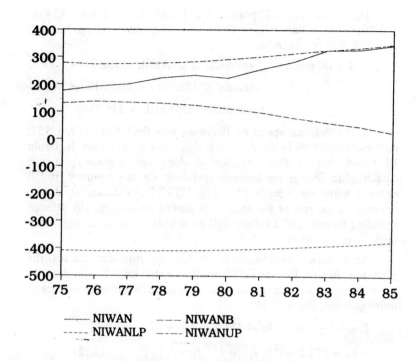

FIG. 21.5 : USSR Welfare Function

(*d*) critical US defence spending for USSR phase change what determines the shape of USSR welfare function is US defence spending. Next we will clarify the critical points of US defence spending that causes the phase change on welfare function of USSR defence spending. For this purpose, we must check the number of real root of $dW_z / dz = 0$. Determinant equation $4a^3 = 27b^2$ decides the critical points. From $4*(0.5257424)^3 = 27*b^2$, we obtain

$b = \pm\ 1467261823$. And we seek US defence spending that satisfies the equation.

$$\pm 0.1467261823 = -\ 0.3165954021 + 0.001222856 * \text{USAMIL} \quad (3.10)$$

Then, the critical points are USAMIL = 138.91187498, 378.88489233 billion dollars. Then we get the phase change formula of USSR defence spending.

USSR welfare :   unimodal if USAMIL < 138.9119   (3.1)

bimodal if 138.9119 < USAMIL < 378.8849

unimodal if USAMIL > 378.8849

If US defence spending becomes less than 138.9119, USSR public opinion turn to the dove in which hawk never exists. It should be noted that in this situation it does not necessarily occur catastrophic change on defence spending. On the contrary, if US defence spending exceeds 378.8849, USSR hawk dominates public opinion. In the rest of the area, both parties coexist. As US defence spending records 256.2 billion dollars at 1989, there is enough room to go over 378.8849.

Next, how much amount of US disarmament causes the predominance of dove in USSR military behavior ? It would easily show the critical disarmament which shifts USSR bimodal welfare to unimodal one. From (2.16).

$$D = 4a^3 - 27b_t^{\ 2} = 4a^3 - 27\ (\ \Sigma\ k_i\ y_{t-i} + f)^2$$

$$= 4a^3 - 27\ [\ \Sigma\ \lambda^{1/2}\ h_i\ y_{t-i} + \lambda^{1/2}\ (g - d_{x,t})\ ]^2 \quad (3.12)$$

$a = 0.5257424\ b = -\ 0.3165954021 + 0.001222856 * \text{USAMIL}$

we have the criterion.

$$\pm\ 0.1467261823 = -\ 0.3165954021 + 0.001222856 * (\text{USAMIL} - d_x)$$

$$(3.13)$$

$$0.001222856 * (\text{USAMIL} - d_x\ ) = 0.4633215844,\ 0.1698692198$$

$$(3.14)$$

Then we can observe the US disarmament criteria for USSR phase change in the sample period.

$$\text{USA} - d_x = 378.88482732,\ 138.91187498 \quad (3.15)$$

$$d_x = \text{USAMIL} - 138.91187498\ \text{dove} \quad (3.16)$$

d_x = USAMIL − 378.88482732: hawk   (3.17)

Disarmament criteria (3.16) indicates the amount of US disarmament shows the predominance of dove of USSR. FIGURE 21.6 shows the US disarmament and its disarmament rate.

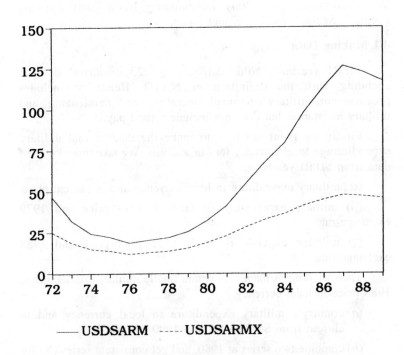

— USDSARM ----- USDSARMX

Fig. 21.6 : US Disarmament for USSR Welfare Phase Shift

From the figure it should be noted that it needs much disarmament inn order to have USSR dove as time passes by. Of course within the sample period, welfare function stayed in bimodal mode.

4. Estimating Main OECD Model and Effect of Disarmament

Disarmament of US/USSR necessarily has an effect on NATO and the other countries' defence spending. In order to disclose the affect, it require to build military spending model for these countries, in which we model international interrelation among these countries. Countries included are ; Canada, Belgium, Denmark, France, former West Germany, Greece, Italy, Luxembourg, Netherlands, Norway, Portugal, Spain, Turkey, UK and Japan.

4.1 Making Data

SIPRI yearbook holds database of '27 countries' defence spending with the definition of NATO. Hence, it includes procurement, military personnel, operations and maintenance, and military assistance, but does not include retired pay.

Firstly we point out how to make the data of real military expenditure in local currency for our analysis. We take three kinds of data from SIPRI yearbook.

(α) military expenditure in local currency and in current price

(β) military expenditure of US$ at 1979 price and 1979 exchange-rate

(γ) military expenditure of US$ at 1988 price and 1988 exchange-rate

Then we process them, and obtain finally military expenditure at 1988 price in local currency.

(a) country's military expenditure in local currency and in current from SIPRI 1982 and 1990 editions

(b) combine two series at 1980, and get consistent series (X) for 1972-1989

(c) military expenditure of US$ at 1979 price and 1979 exchange rate from 1982 edition X$79 military expenditure of US$ at 1988 price and 1988 exchange rate from 1990 edition X$88

(d) exchange rate from National Account (1960-1989) (OECD) e79 and e88 in local currency per dollar

(e) military expenditure at 1979 price X79 = X$79 / e79

military expenditure at 1988 price X88 = X$88 / e88

(f) price deflator at 1979, p79 = Xi / X79

price deflator at 1988, p88 = Xi / X88

(g) combine two series pi, 79 and pi, 88, and get consistent p88

(h) military expenditure at 1988 price in local currency XMIL = X / p88

4.2 Estimating Other Countries' Models

Prior to estimation, we posit a model by expanding the basic Richardsonian equation (2.5) in two directions :

$$x_t = ky_t + (1 - a) x_{t-1} + g \qquad (4.1)$$

$$g = g_0 + g_1 \, GNP_t + g_2 \, p_t \quad g_1 > 0 \quad g_2 < 0$$

where constant term g is dependent upon either real GNP_t or Pt (domestic price level), or on both. (a) Although original Richardson model determines nation's defence spending within international context, this model taken into account domestic factors (GNP_t, p_t) simultaneously. Sign $g_1 > 0$ comes from the economic affluence for the military spending, and the sign $g_2 < 0$ from money illusion. (b) While, in the Richardson model, two rival nations are confronting with each other, the NATO members act for the collective security. Hence ''y'' country has two possibilities : ''y'' country may not only an enemy but also friendly. Therefore we have to distinguish two cases.

(I) enemy country   : k > 0 antagonistic   (4.2)

k < 0 cooperative

(II) friendly country : k > 0 cooperative

k < 0 complimentary

Modified Richardson model is applied to the basic confrontation of the main OECD vs. USSR. The followings are the estimated equations. Countries other than NATO are also included.

Canada :

CANMIL = 2971.55 + 0.3700 * CANMIL t_{-1} + 12.399 * USRMIL   (4.3)

t values : (2.3785) (1.5414) (3.0094)

adj. R^2 = 0.9444   D.W. = 1.9883   SE = 293.4293

CANMIL : Canadian defence spending

USRMIL : USSR defence spending

Simple Richardsonian is valid to Canadian case.

France :

FRAMIL = − 46187.95 + 0.060537 * FRAGDP 85 − 145.5182

* USMIL   (4.4)

t values : (− 4.0435)   (13.9979)   (− 3.01326)

adj. R^2 = 0.9746 D.W. = 2.0141   SE − 3514.28   sample :
1973-1988

FRAMIL : France defence spending

USMIL : US defence spending

FRAGDP85 : French real GDP at 1985 price

Economic condition is the most influential factor in supporting defence spending. Still US defence spending is complimentary with France. Richardson type equation is not valid for French case.

The next equation is simplistic.

FRAMIL = 113854. 3 + 286.18 * USRMIL   (4.5)

(23.9574)   (8.0475)

adj. R^2 = 0.8416 D.W. = 0.4474   SE = 8076.8 sample : 1973
1985

We take the second model.

Former West Germany :

FRGMIL = 27155.8 + 0.2631 * $FRGMIL_{-1}$ + 0.0096226 *
FRGGDP85   (4.6)

t values :   (4.6700)   (1.5125)   (3.3086)

adj. R^2 = 0.9146 D.W. = 1.8691 SE = 696.17 sample : 1973-1989

FRGGDP85 : real GDP of West Germany at 1985 price. Even West Germany takes robust internal influence from domestic economic rather than from the outside.

Italy :

ITAMIL = 2295.04 + 0.71751 * $ITAMIL_{-1}$ 12.7553 * USRMIL
                                                                 (4.7)

t values :   (1.0743)   (3.6861)   (2.0821)

adj. R^2 = 0.9406 D.W. 1.8352 SE = 567.35 sample : 1973-1985

Simple Richardsonian model is valid. As Italy is not directly confronted with USSR, USRMIL variable is a kind of proxy variable. In this case proxy denotes a sort of international uncertainty.

UK :

$$UKMIL = 9941.3057 + 0.6909943 * UKMIL_{-1} + 4871.9168*PUK88$$

$$(4.8)$$

t values :   (3.1795053)   (3.8857236)   (2.887825)

–0.0222939*UKGDP85

(– 2.1326343)

adj. R^2 = 938906 D.W. = 2.275239 SE = 547.4471   sample : 1973-1988

UKMIL : UK defence spending

PUK88 : UK price level

UDGDP85 : UK real GDP at 1985

Purely internal factors determine defence spending under price illusion and under the tade-off with other items of expenditures. (negative sign, UKGDP85)

Japan:

$$JPNMIL = 353.156 + 0.\,66169*JPNMIL_{-1} + 2.3116 * USRMIL$$

$$(4.9)$$

t values :   (1.5866)   (3.49874)   (2.14720)

adj. R^2 = 0.9659 D.W. = 2.22946 SE = 73.7942 sample : 1973-1985 where :

JPNMIL : Japanese defence spending

Simple Richardsonian model describes the Japanese case.

The total system including US/USSR models and these other countries' models shows satisfactory performance for 1980-1993. It is also possible to have an effect of unilateral and mutual disarmament scenario to the other countries' defence spending. Concerning these kinds of simulation, we merely refer that most countries reduce defence spending by disarmament with a few exceptions.

4.3 Effect of Disarmament on the World Economy

Finally, the effect of disarmament of US and former USSR is examined using internationally linked macroeconometric models. After the World War II, the world economy experienced synchronised recessions by the first and the second oil shocks: both are external shock being brought about from political events. In the same meaning the paper would check whether external shock of world disarmament may cause the world recession beginning 1989 in industrial countries. Another possibility of synchronization comes from the integration of international financial market. [B.K. Goodwin 1994] [K.P. Kimbrough 1993] As the check for the latter possibility may be beyond the scope of this paper, we have to provide other kinds of analysis.

Disarmament scenario is put into macroeconometric models in virtue of government expenditure : (1) US making disarmament of 20 billion dollars each year during 1988-89 and disarmament of 15 billion dollars during 1990-1993, and (2) USSR having that of 10 units during 1990-1993. Figure 21.7 and Figure 21.8 shows outcome of US and USSR simultaneous disarmament in both countries respectively, illustrating relatively big effect on both countries' defence spending with repercussion over the other countries'.

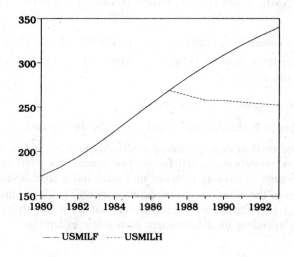

Fig. 21.7 : Arms Race vs Disarmament in USSR

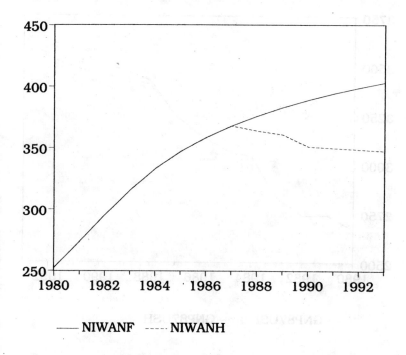

Fig. 21.8 : Arms Race vs Disarmament in USSR

Disarmament will decrease production in a Keynesian framework. Figure 21.9 shows reduction of real GNP in US economy with omitting the other countries' figures : US disarmament decreases economic growth of US economy by about 0.5%, and decreases that of Japanese economy by about 0.3% in our models.

And finally the figure below is illustrating decrease of export of seventeen countries including Summit Seven.

Then we could conclude that the current world recession may be largely caused by some mechanism of internationally integrated financial market.

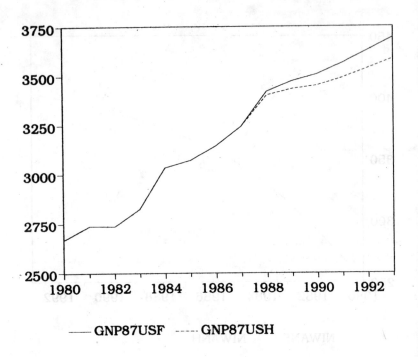

GNP87USF ----- GNP87USH

Fig. 21.9 : Real GNP of US

5. Concluding Remarks

In the last we conclude from the estimated result.

(1)Two types of action-reaction models have successful estimation for the arms race of the Cold War era. This kind of model could be called the conventional model. The process of spending could be explained as a continual process for seeking optimal level of defence spending which is dependent upon international and internal factors. The conventional model still has a wide range of application; we already confirmed the applicability to the Middle East region within international context, and to Latin America within internal context. The Richardson model also proved to be useful for describing interrelationship among defence spending of the Middle East ; action-reaction mechanism also dominates this area. What

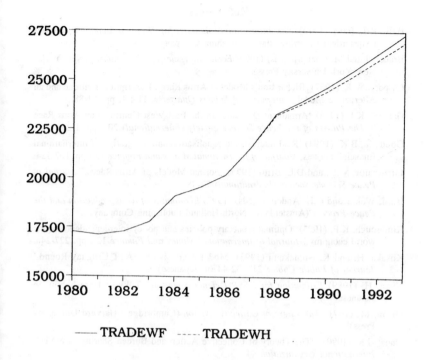

Fig. 21.10 : World Export

interests us in that analysis is arms transfer to the Middle East plays an important role. So, regulation for arms transformer to this area has to be taken into special consideration so as not to break the balance of power. Turning to a possibility of fitting other kind of models, Olson's collective action theory would be worthy of investigation.

(2)We are now in the transition era. Some other kinds of models other than conventional models may be adequately fitted for the military situation after the end of the Cold War. Conventional model has to give way to the new model. The new model has to explain both the Cold War and those after the Cold War.

(3)The power of breaking down the Cold War is the failure of two big economies of US and USSR. This mechanism is not incorporated into our system. Mechanism under which excess resource allocation to defence spending cause collapse of economy could not be explained by Keynesian type economic model.

(4)World disarmament could be possibly a reason for synchronised world recession. However current world recession may be caused by other mechanisms.

References

Ball, R.J. (ed.) (1973) *The International Linkage of National Economic Models* (Amsterdam : North-Holland Publishing Company).

Brauer, J and M. Chatterji (eds) (1992) *Economic Issues of Disarmament* (New York : New York University Press).

Caspary, W.R. (1967) 'Richardson's Model of Arms Race : Description, Critique and an Alternative Model,' *International Studies Quarterly*, 11 : 1, pp. 63-88.

Gantzel, K.J. (1973) 'Armament Dynamics in the East-West Conflict ; an Arms Race ?, *The Papers of the Peace Science Society (International)* 20 : pp. 1-24.

Goodwin, B.K. (1994) 'Real interest rate equalisation and integration of international financial markets,' *Journal of International Money and Finance* 13 : pp. 107-124.

Intriligator, M.D. and D.L. Brito (1975) 'Formal Models of Arms Races,' *Journal of Peace Science Society (International)* 77-88.

Isard, Walter and C.H. Anderton (eds) (1992) *Economics of Arms Reduction and the Peace Process* (Amsterdam : North-Holland Publishing Company).

Kimbrough, K.P. (1993) 'Optimal monetary policies and policy interdependence in the world economy,' *Journal of International Money and Finance* 12 : pp. 227-248.

Kosaka, H. and K. Numakami (1993) 'Model Analysis on GATT Uruguay Round,' *Journal of Public Choice* 21 : 32-44 (in Japanese).

Niwa, H. (1989) *Estimation for USSR Defence Spending* (Tokyo : Harashobou) in Japanese.

Olson, M. (1971) *The Logic of Collective Action* (Cambridge : Harvard University Press).

Oneal, J.R. (1990a) 'The Theory of Collective Action and Burden Sharing in NATO' *International Organization* 44 : 3.

Oneal, J.R. (1990b) 'Testing the Theory of Collective Action - NATO Defence Burdens, 1950-1984' *Journal of Conflict Resolution* 34 : 3.

Richardson, L.F. (1960) *Arms and Insecurity* (Pittsburgh : The Boxwood Press).

Strauss, R.P. (1971) 'An adaptive expectations model of the East-West arms' *Peace Research Society*: Papers XIX, pp 29-34.

Tanaka, A. (1981) *Chinese International Conflict Behaviour : 1949-1978* (MIT, Ph. D. Dissertation).

Wagner, D.L., R.T. Perkins, and R. Taagepera (1975) 'Complete Solution to Richardson's Arms Race Equations,' *Journal of Peace Science* 1:2.

Zeeman, E.C. (1977) *Catastrophe Theory-Selected Papers 1972-1977* (London : Addisson-Wesley Publishing Company, Inc).

Chapter 22

Tri-Polar World and the Strategies for Japan: A Cost Benefit Analysis

Toshitaka Fukiharu

According to L.C.Thurow [1992], *Head to Head*, the world after the collapse of the former Soviet Russia in 1991 will be shifted from the one with two polars: the US and Russia, to the one with three polars: the US, Japan, and EU (EC). The aim of this short paper is to conduct a cost-benefit analysis of the strategies for Japan in such an age of structural changes. In the first place, .his proposition of the prospect of the world with three polars is examined: will Japan be one of the three polars? According to Thurow, HDTV appears to be the only future industry in which Japan has the comparative advantage without question. (p.260) Until recently, this evaluation has been supported. However, brand-new reports changed this evaluation: the digital-system HDTV will be the mainstream in the future, while Japan has advantage solely in obsolete analog-system HDTV. (*Time*, March 7, `94). In the semi-conductor industry, one-time Japanese rising share (p.167) began to shrink.

Final evaluation of the (supposed) Japanese advantageous is on the automobile industry (p.165). We begin with a brief history of the Japanese automobile industry. Before and during the World War II, main outputs of NISSAN and Toyota were trucks for military-use. The U.S. Strategic Bombing Survey (USSBS), however, reported poor quality of automobiles (including trucks) made in Japan. According to USSBS, it was only aircraft industry that Japan reached international standards of product-quality (Thurow also rated high marks on zero-fighters, p.291). Prohibition of the aircraft production in the Post-war years forced the transfer of Japanese manpower from aircraft to automobile industry.

Some may be tempted to use this fact for a refutation of the *Klein hypothesis*: lower military expenditure induces higher economic growth. If Japan were allowed in the post-war years to produce aircraft through higher military expenditure, she might have been able to move into export of commercial, aircrafts, with higher economic growth than the actual one. This proposition may well be refuted again in the following way. True, in the short-run the *Klein hypothesis* may have been refuted. However, in the long-run, the probable Japan as a military super-power have lived in quite an unstable world with a possibility of repeated ruin: lower growth rate. In any case, starting in the post-war days as an *Infant Industry* with protective high tariffs and *Industrial Policy* Japanese auto-makers took off in the 1970s.

The '80s may be the prime time of the Japanese automobile industry. After reaching 30 percent of the US auto market share in the late 1991, the Japanese share began to dip. These downturn facts might be a short-run aberration from the rising trend of Japanese industries, and tentatively we assume that Japan will be one of the three polars of the world economy, following Thurow.

In the second, we examine his prospect of the 21st century as the world dominated by *Managed Trade*, not *Free Trade*. Japan has benefited from *Free Trade*, and she will suffer from *Managed Trade*, unless it becomes *Freer Trade*. Whatever strategy to secure *Free Trade* will be her most important one. Thurow's prospect is directly influenced by the integration of European market in 1993. Indirect influences on him appear to be the following two facts. The first one is: when his book was written, GATT's Uruguay Round was in the deadlock, especially in the field of agriculture (p.92). However, a compromise agreement was reached between the negotiating countries in 1994. The second fact was the *Self-Export-Restraint* by the Japanese automakers (p.88). However, when the *transplant* factories began full operation in the U.S., SER became meaningless, and it is abolished this year 1994. One might argue that the integration of European markets alone is sufficient to depict the 21st century as the world dominated by the *Managed Trade*. Even for this claim a counter-argument is possible. As Thurow explains (p.105), for the creation of *EMS* (European Monetary System), not only the monetary but also budgetary cooperations of each member government are indispensable, and each country is asked to satisfy the two restrained budgetary performances as partial conditions to enter *EMS*. Germany, supposedly the leader of the *EC*, may not be able to satisfy one of them: her "accumulated" budgetary debt may exceed 60 percent of GDP in 1999. (*NIKKEI*, April 30, '94). These counter arguments may also be nothing but a short-run aberration from the long-run trend asserted by Thurow.

Accepting his assertion as a long-run argument, we examine what could lead *Managed Trade* to *Freer Trade*. To conduct this examination, let us review the history of *Free Trade*. Free trade agreements may be said to date back to the British *Commercial Treaty* of 1786 with France. The British negotiator was a Smith's student, W.Pitt, and aimed at reducing tariffs. However, (although imperfect) free trade brought England, in the early stage *Industrial Revolution*, a huge trade surplus, and France, a serve deflation, which, some argued, was one of the causes of *French Revolution* (A.Mathiez [1922, vol.1, Chap.3], *La Revolution Francaise*). Between the *French Revolution* and the *Continental System* by Napoleon, Britain and France were in a state of war. It was only in 1860, when the two countries concluded the *Commercial Treaty* to reduce traiffs. By this time, France also entered the *Industrial Revolution*. We must remember that this `free trade' *Commercial Treaty* between England (negotiator: R.Cobden) and France was criticised by the economists in those days since it stipulated *free trade* only between the two countries and high tariffs were imposed on the imports from the third country. As history reveals, however since then a lot of two-country commercial treaties were concluded, and the *age of free trade* was realized by the network of these treaties.

What was the driving force of the world-wide free trade? As a candidate, the Industrial Revolution can be mentioned. As A. Smith asserted, technological innovations cannot be incorporated into a country without sufficiently large markets. (A.Smith [1950: p.415]. *the Wealth of Nations*.) Indeed, one case to support Smith's assertion is discussed in *Head to Head* (p.207): When *IBM* established its subsidiary in Japan to enter the Japanese market, *IBM* could not but allow Japanese computer-infant-industry to use fundamental licenses possessed by *IBM*. Although Thurow insists that the Japanese government ordered it as a condition for the entry, it would be more accurate to interpret it in such a way that the computer, a great innovation, required huge markets and *IBM* obeyed to the order of the great innovation. From this reasoning it follows that with future sufficient technological innovations which require huge markets, EU (EC) and other *Trading Blocs* cannot but open `their' blocs to others to expand markets for `their' innovations, leading to *Freer Trade*. The lack of innovations may explain why the *Bloc Economy* of the 1930 did not lead to *Freer Trade*. (SEE M.A.Bernstein [1987], *the Great Depression*.) Thus, the strategy to secure *Free Trade* is to promote innovations.

There is a very important qualification, however, for advocating further *Industrial Revolutions*, even if they could dissolve the Trading Blocs into *Freer Trade*. It is the fact that the *Industrial Revolutions* so far have been characterised by mass-production through mass-consumption of resources, and they are the culprits for *oil Crises* in the 1970s and *Environmental Problems* in the 1990s. For both problems the key word is *Energy*.

In what follows, to refine the Japanese strategies, we proceed to the discussion of *Energy Problem*. Faced with the 'crisis' in the 1930s: the severe economic stagnation, Japan invaded Manchuria. This aggressive strategy in 'crisis' brought about the US oil embargo as sanction, where Japan imported 80 percent of its oil consumption from the US. We learned a historical lesson that a *threat* strategy of a super-game does not necessarily achieve a Pareto Optimal solution to the game depending on players' time preferences and imperfect information. It is a well-known fact that is the situation of scant information. Japan regarded the sanction as the virtual declaration of war against her, and knowing there is no hope of victory against the US she adopted *first strike strategy* : Japan had only a 18 month-stock of oil. (See Toland [1970], *the Rising Sun*.) After the devastating defeat in the World War II, Japanese strategy has become *alliance*, whereas she had no other strategies. Her strategy did work superbly.

In *Oil Crunch*, perhaps the greatest crisis after the World War II, Japan could secure uninterrupted oil consumption, thanks to the strategy of *alliance*. Of course, she tried everything for energy conservation, and her economic strength soared during this 'crisis' era, so much so as to create trade frictions with the US in the 1980s. This is the second proof of the characteristic of the Japanese economy: faced with crises inflicted by the outer forces she can manage to overcome them by technological or managerial innovations. It must be noted, however, that these innovations are mainly *Process Innovations*, not

Product Innovations (Head to Head [p.353]. *Iacocca* [1984], Iacocca; an *Autobiography*, insisted that *Toyota System* had been invented in factories of Ford. The point, however, is that *Toyota* brought that system into perfection.)

In spite of her energy conservation, Japan consumed huge amounts of oil, which has been supplied because of the allied countries' confidence that Japan has no intention (or capability) of aggression, even when faced with the 'crisis'. Since the collapse of the former Soviet Union in 1991, however, the Japanese strategy in the field of *energy* appears to have been in difficulty. Since the *Oil Crisis* she tried to solve one of the *security problems* stemming from oil dependence on foreign sources by developing Plutonium processing technology: we call this the strategy of *Energy Self-Reliance* (ESR strategy). While stored Plutonium could solve one of the main *security problems*, it produced a new dilemma for Japan. A lot of countries feared that Japan has intention of possessing *Nuclear Weapons* using stored Plutonium, in spite of the pledges of successive Japanese premiers to sign the renewed *Nuclear Non-Proliferation Treaty*. (Time [Nov. 30`92]) This fear has been multiplied since the North Korea declined the nuclear inspection by the International Atomic Energy Agency in February 1993 and the strategic disadvantages of allied forces between South Korea and the US against North Korea was revealed in Time [April 4, 1994].

In what follows, we assert that Japan has no other choice but abandoning the *ESR* strategy, even if she has no intention of possessing nuclear weapons. As the Japanese experience in the *Oil Crisis* shows she could not only secure uninterrupted energy consumption through political alliance with other countries, but also innovate technologies. We must assume that at the same time a *Trade War* through innovations was created as a byproduct. But the *Trade War* of this type is different from *Trade War* through mutual traiff-imposition, and in a sense the former are necessary wars in capitalism. Without them, capitalism itself might collapse due to mass-unemployment stemming from the lack of innovations-induced investment. Of course, a caution is required here: not all the innovations are conducive to the expansion of employment even though they surely encourage investment demand. Recent computerisation appears to produce *mismatch* in labor market: excess demand for *high-tech* workers and excess supply of *low-tech* workers. Some may interpret this *mismatch* as a *short-run* lack of investment for *education*, and an opportunity for long-run economic growth through expanded public investment on education.

Returning to the original energy problem, we examine three cases of energy shortage. The first two cases are the ones in peace-time. On the one hand, suppose that other countries suffer from energy shortage in the future, while Japan has sufficient energy supply thanks to the successful development of Plutonium processing technology: success of the *ESR* strategy. A natural course of Japan in that case would be military build-up including the production of *Nuclear Weapons* to protect 'affluent Japan' who occupies *one polar* in the world, although, theoretically, Japanese sufficient energy could be supplied to other suffering countries as *economic aids*. Judging from the poor performance in the field of economic aids (although absolute value of Japanese economic aids is No.1 in the world), the former course to the military build-up would have much higher probability than the latter course. This would lead to an *arms race* in Asia: high cost for Japan. On the other hand, when the future energy-shortage evenly hits all the countries including Japan who abandoned the *ESR* strategy, this hardship will be overcome as we experienced in the decade of *Oil Crisis*, so long as the existing *Nuclear Weapons* are redundant for the lack of opportunity to use in actual wars.

Final examination is concerned with the energy-shortage in the war-time. Suppose that Japan is still allied with the US who is about to enter a war with other countries. We must remember the fact that the US nuclear strategy appears to have shifted from the one of preemptive nuclear *first-strikes* to the one of defensive nuclear *counter-attacks*, as exhibited by her intention of abolishing *MIRVs*. In this case, the US will not feel the shortage of *nuclear* weapons, although she feels the shortage of *conventional* weapons. In this case, huge stockpiles of plutonium, made possible by the Japanese *ESR* strategy, would mean rather as additional cost for the US than a *benefit* for her, since either it must be specially protected from the direct theft by the enemy when the alliance countries, or it implies the enemy's expanded nuclear arsenal if the alliance between the US and Japan is dissolved for whatever reasons. This would produce *political frictions* between them an additional cost for Japan. So long as Japan desires the uninterrupted success in economics in the framework of world-wide political stability through the alliance with the US, the *ESR* strategy would produce unbearable cost for Japan.

In this way, the *benefit* of *ESR* strategy could be obtained mainly from the viewpoint of *Nationalism*, whereas *Japanese* nationalism produce high cost through or arms race in Asia. My final evaluation is that cost is greater than the total benefit. Recent Japanese strategy for her seat in the UN Security Council (we call this the *SC* strategy) would be justified only from the viewpoint of *Nationalism*, although the proponents assert that the *SC* strategy is necessary to play a more 'meaningful' role. True; not many *Asian* countries have expressed their support for the Japanese seat. But, they may well have done so in exchange for other purposes: for example to obtain economic aids from Japan *without* any politico-economic conditions. Even the US, who has supported the Japanese seat in the Security Council, appears to have done so in exchange for `Japan as the US financier' who supplies money at the US will. (*NIKKEI*, `94, May 22) Note that economic aids can become an important strategy: by threatening the suspension of economic aids, Japan can demand foreign countries with human-rights or environmental violations to remedy those violations, or she can encourage a country with high birth rate to adopt policies to curtail it by making the adoption a requirement for Japanese economic aids: the *Economic Aids* strategy (the *EA* strategy). Needles to say, the *EA* strategy must be chosen on Japanese own judgement, independently of any other nations including the US. (Note, however, the above-mentioned *Japanese* historical lesson.)

It is apprehended that Japan cannot apply the *EA* strategy effectively if she is forced to make compromises to obtain her seat in the UN Security Council. It is all the more so in the atmosphere of the worldwide criticism that Japan has not liquidated her past yet (Time [Oct.7,`91]). My evaluation is that the *SC* strategy implies more cost than *benefit*, since Japan is too hasty. Japan can obtain the seat in due time without any political compromises by applying the *EA* strategy honestly, while gradually liquidating her past. This would be the only strategy Japan can select to acquire the `credibility to act as an honest broker' (R.Nixon, *Beyond Peace* [1994]). The general public in Japan must decide whether to support *ESR* strategy clearly understanding these cost benefit analyses.

It goes without saying that the expansion of nuclear energy is accompanied by great danger of environmental pollution. In the late 1970s, the US government began subsidising wind-power electricity as an alternate energy, and it became a full-fledged clean energy industry (Time [Japan. 13,`92]) The US government is reported to drive for the export of technologies for environmental protection, estimating that the market for this industry will be doubled to $60 billion in 2000 (*NIKKEI*, Nov. 24,`93) This US strategy , stressed by Vice President Gore, may well be conducive to an `environmental-friendly' *Industrial Revolution*. In this field, however, trade-friction appears to emerge between the US who has comparative advantage and Japan who reportedly tries to protect Japanese *infant* industry. (*NIKKEI*) Japan cannot use the argument of infant industry any more, and the protection of infant industry (even if there exists one) should be shunned. Whereas by the traditional *Industrial Policy* Japan might be able to surpass the US technological level in this expanding 'environment-friendly' energy industry, the expected innovations from such protection would still remain a *Process Innovation*. From the above standpoint it would be inferior to *product Innovation*, since 'labour saving' property accompanying *Process Innovation* is too visible. (A typical example is *Toyota System*.) This is the case, even if it has its own employment-expanding effect through investment, and it has been stimulating (accelerating) the rivals' *Product Innovations*. Note that those rivals boast of the visible image of employment expansion, ignoring the fact that a part of expansion stems from the others' *Process Innovations*. In this way, the Technology *Self-Reliance* strategy (*TSR* strategy) would be a desirable strategy for Japan, and, indeed, she is in the process of pursing that goal. However, a condition must be placed on the *TSR* strategy: 'Japanese' *Product* Innovations must have their origins in purely civil purposes such as 'environment-friendly' energy industry. We cannot deny that many *product* innovations have its origins in military purposes. Indiscriminate pursuit of *Product Innovations* may well lead to a Japanese military build-up, which might cause an *arms race* in Asia. It may be said as a conclusion that whether Japan can truly become one of tri-polars of the 'stable' without military build-ups depends on whether she pursues the *EA* and (conditioned) *TSR* strategies, avoiding the *SC* and *ESR* strategies. If she pursues the *SC* and *ESR* strategies, Japan will be one of tri-polars of the `unstable' world. This may appear a formidable task for Japan. However, we must remember that as explained above, post-war Japan has displayed her strength faced with *formidable* crises.

Chapter 23

Asian Military
Situation—Present and Future

Nagaharu Hayabusa

The military situation of East Asia is beginning to change dramatically with the end of the East-West Cold War in autumn 1989. The declining presence of the two superpowers—the United States and the Soviet Union, the relative strengthening of China's influence, and moves towards a military expansion centering on the ASEAN countries. In the first meeting in July 1994 of the ASEAN Regional Forum (ARF) held in Bangkok, representatives from 18 countries including the major western industrialised countries, the United States and Russia participated to comprehensively discuss the political and military situations of the East Asian region. This was seen as the first step toward building of a security framework in Asia similar to the European CSCE.

Such moves are complex and do not seem to show a consistent direction. It is unclear what kind of framework will drive out of it. However, it is clear that Asia's political and military situation, in which the predominance of the United States and the Soviet Union were even stronger than that in Europe under the East-West Cold War structure, has entered a new phase.

The U.S. military presence in East Asia has consistently declined since the end of the Vietnam War in 1975. The US withdrawal after the end of the Cold War from the Philippines which was the main defence base for the Southeast Asian region was epoch-making. The US Air

Force, after its withdrawal from the Philippines, secured a small-scale base in Singapore, but its weight is small from an overall view. Troop deployment in 1992 is as follows:

— US Forces in Japan (mainly in the main island of Okinawa): 24000 soldiers, tactical aircraft 190.

— US Forces in Korea: 26000 and tactical aircraft 80

— Seventh Fleet: 60 65000-ton-class warships, 140 carrier-dorne aircraft

Of these, the US Forces in Japan and Korea are based in order to fulfill US obligations of its bilateral military pacts with the two countries, but with unprecedented thaw between the two superpowers as a result of the breakdown of communism in the former Soviet Union, the US presence is no longer significant in reality, and the only role it is playing is deterring the military threat North Korea poses on a limited basis.

Troop deployment of Russia in the Far East after the end of the Cold War remains Unchanged as Follows:

— Military personnel: 290000 (33 brigades)

— 750 1.92 million-ton-class warships

— Tactical aircraft 1430

This scale is substantially larger than that of the US army and naval forces. However, its military capability is seen to be extremely low. Due to the intense fiscal difficulty, Russia is drastically cutting its military expenditure, forcing its equipment maintenance to be idle, thereby sharply declining the frequency of personnel training. One high ranking officer of the US armed forces declared that the Russian forces in the Far East is no longer an enemy of the US Forces which are deployed in Asia.

Under the Cold War structure, the Cam Ranh Bay in Vietnam was the military base besides Siberia for the former Soviet Union. The Russian forces virtually withdrew from this base in 1990. Although some supplementary warships are still stationed in response to the wishes of the Vietnamese government, there are no battleships, and there are virtually no fighter planes.

With decline of the military presence of the two superpowers, a military vacuum has been created in the East Asian region, particularly in Southeast Asia and its periphery. The ASEAN countries have expressed a strong apprehension toward this trend. The Vietnamese invasion into Cambodia in December 1989, and the military expansion trend of China since the latter half of the 1980s have also aroused the anxiety of the ASEAN countries. To alleviate this anticipation, these countries resorted to arms expansion. Besides the military vacuum, the three causes which forced the ASEAN countries to resort to military expansion are as follows:

(1) The dynamic improvement in their governments' economic and fiscal positions accompanying their vigorous growth;

(2) The situation in the past where the level of military capability was extremely low and their defense capability was virtually nil.

(3) The strong deep rooted national sentiment to capture China as a potential military threat. Particularly, the recent Chinese proclamation of their law on 'territorial waters' and strengthening of action base in the Spratley Islands and so on have caused the apprehension of the ASEAN (including Vietnam) to exacerbate.

Below are the major moves which show the recent changes in the Chinese military situation.

(1) Reducing the military personnel by more than 1 million for modernization purposes on the one hand, and on the other reforming their organization and composition. Promoting dramatic modernization in equipments centering on the naval and air forces such as improving and diversifying nuclear arms, building of new warships, and purchasing SU-27 fighter jets from Russia.

(2) Moves to expand the scope of its activity in the seas such as expansion of fishing and natural resources exploration in the Spratley Islands, strengthening of its military bases, promulgation of its law on territorial waters and its effectuation, clarifying in its report to the 14th National People's Congress of the Chinese Communist Party the defense of its vested rights in seas.

(3) The specific content of the Chinese military modernization are as follows: It seems that China maintained its own independent development efforts on nuclear arms and deployed a new type of Intermediate Range Ballistic Missiles (CSS-5). In navy, they seem to have built a naval

vessel that can carry helicopters, in addition loaded warships with new type of missiles. In air force, China introduced the SU-27 fighter jets and promoted the development and deployment of F8-II fighters jets.

It is extremely difficult how to assess such a military modernization. From the growth of defence expenditures (Table 23.1), vigorous promotion of modernisation is clearly seen. However, the level of military strength is as shown in Table 23.2, and taking into consideration the largeness of its national land which is 26 times that of Japan and its population more than 10 times, the level is considered to be not as high.

Table 23.1 : *Trend of China's Defense Expenditure*

Unit: billion Yuan

	1987	1988	1989	1990	1991	1992	1993	1994
Defence	20.38	21.53	24.55	28.97	32.51	37.0	42.5	52.04
Growth %	1.8	5.6	14.0	18.0	12.2	13.8	14.9	22.4

Note :   Defence expenditure is in according with the report of the financial Department Head at the People's Congress

Table 23.2 : *China's Military Strength*

		1985	1993
Nuclear Arms	ICBM	A few	A few
	IRBM/MRBM	More than 100	About 100
	SLBM	Under	Under
		development	development
	Medium Range	About 120	About 120
	Bombers		
Army personnel		About 3.16 million	About 2.3 million
Naval vessels		About 1,740	About 1,060
	(of which submarines)	(about 110)	(About 90)
	Tactical aircraft	About 760	About 880
Air Force tactical aircraft		About 5,250	About 5,290

The characteristics of military modernisation trend of ASEAN member countries are as follows:

(1) The present status of defence capabilities of the six countries are limited, and although these countries are endeavouring to modernise their military capabilities, their level is substantially lower than those of industrialised countries.

(2) Defence expenditures for the five countries (excluding Brunei) show that Thailand spent about $2.7 billion in 1992, while those for other countries were between $1 and $2 billion. (Japan's defense expenditure for 1992 was $33 billion, ranking fifth or sixth largest in the world. The United States spent about 14 times that of Japan the same year.)

(3) The major motive for modernisation is to renew old weapons. But, it is clear that apprehension for uncertainty in the region and the potential threat of China are also causes and these are common in virtually all the countries in the ASEAN.

In the purchase of new equipment by the six countries, Indonesia's purchase of 39 East German warships, the 5-10 year arms purchase agreement totalling to some billion sterling pounds signed between Malaysia and Britain in September 1988, and Thailand's introduction of Spanish made helicopter carriers, all should be noted, but will not bring about a major change in the military balance of Southeast Asia.

The military balance of East Asia in Table 23.3 and the trend of military expenditure of Asian countries in Table 23.4 will follow. As for the military balance, it can be said that the level of naval and air force in the six ASEAN members in future will continue to be upgraded steadily, and the armed forces in Vietnam will dramatically decline. However, even of military expansion was achieved at a high pace, the gap in military compared with those of the United States, Japan and China will not easily narrow. In the unlikely event that the six ASEAN countries, the three Indochinese countries (a decision was made to include Vietnam as a member of the ASEAN in July 1994) formed an unified army, it is unthinkable for them to become a threat to Japan or China.

In examining the future Asian situation from a long-term perspective, in the 21st century it is certain that after the year 2010, China will surpass Japan in the economic field and will hold an overwhelming influence over Asian countries. The attitude of the ASEAN countries (the three Indochinese countries will have long

Table 23.3: Military Balance In East Asia

Country	Army	Navy	Air Force	Country	Army	Navy	Air Force
Thailand	190,000	About 300 (57,000 T)	Tactical About 185	Vietnam	700,000	About 100 (33,000 T)	Tactical About 240
Philippines	68,000	About 70 (45,000 T)	Tactical About 60	Laos	33,000	Unknown	Tactical About 30
Indonesia	203,000	About 160 (167,000 T)	Tactical About 100	Cambodia	50,000	About 16 (2,000 T)	Tactical About 20
Malaysia	90,000	About 90 (30,000 T)	Tactical About 80	South Korea	550,000	About 220 (130,000T)	Tactical About 470
Singapore	45,000	About 160 (33,000 T)	Tactical About 190	North Korea	1,000,000	About 620 (90,000 T)	Tactical About 890
Brunei	3,000	About 13 (About 800 T)	Tactical About 4	Taiwan	310,000	About 520 (240,000 T)	Tactical About 520
				Japan	145,000	About 260 (378,000 T)	Tactical About 530

Source   Naval statistics from the Japan yearbook (1993-94) Others from Military Balance (1993-94)

Note   Include Naval tactical aircraft

Table 23.4: Trend of Asian Countries' Military Expenditure

Unit: billion

	1989	1990	1991	1992	1993
China (Yuan)	24.55	28.97	32.51	37.0	42.5
Growth (%)	14.0	18.0	12.2	13.8	14.9
South Korea (Won)	6,014.8	6,637.8	7,452.4	8,410.0	9,215.4
Growth (%)	9.0	10.4	12.3	12.9	9.6
North Korea (Won)	4.1	5.2	5.2	4.48	4.70
Growth (%)	5.4	26.8	0	Δ*13.8	4.9
Taiwan (Yuan)	212.8	233.4	250.5	262.3	274.9
Growth (%)	11.2	9.7	7.3	4.7	4.8
Thailand (Bhats)	46.30	52.60	60.58	68.81	79.60
Growth (%)	0	13.6	15.2	13.6	15.7
Indonesia (Rupee)	—	2,674.0	3,015.0	3,596.0	4,040.0
Growth (%)	—	—	12.8	19.3	12.3
Malaysia (Ringgit)	3.75	4.23	4.30	5.0	5.06
Growth (%)	Δ*11.1	12.8	11.8	5.7	1.2
Singapore (S.Dollar)	2.92	3.08	3.68		
Growth (%)	11.5	5.5	19.5		
Philippines (Pesos)	27.34	23.80	27.12	29.1	29.9
Growth (%)	150.8	Δ*12.9	13.9	7.3	2.7

Note 1   China's defense expenditure is based on the report to the People's National Congress by the Financial Department Head.

Note 2   South Korea's defense expenditure is based on its Defense Report.

Note 3   Defense expenditures for other countries are based on the "Military Balance."

become members of the ASEAN then) whose military strength will be dramatically enhanced then to grasp China as a potential threat will be certainly stronger. However, Japan, whose territory is not be connected by land will virtually not feel this threat. The common view of the Japanese Self-Defence Force leaders is that if Chinese military continue to expand at the present pace for the next 25 years, then Japan might feel threatened.

China's defences policies maintain a certain degree of rationality and its self-control is functioning. However, there is a tendency for China to take belligerent and tyrannic attitude at times, and thus there is a possibility of China resorting to its military strength to protect its classic vested interests.

Therefore, the essential point security in the 21st century, will be how to prevent China's dogmatic action. Russia, India and the Korean peninsula after unification could become a disturbance factor and a possibility of regional conflicts emerging in various corners cannot be excluded. However, compared with China's compulsive action in any form, that implication is small.

United States and Japan are the two countries which can prompt China to maintain its self-control even if China became a major military power. If the United States and Japan can cooperate to evoke China's need for self-control, any provocative action can be prevented. If ASEAN countries can also cooperate, that certainly will further increase.

The problem is that it is unclear whether the Clinton Administration is fully aware of the indispensability of continued presence of the US forces in the 21st century to prevent a possible Chinese dogmatic action (not only military). If the United States has virtually withdrawn its military presence from Asia after 2010 as a result of financial difficulty or mounting isolationism, the role to prevent China from any haughty action will be burdened on Japan. In that case, the only direction that Japan can take politically will be an even closer cooperation with the ASEAN countries, and militarily will be a speedy arms expansion that can counter the Chinese strength.

Japan's defence expenditure in fiscal 1994 is ¥4.68 trillion, which is about 0.9 percent of the projected GNP. After Japan lost World War II, it has consistently maintained a stance to have a `light armament policy' (see Table 23.5). This policy in principle remains unchanged and will not change in the near future. Review of Japan's defence policies were made in the three cabinets following the Hosokawa

Table 23.5 : Military Capability of Major Countries/Regions

Army		Navy			Air Force	
Country	Soldiers (1,000)	Country	Tonnage (1,000)	Vessels	Country	Tactical Aircraft
China	2,300	former Soviet Union	6,874	2,460	former Soviet Union	7,820
former Soviet Union	1,500	U.S.	5,792	1,130	China	6,170
India	1,100	China	909	1,060	U.S.	5,040
N.Korea	1,000	U.K.	883	350	France	920
Vietnam	700	France	454	350	N.Korea	810
U.S.	601	Peru	262	50	Germany	770
S.Korea	550	India	247	170	India	730
Pakistan	515	Taiwan	238	520	Israel	700
Turkey	450	Turkey	208	230	Syria	640
Iraq	350	Germany	196	220	Turkey	570
Germany	316	Italy	175	200	U.K.	540
Taiwan	312	Brazil	169	180	Taiwan	520
Iran	305	Indonesia	167	150	Egypt	510
Syria	300	Spain	163	210	Poland	500
Egypt	290	Canada	158	80	Sweden	500
Japan	150	Japan	331	160	Japan	490

Note : Statistics of the Army and Air Force are based on the "Military Balance" (1992-93), and for the Navy on the Jane yearbook (1993-94).

Cabinet initiated in the summer of 1993, and these efforts focussed on how to fully participate in the U.N. Peace-Keeping Operations by consolidating the restrictions of the Japanese Constitution. There is no possibility of converting to military expansion. Although the Cabinet of Prime Minister Tomiichi Muramaya is setting forth a policy toward arms reduction, the possibility of a dramatic reduction is narrow under the present situation.

A change in the present Japanese policy will come only when China's potential military threat has increased in the 21st century as I previously mentioned. As such, to prevent an unhappy case, hopes are pinned on maintaining the US military presence, or establishing as security network by the Asian region as a whole such as by utilizing the ASEAN Regional Forum or the APEC to cover the vacuum.

Chapter 24

Military Expenditure and Developing Nations

Manas Chatterji

1. Introduction

One of the most important problems facing humanity today is the immense disparity in the level of income and, consequently, in the standard of living among the people in different parts of the world. What is more disheartening is that the growth rate of income of the poor countries is consistently far below the rate of economically advanced countries. Based on the resulting difference of income, countries are generally divided into two groups : developed and underdeveloped.

The term underdeveloped has been used with different connotations in the literature of economic development. Sometimes it meas a low ratio of population to area, sometimes a scarcity of capital as indicated by the prevalence of high interest rates, and sometimes a low ratio of industrial output to total output. A reasonably satisfactory definition of an underdeveloped country seems to be a country that has good potential prospects for using more capital and labor and/or more available natural resources to support its present population at a higher level of living ; or, if its per capita income level is already fairly high, to support a larger population at a level of living no lower than at present. The reasons for underdevelopment include poor natural environments, severe climate conditions, niggardly endowment of resources, past social and cultural development, and restrictive religious practices. All these factors confine production far within the maximum possible frontier, leading to low output. Frequently, the result is a vicious circle of low output, high propensity to consume, and low levels of savings and capital accumulation.

Another dimension of the problem of economic development is regional or spatial. Unfortunately, insufficient attention was paid to this area until recently. Considerable theoretical and empirical work has been done on the problem of allocation of resources over time, but not over space. Clearly, the spatial aspect is also important. This is particularly true for a country like India, which is striving for economic growth despite extreme regional variations within a framework of a federal democratic system of government. The situation is basically the same in all the so-called poor countries, especially those that were under foreign rule for a long time. Examples are Indonesia (Java versus the other islands), Burma and Thailand (capital cities and rice growing areas versus up-country), and Brazil (Northeast versus Central-South area).

The formation of Bangladesh from former East Pakistan is a classic example of regional divisiveness. The general pattern is one of economic disparity between the capital or port city and the rest of the country. Look at the economic history of any of these countries. Typically, industrialization has started at a few focal points, mostly port cities. These were points deemed convenient to the rulers, not necessarily optimum locations. In addition to economic development, these points got an earlier start with respect to education, health care facilities, and all the other benefits and drawbacks of Western civilization. The cases of Calcutta, Bombay, and Madras in India are good illustrations.

This regional variation can also be observed in regions of developed countries, such as the United States, the United Kingdom, Western Europe, the USSR, and the like. In the United States, for example, the central cities, Appalachia, the northern Michigan and Minnesota regions, and other areas provide clear examples of poverty and underdevelopment in the midst of affluence and prosperity. Although on the absolute level there is no comparison between this underdevelopment and that of the developing countries, in relative terms it is nevertheless very striking.

There are many reasons for the growth and decay of a region : economic, social, and political. One factor that has become prominent in recent years is the availability of resources, particularly energy.

Scarcity of energy, its high price, and the development of new sources are going to have a profound impact on regional growth and decline. For example, the northeastern United States is in decline and is in need of reindustrialization, whereas the South and Southwest are growing and need management of this growth.

The spatial dimension is also highlighted by the continuous confrontation between developing and developed countries (North versus South) ; between ethnic, racial, linguistic, and religious groups ; and, more recently, between ecological and development goals. Historically, this phenomenon of conflict and its resolution has not been integrated into social science theory. Consequently, the solutions obtained from such an analysis, though optimal, in an economic, social, or political perspective, have not been found to be practical in terms of policy.

Due to the existence of internal conflict and international relations poor countries are spending considerable amounts of scarce resources on military spending. Table 24.1 gives some aggregate information about the magnitude of this spending. Table 24.2 gives some information of other related variables for some countries. It is clear from these figures, that although superpowers are cooperating in reducing their military spending, there is no such change visible in the case of the developing countries. In fact, the scene has moved to regional conflict from superpower confrontation. This unabated trend in military spending has implications for economic development. Some scholars such as Benoit (1973) found a positive relationship between defense spending and economic development while others have severely criticized his findings. I have summarized elsewhere in a descriptive fashion the views presented by scholars in the two contesting groups, Chatterji (1991b).

In this paper, I shall present some analytical techniques which have been used by those scholars in this context and indicate how their techniques can be used for testing and estimating the relationship between military spending and development.

There are usually three types of analytical tools used in this area, namely :

1. Regression and Econometric Analysis
2. Input-Output Type Analysis
3. Graph Theory Approach

Due to the limitation of space, I shall only discuss the application of the first technique. Interested readers are referred to Chatterji (1991a) for the other techniques.

2. Regression and Econometric Analysis of Military Spending and Economic Growth.

Economic growth can be defined in various ways. One way is to measure the rate of growth of real GDP (Gross Domestic Product). However, economic growth is not the same thing as economic development, since it avoids the question of income distribution. Economic growth can take place when total savings leads to capital accumulation which in turn helps to produce more goods. Increased production leads to industrialization and urbanization with all their positive and negative externalities ; efforts on the part of the governments to achieve the targets may reduce income inequality and lead to social justice. The question then, is to decide the amount of optimum savings and how to allocate. Military spending does not contribute to economic growth ; it produces goods which in themselves are not capital goods since they do not lead to further production. In that context, growth itself is not the end, but is a means to an end. As A. Sen (1983) puts it :

I believe the real limitations of traditional development econo-mics arose not from the choice of means to the end of economic growth, but it the insufficient recognition that economic growth was no more than a means to some other objectives. Thee point is not the same as saying that growth does not matter. It may matter a great deal, if it does, this is because of some associated benefits that are realized in the process of economic growth. Entitlement refers to the set of alterative commodity bundles that a person can command in a society using the totality of rights and opportunities that he or she faces.

On the basis of this entitlement, a person can acquire some capabilities, *i.e.*, the ability to do this or that (*e.g.*, be well nourished), and fail to acquire some other capabilities. The process of economic development can be seen as a process of expanding the capabilities or people. Given the functional relation between entitlements of persons over goods and their capabilities, a useful—if derivative— characterization of economic development is in terms of expansion of entitlements.

A considerable amount of material on growth theory is available. Growth theory, including studies of such topics as growth models starting from the Harrod-Domar theory, the Turnpike theorems, and others. However, in these multisector, multi-period models, defense spending traditionally has not been treated as a separate variable. In some instances investment funds have been divided into two parts as consumption and investment goods producing. Mahalanobis's (1963) model is based on that principle. And when the growth models have been linked with the macroeconomic models, military expenditure as an explicit variable in a macroeconomic theory has not been taken into consideration. One such example is by Smith (1980).

The welfare function is dependent on security variable (S) and output of civilian sector (C).

$$W = W\,(S.C) \qquad (1)$$

Although it may not be appropriate to assume, we can set

$$S = S\,(M.E.) \qquad (2)$$

i.e., security is a function of military expenditure M and strategic environment E. E may be determined from the Richardsonian equation of Arms Race where one country's military spending depends on the strength of its adversary. Total output is made up to civilian and military expenditure. *i.e.*,

$$Y = pC + qM \qquad (3)$$

Where p and q are prices relative to the price of total output.

The objective will be to maximize (1) subject to (2) and (3). The first order condition for that is :

$$W_S\,SM = (q/p)\,W_C \qquad (4)$$

W_S is the partial derivative of the welfare function with respect to S and the same is true for S_M and W_C.

If the form of the welfare function and the production function are respectively

$$W = A \left[dC^{-a} + (1-d) S^{-a} \right]^{-1/a} \qquad (5)$$

(CES)

$$S = BM^b E^c \qquad (6)$$

(Cob-Douglas)

The first order condition in logarithm becomes

$$\log m = A_0 + A_1 \log C + A_2 \log(\mathcal{V}_p) + A_3 \log E \qquad (7)$$

where $A_0 = \{ \log [(1 - d/d] + \log b - a \log B \} / (a + ab)$

$A_1 = (1 + a)/(1 + ab)$

$A_2 = -1/(1 + ab)$

$A_3 = -ac/(1 + ab)$

Thus the elasticity of substitution between security and civilian output and the parameters for the security and production function can be estimated empirically. Although the estimation of this equation was made for England, similar formulations can be made for other countries provided we can make a realistic security production function and the strategic environment E.

Of course, the strategic environment can be defined in terms of alternative formulations. Others try to relate military expenditure for a country such as the U.S. to the structural instability of capitalist economy, monopoly power, under-consumption, unemployment, income concentration, union power and political variables. Griffin et al. (1982) conclude, by analyzing empirical observation, that military expenditure in the U.S. is related to cyclical stagnation and that geo-political influences are secondary.

Smith (1980) tested the hypothesis that reduced investment has been a major opportunity cost of military expenditure. He made a regression relation of potential output with the share of military expenditure, growth rate, and demand pressure. He found a significant regression coefficient of -1 between military spending and investment.

Working with national accounting identity

$$Q - W = Y = C + I + M + B \qquad (8)$$

where   Q = potential output

W = the gap between actual and potential output

C = consumption (private + public)

I = investment (private + public)

M = military expenditure

B = current account balance of payments

and using lower case symbols denoting shares (8) can be written as :

$$i = 1 - w - c - m - b \qquad (9)$$

The share of consumption can be postulated as :

$$c = \alpha_0 - \alpha_1 u - \alpha_2 g \qquad (10)$$

where u is the unemployment rate and "g" growth in actual output. Assuming $(w+b) = \beta u$, he reformulates (9) and estimates :

$$i = (1 - \alpha_0) - (\beta - \alpha_1) u + \alpha_2 g - m \qquad (11)$$

It is often suggested that, to be most effective, the theoretical models must include social, political, and other non-economic variables (in addition to economic variables). Notwithstanding the problem of measurement, social scientists are increasingly employing such techniques as factor analysis. However, some of these data-oriented approaches ignore a priori tested hypotheses about socio-economic determinants of growth. Scholing and Timmermann (1988) propose a path analysis to correct this deficiency. In this path model they use latent variables (like socio-economic status, political participation, human capital, technical progress, international competitiveness, etc.).

This model with latent variables is defined by two linear equation systems, *i.e.*, the inner model and the outer model.

The inner model describes the core theory containing the theoretical relationship of latent variables. The structural relations are :

$$Y = BY + \Gamma X + \delta. \qquad (12)$$

where $Y = (Y_1, Y_2, ...Y_n)$ denotes the vector endogenous inner variables

$X = (X_1, ... X_m)$ denotes the vector of exogenous variables, $\delta = (\delta_1, ... \delta_n)$ denote the residual vector, and B_{nxn} and $\Gamma_{v\xi\mu}$ are matrices of parameters.

The outer model describes the relationship between the observable (manifest) variables and the latent variables they measure. The structional equation of the outer model is :

$$Z = \pi^1 X + \varepsilon_1 \tag{13}$$

$$R = \pi^2 Y + \varepsilon_2 \tag{14}$$

where $Z = (Z_1, \ldots Z_p)$ is the vector of outer variables associated with exogenous inner variables (X) and $R = (R_1, \ldots R_q)$ is the vector of outer variables associated with the endogenous inner variables (Y). x^1_{pxm} and π^3_{qxn} are regression matrices with loading factors at the diagonals and zeros in off-diagonals.

The authors, using Wold's method of partial least squares (PLS), estimate the above equations taking nineteen inner variables, and 118 associated outer variables, for seventy developing countries. They found that non-economic factors such as climate, ethnic homogeneity, intensity of government control, infrastructure, etc. have significant direct and indirect relationships. There are some scholars, such as Neuman (1978), who believe that too much emphasis should not be placed on macrostatistical analysis :

> Apparently, secondary manipulation of macrostatistical indicators, by academic and policy making centers thousands of miles from the areas being studied, misses and essence of what is actually going on in the countries under study. There may be no substitute for field research on this issue, at least until we have discovered which are the relevant variables and have collected sufficient information about them.''

Palma (1978) for example, proposes Dependency Theory as a formal theory of underdevelopment. In his estimation, such approaches as stages of growth, modern-traditional sociological typologies, dualism, functionalisms, etc., do not integrate into their analysis the socio-political context in which development takes place. The mechanical formal nature renders them both static and unhistorical. Some radical analysts such as Sunkel (1973) blame the neo-mercantilist system of domination by transnational conglomerate (TRANCO) for hindering the growth potential of developing countries emanating from the demand side as against foreign aid and investment.

The relationship of military spending to economic growth is a part of a broader consideration of causal link between gross domestic product (income) and government (public) expenditure. Some argue that the causal flow runs from the level of economic development to government expenditure. Others treat government expenditure as exogenous which determine the economic growth. Ram (1986) uses the Granger-Sargent procedure to determine the direction of the flow. It is given by :

$$Y_t = \sum_{j=1}^{m} a_j \, Y_{t-j} + \sum_{j=1}^{m} b_j \, X_{t-j} + \alpha + \beta \, T + U \qquad (15)$$

Where Y and X are variables for which the direction of the causality flow needs to be determined, T is trend, and U is random disturbance.

The Granger-Causality rests on the proposition that X causes Y, if and only if Y is better predicted by using the past history of X than by not doing so, with past of Y being used in either case. Thus, testing for the absence of Granger-casual flow from X to Y is equivalent to testing the hypothesis $b_j = 0$ for all j. In this case, Y denotes income and X denotes government expenditure. Ram selected 63 countries from 1950—1980 and used a dummy variable (1973 energy shock) and different lags (m = 2, 3, 4) and found no significant causality between GDP and government expenditure.

Lotz (1970) discussed the same relationship of thirty-seven developing countries through a factor analysis study. His variables are :

Yp   Per capita income (in U.S. dollars).

U   Per cent. of population living in urban areas.

MO   Exports of minerals and oil as a per cent. of total exports.

X/Y   Exports as a per cent of GNP.

Li   Literacy rate of adults.

Qp   Notes and coins per capita (in U.S. dollars).

W/Y   Expenditures on social welfare as a per cent of GNP.

(E+H)/Y   Expenditures on education and health as a per cent of GNP.

D/Y   Expenditures on defense as a per cent of GNP.

ES/Y   Expenditures on economic services as a per cent of GNP.

EXP/Y   Total government expenditures as a per cent of GNP.

His conclusion is that government spending is related to not only the level, but also to the character, of economic development. Where development has penetrated all sectors of the economy, welfare spending appears to have increased. Further defense and economic services are not closely related · to the stages of economic development.

Landau (1983) investigated the relationship between total government expenditure and the rate of growth of per capita output based on ninety-six non-communist countries. The growth rates examined were for the periods 1961-70, 1961-72, 1961-74, 1961-76 and 1961-68 and 1970-76. He also considered total investment in education, weather and energy consumption. His conclusion is that the effect of an increase in government spending reduces the growth rate of income, for the full sample of countries, weighted or unweighted by population for all six time periods covered excluding or including the major oil producers.

Landau (1986) extended the analysis by disaggregating the Government expenditure variable (including national, state, and local governments). The classifications are : (1) consumption other than defence or education ; (2) education ; (3) defense ; (4) transfer ; and (5) capital expenditures. The revenue source are current revenue, the deficit and foreign aid. To measure regulatory impact, he took the rate of change of money supply, inflation rate, an index of real exchange rate and the real interest rate. He also took many variables to measure international economic conditions ; human and physical capital ; population, industrial shares in GDP ; and historical-political factor including conflict variables. He found a strong negative relation between the level of per capita expenditure and the growth rate. The Military did not have an impact on economic growth, government expenditure as such has no relation with growth, and foreign aid showed no impact on growth.

It is to be emphasised again the growth in GDP does not necessarily imply development since the distribution of income (say as measured by GINI coefficients) can be quite asymmetric ; and besides, income inequality may not decrease as a consequence of economic growth. Rubinson (1976) showed through a cross-section regression analysis that it is the economic influence and dominance of a state in relation to the world economy which influences inequality of income within that state. His regression measured the effect of kilowatts-hours ; per capita, government reserve, exports ; and the value of and index of social security insurance.

Deger (1981) has shown that for fifty developing countries over the period 1965-73 there was a statistically negative relation between the share of military expenditure in GDP and the share of public education expenditure in GNP. The same was true between military expenditure and average propensity to save.

Maizels and Nissanke (1986) investigated, through regression analysis using dummy and qualitative variables, the determinants of military spending. They found the following statistically significant factors : (1) war/tensions with neighboring countries ; (2) domestic factors, namely repressing internal opposition groups ; (3) global power block and most important share of central government budget in GDP. Investment in developing countries by transnational corporations was not a determinant of military expenditure.

Frederikson and Looney (1983) divided Benoit's list of countries into two groups. Group I consisted of twenty-four countries characterized by a relative abundance of financial resource. Group II consisted of nine countries which were relatively resource-constrained. Using Benoit's methodology, data and time frame, the same regression equation (as follows) was computed :

$$DIVGDP = f(INV, AID, DEFN)$$

where GIVGDP = real growth in GDP minus real growth in defense expenditure

INV = gross capital formulation or a percentage of GDP

AID = receipts of bilateral aid or a percentage of GDP

DEFN = defense expenditure or a percentage of GDP

The authors found that the coefficients of defense in the resource abundant group is significantly positive, whereas in the resource constraint group, it is negative. It means that in poor countries, defense spending retards growth, whereas in non-resource-constrained countries defense expenditure contributes growth. Specifically, the equations are :

$$CIVGDP = 1.77 + \cdot 161NVEST + \cdot 12AID \qquad (18)$$
$$(6.11) \qquad (3.070)$$

$$+ \cdot 22DEFN \qquad R^2 = .89$$
$$(3.77)$$

For the resource abundant group and

$$CIVGDP = 4.72 + \cdot 15INVEST + \cdot 19AID \qquad (19)$$
$$(1.92) \qquad\qquad (1.46)$$

$$- 1.22DEFN$$
$$(-3.52)$$

$$R^2 = .76$$

The regression equation of Benoit's fourty-four countries is

$$CIVGDP = 1.14 + .21INVEST + \cdot 13AID \qquad (20)$$
$$(5.57) \qquad\qquad (2.30)$$

$$+ \cdot 23DEFN$$
$$(1.34)$$

$$R^2 = .61$$

Ball (1985), however, thinks that Frederiksen and Looney have obscured more than they have explained regarding the true nature of the growth-military expenditure. According to her, they have presented an unrealistically simplified view of the relationship since other factors such as civilian expenses, term of trade, etc. are more important. According to her, this complex relationship cannot be expressed by a few equations estimated by weak data. Also, their suggestion of a positive relation for resource rich country can be illusory. Again, the difficulty in inferring causality on the basis of statistical analysis remains.

Consistent with the hypothesis, he found that the signs of the path coefficients are correct in showing that, (*a*) DEF has a positive relation with both GDP-manufacturing and social development factor (DEF) ; (*b*) SDF has a positive relation with GDP-manufacturing ; (*c*) military expenditure has negative relation with both EDF and SDF rejecting Benoit's hypothesis.

In a recent article, Adams et al. (1989) addressed Benoit's hypothesis for ninety-five countries for the period 1974-86. The divided these countries into "low income," "middle income" and "industrial market economies." Their initial conclusion, based on the extension of Feder's (1982) analysis, is that there is a positive correlation between government spending (and even military spending) and economic growth once the other growth determining factors have been taken into account if we exclude "warring" countries. Including the warring countries, however, the relationship is negative.

Biswa and Ram (1986) used the augmented Feder model

$$\dot{Y} = \alpha\,(I/Y) + \beta\,(\dot{L}) \left[\frac{\delta}{1+\delta} + C_M \right] \frac{\dot{M}}{M/Y} \tag{21}$$

$$\dot{Y} = \alpha\,(I/Y) + \beta L) + \left(\frac{\delta}{1+\delta} - \theta \right) [\dot{M}/\,(M/y)\,] + \theta M \tag{22}$$

where $\dot{Y} =$ annual rate of growth of total output (GDP)

$I/Y =$ investment-output ratio

$\dot{L} =$ annual rate of growth of labor force

$\dot{M} =$ annual growth of military expenditure

$\theta = C_M\,[M/Y - M]$

and C_m represent the externality effect of military output of the civilian sector (obtained from the production function of C, the civilian output). If $C_M > 0$ and $\theta > 0$, increase military output will imply a higher rate of growth of total output Y.

Deger and Sen (1983) reputed Benoit's econometric evidence for India indicating that the positive effect of military expenditure on economic growth is exaggerated. Their intertemporal discounted welfare function to maximize is

$$W = \int_0^\alpha e^{-\rho t}\, u(1-m, s, \theta)\, dt \tag{23}$$

subject to : $s = m - (\delta + \beta)s = m - \alpha s$ (24)

where ρ is rate of time preference

m, s are average coefficients M/Y and S/Y respectively and is a threat perception variable. Based on this optimizing of the specification, the authors prove that a stable steady-state equilibrium will hold which can be brought to a higher level by an increased perception of threat θ. They justify the validity of the model on the basis of India-Pakistan relations. Their point of contention is that defense is related to non-economic factors such as security and counteract threat rather than economic variables. To test whether there is spinoff or economic impact they estimate the following equation :

$$X_{it} = \alpha_{0i} + \alpha_{1i} M_{1i} + \alpha_{2i} V_i + k_{it} \qquad (25)$$

where V_i is the value added in ith industry whose output is X_{it}, and M is military expenditure. Based on the estimated equations for the sectors (*i*) basic metals, (*ii*) metal products, (*iii*) machinery not electrical (*iv*) electrical machinery, and (*v*) transport equipment, they find that increases in military expenditure have insignificant effects of industry output.

To examine the interaction between growth, investment and military expenditure, they specify the following model and estimate it with 1965-73 time series averages of cross section of 50 LDC's. (Here, g is average annual growth rate of GDP, i is investment share in GDP, m is the defense burden, y is income, n is the growth rate of population, a is the net foreign capital transfer, y is the change in national income, D is the difference between per capita incomes measured at purchasing power parity at official exchange rates, N is the total population, DI is the dummy variable for oil producing countries with balance of payments surplus, and D2 the same for war economy).

The three stage least squares estimates are (t-values in parentheses)

$$g = -9.63 + 0.83i + 0.20m - 0.13y + 0.39n - 0.28a$$
$$\;\;\;\;\;(-1.50)\;\;\;(1.84)\;\;\;\;(0.98)\;\;\;(-0.76)\;\;\;(0.63)\;\;(-2.09)$$

$$(26)$$

$$i = 14.17 + 0.54g + 0.03y - 0.35m + 0.33a$$
$$\;\;\;\;\;(10.8)\;\;\;\;(2.75)\;\;\;\;(4.42)\;\;\;(-2.75)\;\;\;(3.83)$$

$$m = 3.69 + 0.18y - 0.30D + 0.00IN + 4.99DI + 13.33D2.$$
$$\;\;\;\;(5.59)\;\;\;(3.25)\;\;(-2.82)\;\;\;\;(0.37)\;\;\;\;\;(4.19)\;\;\;\;\;\;(15.63)$$

The multiplier of military burden on growth can be easily calculated as

$$\delta g / \delta m = (\alpha_2 + \alpha_1 \beta_3)(1 - \alpha_1 \beta_1) = -0.1633.$$

Clearly, when taking *all* interdependent effects together, the effect of an increased military burden is to reduce the growth rate.

Deger and Smith (1983) based their results on cross-sectional data of 50 LDC's and found that military expenditure has a small positive effect on growth through modernization and a larger negative effect through savings. But the net effect of the growth was negative. These equations estimated by three stage least square are :

$$g = -8.93 + .92s + .35m - .49p + .59a - .26y + .16r$$
$$(-2.43) \quad (3.78) \quad (2.77) \quad (-1.08) \quad (2.89) \quad (-2.44) \quad (1.42)$$

$$R^2 = 0.2260$$

$$s = 14.54 - .43m + .48g + .037gy - .67a - .75p \quad (27)$$
$$(9.70) \quad (-3.16) \quad (1.92) \quad (4.55) \quad (-7.63)(-.03)$$

$$R^2 = 0.8651$$

$$m = 3.98 + .19y - .30(q - y) - .02N + 4.67D1 + 11.31D2$$
$$(4.53) \quad (2.61) \quad (-2.09) \quad (-4.29) \quad (3.65) \quad (10.83)$$

$$R^2 = 0.7808$$

(t ratios in parentheses)

The variables are :

g : average annual growth rate of real GDP

s : national savings ratio

a : net external capital flows as a percentage of GDP

m : share of military expenditure in GDP

p : rate of growth of population

y : 1970 per capita income at official exchange rates

r : average annual growth rate of agricultural product

p : rate of change of aggregate price level per annum

q : 1970 per capita income at purchasing power parity

N : total population

D2 : dummy for Israel, Jordan, South Vietnam, Egypt, Syria, India, Pakistan

D1 : dummy for Iran, Iraq, Libya, Saudi Arabia

Chan (1985) sums up the research already done and its potential avenues : "We have probably reached a point of diminishing returns in relying on aggregate cross-national studies to inform us about the economic impact of defense spending. Instead, it appears that future research will profit more from discriminating diachronic studies of individual countries. As some analysts have already noted, the search for universal pattern applicable to all places and times is likely to be disappointing. The claims to generality based on the results of such searches tend to entail substantial costs in empirical sensitivity and specificity. Moreover, the discrepancies in these results are not easily reconciled, as they are often based on different measures and samples. An alternative and perhaps more fruitful approach would be to eschew claims of generality at the present, while recognizing the complexity of the problem.''

There are a number of basic criticisms made by contributors to this subject matter. The first is the question of the use of cross section data. Regression equation can be estimated on the basis of cross section data. The data refers to a particular time period, say 1960, for all countries.

Alternatively, we can have a particular country, say India, for which we may have data for the time period 1950-90. It will be a time series data. A pooled time series cross section data exists when we have information for a number of countries for a number of years. Benoit used pooled data, except that he took the *average* for a number of years for the countries. He also conducted time series analysis for some countries such as India. The major difficulty in the time series analysis is that there is a frequent occurrence of collinearity which is, however, not so server in the case of cross section data. The use of cross section data is appealing since micro information is quite important and many macro hypotheses are nothing but the aggregate of micro behavior. If ε_{jt} denotes the error term of Jth country for t^{th} period, it is composed of two parts, i.e.,

$$\varepsilon_{jt} = \gamma_j + \gamma_{jt}$$

where γ_j represents individual firm effect and γ_{jt} is the residual random variable for j^{th} country at time period 't'. If the data for some country for every observation period is available, we can compute the average residual for a particular country over the years. If we wish to compare the time series regression coefficient with that of the cross section, we have to first average the regression coefficients of a firm over time and average the times series coefficients over countries. Obviously the two regression coefficients will not be equal. The cross section will show long-run adjustments whereas times series will reflect short term fluctuations. So cross section coefficients will show adjusted long run coefficients. Thus, "cross section cannot be used successfully to make time series predictions unless a systematic relationship between the cross section and time series estimates has been firmly established" (Kuh, 1959). For this purpose, a pooled or rectangular data structure is necessary ; otherwise the analysis will be structurally incomplete.

The second point which is directed against Benoit's analysis is that his arms race equation is not based on action-response Richardsonian type analysis. A third argument is the absence of any separate equation of international trade, separate manufacturing sector and technical and educational infrastructure sector. Also, it requires an appropriate lag in time. I have been working on a model to included all the above considerations for India during the period 1950-86 and estimating it by three stage least square. The following is an outline of the progress report.

3. Towards a Socio-Econometric Model of Arms Race in the Indian Sub Continent

There has been considerable development in econometrics as models have been developed on the international, national and regional levels. However, in such models, social and political factors are seldom taken into account. Without these considerations, such models in the area of military spending will be of limited value. It will be worthwhile to integrate some conflict models with econometric models. I shall outline such an attempt. Let me first present Richardsonian equations of action and reaction models for India and Pakistan.

There has been recently a great increase in the literature of conflict resolution models. More and more often, realities of the world are being taken into account and restrictive assumptions are being relaxed. Testing these models is becoming easier because of the development of new concepts and methods to measure qualitative variables. The scope of these models, however, can be enlarged in at least two respects. First of all, spatial relations of the contesting parties, i.e., geopolitical aspects of the conflict, can be considered. For example, when we look at the map of the Indian subcontinent, we find that the geographical boundaries of two other major powers, namely China and Russia, meet the boundaries of India and Pakistan. In considering political relations between India and Pakistan, this factor is of crucial importance. The second factor is the diverse forces that are acting *within* the contending parties. The rate of economic growth, population growth rates, and internal peace and stability are a few of many factors which greatly influence the foreign relations of a country. This is particularly true in Indo-Pakistani relations.

The case of India and Pakistan offers an excellent field of study to which we can apply the modern game theoretical approaches for analyzing a mutual relationship. Here we have two countries which are independent from geographic, economic, political, and social points of view. They can gain much by cooperating or lose much by quarrelling, while the great powers, through cooperation, can act as moderating influences. This study casts some light on these aspects and is intended as a beginning toward a more generalized study to be undertaken in the future.

The emergency of India and Pakistan as two nations in the Indian subcontinent has ushered in a new phenomenon is Asia. Before 1947, they were a single country whose people struggled together for freedom from British rule. When the British decided to leave, the Moslem minority, apprehensive of the Hindu majority, demanded a separate state, which they obtained after a bloodbath, the consequence of religious riots (East Pakistan, a former part of Pakistan, is now Bangladesh). The division of the country resulted in a complete breakdown in the social, economic, and political systems of the country.

India now has about 900 million people, of which 10 percent are Moslem. Of Pakistan's 150 million people, there are very few Hindus. The enmity between Hindus and Moslems in India has not ended as a result of the division of the country. These two groups have now become arch-enemies, spending millions of rupees in defense preparation. Already they have fought significant wars resulting in loss of lives and resources.

The very existence of these two countries depends on their mutual cooperation and friendship. There are many ways through which this friendship can be built. This has to be brought about on a government level, on personal levels, and also through the auspices of other countries. One significant step in the right direction would be in the field of disarmament. This paper throws some light on this aspect. It is not intended to provide an easy solution. since conflict between nations is too complicated a matter to be solved easily. This is just a simple approach to conflict resolution taking into consideration the realities as far as possible.

Following Richardson (1960), we assume that there are three factors related to the arms race : namely, mutual suspicion and mistrust, cost of military expenditure, and grievances. Let us consider these factors for India and Pakistan in the context of their relationship and other internal and external variables.

4. India

So far as India is concerned, there are two fronts to guard, namely, her borders with Pakistan and China. India is suspicious of both these countries. Thus, it can be assumed that the rate of change if its military expenditure will depend upon the military expenditure of Pakistan and China in the previous period. The lag in time period is appropriate since responses are never instantaneous and there is always a time lag in intelligence reports. A time lag of one year is assumed. So we have the following relations :

$$\frac{dM_{1\,t}}{dt} \quad kM_{2\,(t-1)} + nM_{3\,(t-1)} \qquad (28)$$

where

M_t = military expenditure (rupees in millions) of India for the time period t

$M_{2\,(t-1)}$ = military expenditure (rupees in millions) of Pakistan in the time period (t–1)

$M_{3\,(t-1)}$ = military expenditure (rupees in millions or an index number) of Communist China in the period (t–1)

The L.H.S. of (28) denotes the rate of change of Indian military expenditure. k and n are positive constant which, following Richardson's terminology, can be called "defense coefficients." It is difficult to obtain data for China's defense expenditure ; when not available we can try to employ some index numbers or use some proxy variables. This equation (28) represents the mistrust and suspicion on the part of India against Pakistan and China. It is true that mistrust is a qualitative aspect of a state of mind. However, we assume that military expenditure is a satisfactory yardstick for measuring it.

The second factor involved in the arms race is the cost of keeping up defenses. It can be assumed that a significant portion of resources devoted to military efforts is a complete waste, while other military expenditures, such as road building, etc., may have some economic value. It is the purpose of this paper to identify those items as useful and useless, nor do we want to make any estimate of them at this stage. Secondly, since the investment resources of India are limited, military expenditure implies sacrificing economic development (in terms of increase in the Gross National Product) that would have been achieved if the resources were used for national economic development.

The third factor is the amount of benefit that would have accrued to India if these two countries become friendly and cooperated in the field of international trade. In regard to many commodities, such as jute, tea, textiles, etc., the foreign exchange income of both countries could increase appreciably if they cooperated in production and distribution. These three factors have been combined in the following equation :

$$O_{it} = a_1 M_{1t} + b_1 Y_{1t} + c_1 T_{1t} \tag{29}$$

O_{1t} = represents the cost of defense

y_{1t} = Gross National Product of India at the time
  t (GNP in short)

T_{1t} = is the foreign exchange income of India at the time period
  t

a_1 = constant showing what percentage of the military expenditure can be treated as waste

b_1 = the percentage of Indian GNP that could have increased if there were a friendly relationship between the two countries. In the extreme case, b_1 may be taken as the percentage of Indian GNP that could have increased if there is no military spending. Since such a utopian situation is nowhere in sight in the near future, we shall take b_1 as the percentage of GNP that could be increased if the defense expenditure is kept at "normal" level. We shall not attempt a comprehensive definition of the term "normal" and shall assume that such a figure has been agreed upon.

c_1 = a constant which shows the percentage of foreign exchange income that could be added with mutual cooperation between these countries.

The third factor related to conflict is grievances. For the sake of simplicity, let us assume that India's grievances against Pakistan consist of two parts (1) alleged collusion with China, and (2) alleged Pakistani inspired rebellions in tribal areas of northeastern India. Again, we do not want to verify whether these allegations are true or not. These two sources of grievances are assumed to be measurable in monetary terms. Some portion of them – say, military help from China to Pakistan – is measurable. The following equation then represents the third factor :

$$Q_{1t} = d_1 A_{2t}^c + e_1 I_{2t} \qquad (30)$$

where

Q_{1t} =   is the grievance of India against Pakistan at time period t

A_{2t}^c =   military help from China to Pakistan in the time period "t" (Rupees in millions)

I_{2t} =   "Help" (assumed to be expressed terms of monetary value) by Pakistan to the tribal rebels in northeastern India (rupees in millions).

d_1 and e_1 are respective weights given by the Indian government to match these threats. These constants can also show what amount of money has to be spent by India to face these dual fronts. The grievance equation involves some qualitative aspects which are assumed to be expressed in quantitative terms. It is easy to identify other "cost" and "grievances," but for the sake of simplicity we assume that all these items can be expressed through the variables we have considered.

Combining the three factors together, we have

$$\frac{dm_{1t}}{dt} = km_{2(t-1)} + nm_{3(t-1)} - O_{1(t-1)} - Q_{1(t-1)} \qquad (31)$$

Based on equation (28) to (31), I have developed a time path of equilibrium military expenditure of the two countries, Chatterji (1991a).

My long term goal is to integrate equations of the type 1 to 4 in an econometric model. Towards that end in this paper, I shall present some correlations (Table 24.5) between some variables listed in Table 24.4 and some regression equations, given in Table 24.6. I admit that the equations are far from definitive. A considerable need exists to refine the data and include some other variables.

It is possible to intend the above approach for a pa time series cross section study of a larger group of developing countries using the same   variables as in Table 24.3 and appropriate lags. Another variable which probably explain arms spending of a country is the extent of the country's degree of conflict with other country's. The data for this variable can be obtained from the conflict and Peace Data Bank (COPDAB). This data bank gives information about symbolic political relations, economic relations, military relations, cultural and scientific relations, physical environment and natural resource relations, and human environment, demographic, and ethnic factors. These variables and the index of conflict situation for each country can be used as independent variables. I hope such more sophisticated studies will be forthcoming in the future.

References

Adamas, F. Gerard, Jere R. Behrman and Michael Boldin, "Defense Expenditures and Economic Growth in the Less Developed Countries: Reconciling Theory and Empirical Results, "in *Disarmament, Economic Conversion and Management of Peace*, edited by M.Chatterji and L.Forcey, Praeger Publishers, New York.

Adekson J. Bayo. "On the Theory of Modernizing Soldier: A Critique." *Current Research on Peace and Violence* 1(1) 1978: 28-40.

Albrecht, Ulrich. "Technology and Militarization of Third World Countries in Theoretical Perspective." *Bulletin of Peace Proposals* 8 (2) 1977: 124-136.

Albrecht, Urich, et al. "Militarization, Arms Transfer and Arms Production in Peripheral Countries." *Journal of Peace Research* 12(3) 1975: 195-212.

Ames, Barry and Ed Goff. "Education and Defense Expenditures in Latin America: 1948-1968." *Comparative Public Policy: Issues, Theories and Methods*. Ed.Craig Liske, et al. Beverly Hills: Sage, 1975.

Arlinghaus, Bruce E. Military Development in Africa: *The Political and Economic Risks of Arms Transfers*. Boulder, Co: Westview Press, 1984, 152.

Askari, Hossein and Vittorio Corbo. "Economic Implications of Military Expenditures in the Middle East." *Journal of Peace Resolution* 11 1974 : 341-343.

Ayres, Ron."Arms Production as a Form of Import—Substituting Industrialization: The Turkish Case." World Development 11(9) 1983 : 813-823.

Azar, E.E., "The Conflict and Peace Data Bank (COPDAB) Project," in *Journal of Conflict Resolution*, Vol. 24 (1980).

Ball, Nicole, "Defence and Development: A Critique of the Benoit Study." *Economic Development and Cultural Change*, Vol. 31 (1983).

Ball, N."Defense Expenditures and Economic Growth: A comment,: *Armed Forces and Society*, Vol.11, No. 2 (Winter 1985).

Ball, Nicole. "The Growth of Arms Production in the Third World." *National Forum* 66 Fall 1986: 24-27.

-------. *Security and Economy in the Third World*. Princeton, New Jersey: Princeton University Press, 1988.

-------. "Defense and Development: A Critique of the Benoit Study." *Economic Development and Cultural Change* 31 (3) April 1983: 507-524.

Ball, Nicole and Milton Leitenberg. *The Structure of the Defense Industry: An International Survey.* New York: St Martin's Press, 1983.

Benoti, Emile. *Defense and Economic Growth in Developing Countries.* Toronto: Lexington Books, 1973.

-------."Growth and Defense in Developing Countries." *Economic Development and Cultural Change* 26 (2) January 1978 : 271-280

Berg, Elliot and Jeremy Foltz. "The Non-military Forces in Africa." Santa Monica, CA: The Rand Corporation, 1987

Biswa, Basudeb and Rati Ram. "Military Expenditure and Economic Growth in LDCs: An Augmented Model and Further Evidence." *Economic Development and Cultural Change* 34 (2) January 1986: 361-372

Boulding, K.E., "Defense Spending: Burden of Boon." in *War/Peace Report.* Vol.13, No.1(1974).

Brauer, Jurgen. "Military Expenditures, Arms Production, and the Economic Performance of Developing Nationas." Ph.D dissertation, Department of Economics, Notre Dame University: (Unpublished), 1988.

Brigagao, Clovis, "The Brazilian Arms Industry." *Journal of International Affairs* 40 (1) Summer 1986: 101-114.

Brzoska, Michael. "Research Communication: The Military-Related External Debt of Third World Countries." *Journal of Peace Research* 20(3) 1983.

Brzoska, Michael. "Current Trends in Arms Transfers." *Defence Security and Development.* Eds., Saadet Deger and Robert West. New York : St. Martin's Press, 1987. 161-179.

-------."The Impact of Arms Production in the Third World." *Armed Forces and Security* 15 (4) Summer 1989: 507-530.

Brzoska, Michael and Tomas Ohlson, Eds. *Arms Production in the Thrid World (SIPRI).* London: Taylor & Francis, 1986.

Chan, Steve. "The Impact of Defense Spending on Economic Performance : A Survey of Evidence and Problems." *Orbis* 29 (2) Summer 1985: 403-434.

Choucri, N. and R. North. *Nations in Conflict: National Growth and International Violence.* San Francisco : Freeman, 1975.

Clare, Joseph. "Whither the Third World Arms Producers?" *World Military Expenditures and Arms Transfers.* Washington, D.C.: U.S.Government Printing Office, 1987. 23-28.

Debelko, David and James M.Mccormick. "Opportuntiy Costs of Defense: Some Cross-National Evidence." *Journal of Peace Research* 14 (2) 1977: 145-154.

Davies, David R."The Security-Welfare Relationship: Longitudinal Evidence from Taiwan." *Journal of Peace Research* 27(1) January 1990: 87-100.

Deger, Saadet, *Human Resources, Government Education Expenditure and the Military Burden in Less Developed Countries,* Discussion paper No.109 (London: Birbeck College, 1981).

Deger, Saadet. "Does Defence Expenditure Mobilize Resources in LDC's?" *Journal of Economic Studies* 12 (4) 1985: 15-29.

--------."Economic Development and Defense Expenditures." *Economic Development and Cultural Change* 35 (1) October 1986: 179-196.

--------.Military Expenditure in Third World Countries. London: Routledge & Kegan Paul, 1986A.

Deger, Saadet and Somnath Sen. "Military Expenditure, Spin-off and Economic Development." *Journal of Development Economics* 13 August-October 1983:67-83.

Deger, Saadet and Somnath Sen."Technology Transfer and Arms Production in Developing Countries. "*Industry and Development*. New York: United Nations, 1985. Vol.15: 1-19.

Deger, Saadet and Somnath Sen. *Military Expenditure: The Political Economy of International Security,* Oxford University Press, 1990.

Deger, Saadet and Ron Smit. "Military Expenditure and Growth in Less Developed Countries." Journal of Conflict Resolution 27(2) June 1983: 335-353.

Deger, Saadet and Robert West. "Introduction." *Defence, Security and Development.* New York: St.Martin's Press, 1987.1-16.

Dickon, Thomas Jr."An Economic Out put and Impact Analysis of Civilian and Military Regimes in Latin South America." *Development and Change* 8 1977: 325-285.

Dreyer, June T."Deng Xiaopeng and Modernization of the Chinese Military." *Armed Forces and Society* 14 (2) Winter 1988: 215-231.

Evans, Carol. "Reappraising Third World Arms Production." *Survival* 28 (2) March/April 1986: 99-118.

Faini, Riccardo, Patricia Annez, and Lance Taylor. "Defense Spending, Economic Structure, and Growth: Evidence Among Countires and Over Time." *Economic Development and Cultural Change* 32 (3) April 1984:487-498.

Faini, Riccardo, Patricia Annez, and Lance Taylor. "Defense Spending, Economic Structure, and Growth: Evidence Among Countries and Over Time." *Economic Development and Cultural Change* Vol. 32 No.3 (1984).

Feder, Gershon, "On Exports and Economic Growth," in *Journal of Development Economics,* Vol. 12 (1982)

Ferejohn, J."On the Effect of Aid to Nationas in Arms Race." *Mathematical Models in International Relations: A Perspective and a Critical Appraisal.* Dina A. Zinnes and John V.Gillespie, Eds. New York : Praeger Publishers, 1976.

Fontanel. Jacques and Saraiva Drumont. "Les Industries d'Armement comme Vecteur due Development Economique des Pays due Tiers-Monde." *Etudes Polemologiques* 40 1986 : 27-41.

Frank, Andre Gunder. "The Arms Economy and Warfare in the Third World." *Contemporary Crises* 4 (2) April 1980 : 161-193.

Frank, I., *Foreign Enterprise in Developing Countries* (Baltimore: Johns Hopkins University Press, 1980).

Frederiksen, P.C., and Robert E. Looney, "Defense Expenditures and Economic Growth in Developing Countries," *Armed Forces and Society*, Vol.9, No.4, pp. 663-645 (Summer 1983).

Frederiksen, Peter C. and Robert E. Looney. "Defense Expenditures and Economic Growth in Developing Countries; Some Further Empirical Evidence." *Journal of Economic Development* July 1982: 113-25.

-----."Defense Expenditures and Economic Growth in Developing Countries." *Armed Force Society* 9 (4) Summer 1983: 633-645.

------."Another Look are the Defense Spending and Development Hypothesis." *Defense Analysis* September 1985.

Gandhi, Ved p. "India's Self-Inflicted Defence Burden." *Economic and Political Weekly* 9 (I35) August 31 1974: 1485-1494.

Gillis, Malcolm, et al. *Economics of Development.* New York:W.W Norton & Company, 1987.

Glick, Edward B. *Peaceful Conflict:The Non-Military Uses of the Military.* New York: Stackpole Books, 1967.

Gottheil, Fred M. "An Economic Assessment of the Military Burden in the Middle East." *Journal of Conflict Resolution* 18(3) September 1974: 502-513.

Griffin, L.J., M. Wallace, and J.Devine, "The Political Economy of Military Spending: Evidence from the United States,"*Cambridge Journal of Economics,* No.6, Academic Press Inc., London, pp.1-14 (1982).

Grindle, Merilee. "Civil-Military Relations and Budgetary Politics in Latin America." *Armed Forces and Society* 13 (2) Winter 1987.

Grobar, Lisa M. and Richard C. Porter. "Benoit Revisited: Defense Spending and Economic Growth in LDCs." *Journal of Conflict Resolution* 33 (2) June 1989 : 318-345.

Gyimah-Brempong, Kwabena. "Deffense and Development: A Review of the Statistical and Econometric Literature." *Rand Note No-2650-USDP.* Santa Monica, CA: The Rand Corporation, 1987.

------."Patterns of Military Expenditure in Africa, 1965-1986." *Draft prepared for Rand Corp. and Department of Defense.* Santa Monica, CA: The Rand Corporation, 1987.

-------."Defense Spending and Economic Growth in Sub-Saharan Africa: An Econometric Investigation." *Journal of Peace Research* 26(1) 1989: 79-90.

Hanning, Hugh. *The peaceful Uses of Military Forces.* New York: Frederick A. Prager, 1967.

Harkavy, Robert E. *The Arms Trade and International Systems.* Cambridge, MA: Bellinger Publishing Company.

Harris, G "The Determinants of Defence Expenditures in the ASEAN Region." *The Journal of Peace Research* 23, 1986.

Harris, G.T. "Economic Aspects of Military Expenditures in Developing Countries." *Contemporary Southeast Asia* 10(1) June 1988: 82-102.

Harris, Geoffrey, Mark Kelly, and Pranowo. "Trade-offs Between Defense and Education/Health Expenditures in Developing Countries." *Journal of Peace Research* 25 (2) June 1988: 165-177.

Hayes, Margaret D."Policy Consequences of Military Participation in Politics: An Analysis of Tradeoffs in Brazilian Federal Expenditures." *Comparative Public Policy: Issues, Theories, and Methods.* Eds.C. Liske, W. Loehr and M. Mccamant. Beverly Hills: Sage, 1975. 21-52.

Heller, Peter S. and Jack Diamond. *International Comparisons of Governmental Expenditures Revisited : The Developing Countries, 1975-1986.* IMF Occasional Paper No.69. Washington, D.C.: International Monetary Fund, 1990.

Hess, Peter and Brendan Mullan. "The Military Burden and Public Education Expenditures in Contemporary Developing Nations." *The Journal of Developing Nations* 22(4) July 1988: 497-514.

Hill, Kim Q. "Military Role v. Military Rule." *Comparative Politics* 11 (3) April 1979: 371-378.

Hollist, L. "Alternative Explanations of Competitive Arms Processes : Tests on Four Pairs of Nations." *American Journal of Political Science* 22 1977: 313-340.

Intriligator, Michael D. "Research on Conflict Theory: Analystic Approaches and Areas of Application." *Journal of Conflict Resolution* 26(2) June 1982: 307-327.

Joerding, Wayne. "Economic Growth and Defense Spending: Granger Causality." *Journal of Development Economics* 21 April 1986: 35-40.

Kaldor, Mary H. "The Military in Development." *World Development* 4 (6) June 1976: 459-482.

--------."Military Technology and Social Structure." *Bulletin of the Atomic Scientists* 33 (5) June 1977: 49-53.

Katz, James E. "Understanding Arms Production in Developing Countries." *Arms Production in Developing Countries: An Analysis of Decision Making.* Toronto: Lexington Books, 1984. 3-13.

-------. *The Implications of Third World Military Industrialization: Sowing the Serpent's Teeth.* Tornto: Lexington Books, 1986.

Kende, I., "Local Wars, 1945-76," in E.Eide and M. Thee (Eds.), *Problems of Contemporary Militarism* (London: Croom Helm, 1980).

Kennedy, Gavin. *The Military in the Third World.* New York: Charles Scribner's & Sons, 1974.

Kohler, Daniel. *The Effects of Defense and Security on Capital Formation in Africa.* Santa Monica, CA: The Rand Corporation. 1988.

Korea University, Labour Education and Research Institute. Economic Development and *Military Technical Manpower of Korea.* Ed. Hyung-Bae Kim. Seoul: Korea University Press, 1976.

Kuh, Edwin, "The Validity of Cross-Sectionally Estimated Behavior Equations in Time Series Applications," in Econometrica 27, pp. 197-214, (1959).

Landau, Daniel. "Government and Economic Growth in Less Developed Countries: An Empirical Study for 1960-1980." *Economic Development and Cultural Change* 35 (1) October 1986: 35-75.

Landau, Daniel, "Government Expenditure and Economic Growth: A Cross-Country Study, *Southern Economic Journal,* 49, pp.783-792 (1983)

Lebovic, James H. and Ashfag Ishaq. "Military Burden, Security Needs, and Economic Growth in the Middle East." *Journal of Conflict Resolution* 31 (1) March 1987: 106-138.

Leontief, W. and F. Duchin, "Military Spending: Facts and figures," in *Worldwide Implications and Future Out look,* Oxford University Press, New York, (1983),

Lerner, Daniel and Richard D.Robinson."Swords and Plowshares: The Turkish Army as a Moderizing Force." *World Politics* 13 (1) October 1960: 19-44.

Lim, Davikd. "Another Look at Growth and Defense in Less Developed Countries." *Economic Development and Cultural Change* 31 (2) January 1983: 377-384.

Looney, Rebort E., "Recent Patterns of Defense Expenditures and Socio-Economic Development in the Middle East and South Asia," (mimeo 1990).

Looney, Rebort E. The Political Economy of Latin American Defense Expenditures: Case Studies of Venezuela and Argentina. Ndw York : Luxington, 1986.

------."Financial Constraints on Potential Latin American Arms Producers." *Current Research on Peace and Violence* 10(4) 1987: 159-168.

------."Socio-Economic Budgetary Contrasts in Developing Countries: The Effect of Alternative Political Regimes." *Journal of Social, Political and Economic Studies* 13 (2) Summer 1988: 195-218.

-------."The Political-Economy of Third World Military Expenditures: Impact of Regime Type on the Defense Allocation Process." *Journal of Political and Military Sociology* 16 (1) Spring 1988A: 21-29.

-------.*Third World Military Expenditure and Arms Production*, New York: St.Martin's Press, 1988B. Cited: IIc, IId, IIIb, IIIc, IIIf.

-------."Conventional Wisdom vs. Empirical Reality: The Case of Third World Defense Expenditures and Arms Production." Unpubished paper 1988C.

-------."Internal and External Factors in Effecting Third World Military Expenditures." *Journal of Peace Research* 26(1) 1989: 33-46.

--------."Military Keynesianism in the Third World: An Assessment of Non-Military Motivations for Arms Production." *Journal of Political and Military Sociology* 17 (1) Spring 1989A: 43-63.

Looney, Robert and Peter C. Frederiksen. "Defense Expenditures, External Public Debt and Growth in Developing Countries." *Journal of Peace Research* 23 (4) 1986.

Looney, Robert E. and Peter C. Frederiksen. "Consequences of Military and Civilian Rule in Argentina: An Analysis of Central Government Budgetary Trade-Offs." *Comparative Political Studies* 20 April 1987: 34-46

Looney, Robert E. and Peter C. Frederiksen." Economic Determinants of Latin American Defense Expenditures." *Armed Forces and Society* 14 (3) Spring 1988.

Lotz, Joergen R. "Patterns of Government Spending in Developing Countries." *The Manchester School of Economic and Social Studies* 38 (2) 1970: 119-144.

Lotz, Joergen R., "Patterns of Government Spending in Developing Countries." *The Manchester School*, Vol. 38 No.2 (1970)

Louscher, David J. and Michael Salomone. *Technology Transfer and U.S. Security Assistance.* Boulder, Co: Westview Press, 1987.

Luchsinger, Vincent P. and John Van Blois."Spin-Offs from Military Technology Manangement." *International Journal of Technology Management* 4(1) 1989 : 21-29.

Lucier, Charles."Changes in the Values of Arms Race Paraeters." *Journal of Conflict Resolution* 23 (1) March 1979 : 17-41.

Luttwak, E., *Coup d'etat* London: Allen Lane, 1968.

Lyttkens, Carl H. and Claudio Vedorato. "Opportunity Costs of Defense: A Comment on Dabelko and Mccormick." *Journal of Peace Research* 21 (4) 1984: 389-394.

Maholanobis, P.C., *The Approach of Operations Research to Planning in India*, Asia Publishing House, Bombay, India, 1963.

Maizels, Alfred and Machiko Nissanke. "The Determinants of Military Expenditures in Developing Countries." *World Development* 14 (9) 1986: 1125-1140.

Mangum, Steven L. and David E.Ball. "Military Skill Training: Some Evidence of Transferability." *Armed Forces and Society* 13 (3) Spring 1987: 425-441.

McKinlay, Robert D. and A.S.Cohen."A Comparative Analysis of the Political and Economic Performance of Military and Civilian Regimes." Comparative Politics 8 (1) October 1975: 1-30.

McKinlay, Robert D. *Third World Military Expenditures: Determinants and Implications.* England: Pinter Publishers, 1989.

Mintz, Alex, "Military-Industrial Linkages in Israel," *Armed Forces & Society* 12 (1) Fall 1985: 9-27.

-------."The Evoluation of Israel's Military Expenditures: 1960-1983." *Western Political Quarterly* 41 (3) September 1988: 489-507.

Moll, Kendall and Gregory Luebbert. "Arms Race and Military Expenditure Models," *Journal of Conflict Resolution* 24 (1) March 1980: 153-185.

Moodie, Michael. "Defense Industries in the Third World: Problems and Promises." *Arms Transfers in the Modern World.* Eds. S. Neuman and R. Harkavy. New York: Praeger, 1979. 294-312.

Mushkat, Mario. "The Socio-Economic Malaise of Developing Countries as a Function on Military Expenditures." *Coexistence* 15 (2) October 1978: 135-145.

Nabe, Ouma. "Military Expenditures and Industrialization in Africa." *Journal of Economic Issues* 18 (2) 1983: 525-587.

Nawaz, Shuja, "Economic Impact of Defense Expenditures," Finance and *Development* 20 (1) March 1983.

Neuman, Stephanie G."Security, Military Expenditures and Socioeconomic Development: Reflections on Iran." *Orbis* 22 (3) Fall 1978: 569-594.

-------."Arms Transfer and Economic Development: Some Research and Policy Issues." *Arms Transfer in the Modern World.* Ed. R. Harkavy. New York: Frederick A. Praeger, 1979. 219-245.

-------. "Third World Arms Production and the Global Arms Transfer System." *Arms Production in Developing Countries: An Analysis of Decision Making.* Ed. James E. Katz. Toronto: Lexington Books, 1984. 15-37.

--------."International Stratification and Third World Military Industries." *International Organization* 38 (1) winter 1984A: 167-197.

--------."Coproduction, Barter, and Countertrade: Offsets in the International Arms Market." *Orbis* 29 (1) Spring 1985: 183-213.

--------."Arms, Aid and the Superpower." *Foreign Affairs* 66 (5) Summer 1988: 1044-1066.

-------."The Arms Race: Who's on Top?" Orbis 33 (4) Fall 1989: 509-521.

Nordlinger, Eric A."Soldiers in Mufti:The Impact of Military Rule upon Economic and Social Change in Non-Western States." *American Political Science Review* 64 (4) December 1970: 1131-1148.

Nzimiro, Ikenna. "Militarization in Nigeria: Its Economic and Social Consequences." *International Social Science Journal* 35 (1) 1983: 125-140.

O'Leary, M.and W.Coplin. *Quantitative Techniques in Foreign Policy Analysis and Forecasting.* New York: Praeger, 1975.

Ostrom, Charles W., Jr. "Evaluating Alternative Foreign Policy Decision Making Models: An Empirical Test Between an Arms Race Model and an Organizational Politics Model," *Journal of Conflict Resolution* 21 (2) June 1977:235-266.

-------."A Reactive Linkage Model of the U.S. Defense Expenditure Policy-Making Process."*American Political Science Review* 72 September 1978:941-957.

Palma, Gabriel, "Dependency: A Formal Theory of Underdevelopment or a Methodology for the Analysis of Concrete Situations of Underdevelopment?" *World Development*, Vol. 6, No.7-8, pp.881-924 (1978).

Peacock, Alan T. and Jack Wiseman. *The Growth of Public Expenditure in the United Kingdom.* Princeton, NJ: Princeton University Press, 1961.

-------."Approaches to the Analysis of Government Expenditure Growth." *Public Finance Quarterly* 7(1) January 1979: 3-23.

Peleg, Ilan. "Military Production in Third World Countries: A Political Study." *Threats, Weapons and Foreign Policy.* Ed. P.McGowen and C. Kegley. London: Sage Publications, 1980. 209-230.

Peroff, Kathleen and Margaret Podolak-Warren. "Does Spending on Defense Cut Spending on Health? A Time Series Analysis of the U.S. Economy 1929-1974." *British Journal of Political Science* 9(1) 1979: 21-39.

Pinch, Franklin C. "Military Manpower and Social Change: Assessing the Institutional Fit."·*Armed Forces and Society* 8 (4) Summer 1982: 575-600.

Pluta, Joseph. "The Performance of South America Civilian and Military Governments from a Socio-Economic Perspective." *Development and Change* 10 1979: 461-483.

Pye, Lucien W."The Role of the Military in Underdeveloped Armies in the Process of Political Modernization." *The Role of the Military in Underdeveloped Countries.* Ed. John J. Johnson. Santa Monica, CA: The Rand Corporation, 1962. 69-90.

Ram, Rati, "Casuality Between Income and Government Expenditure : A Broad International Perspective," in *Public Finance/Finance Publiques*, Vol. XXXXI/XXXXI i me Ann e, No. 3., pp. 393-410 (1986).

Rattinger, H."Econometrics and Arms Races : A Critical Review and Some Extensions." *European Journal of Political Research* 4 1976: 421-439.

Ravenhill, John. "Comparing Regime Performance in Africa: The Limitations of Cross-National Aggregate Analysis." *The Journal of Modern African Studies* 18 (1) March 1980: 99-126.

Richardson, L. Arms and Insecurity: *A Mathematical Study of the Causes and Origins of War.* Chicago: Quadrangle, 1960.

Rosh, Robert M. "Third World Militarization: Security Webs and the States they Ensnara." *Journal of Conflict Resolution* 32 (4) December 1988: 671-678.

------."Third World Arms Production and the Evolving Insterstate System." *Journal of Conflict Resolution* 34 (1) March 1990: 57-73.

Ross, Andrew L. "World Order and Arms Production in the Third World." *The Implications of Third World Military Industrialization.* Ed. James E. Katz. Toronto: Lexington Books, 1986. 272-292.

------."Dimensions of Militarization in the Third World." *Armed Forces and Society* 13 (4) Summer 1987.

Rothschild, Kurt W. "Millitary Expenditure, Exports and Growth." *Kyklos* 26 (4) December 1973: 804-814.

Rubinson, Richard, "The World-Economy and the Distribution of Income Within States: A Croos-National Study," *American Sociological Review*, Vol.41, pp.638-659 (1976).

Russett, B. *The Revolt of the Masses: Public Opinion Toward Military Expenditure.* Beverly Hills, CA: Sage, 1972.

Schmitter, Philippe C. "Foregin Military Assistance, National Military Spending and Military Rule in Latin America." *Military Rule in Latin America: Function, Consequences and Perspectives.* Ed. Phillipe C. Schmitter. London: Sage Publications, 1973. 117-187.

Scholing, Eberhard and Vincenz Timmermann, "Why LDC Growth Rates Differ: Measuring 'Unmeasurable' Influences" in *World Development*, Vol. 16, No.11, pp.1271-1294, Pergamon Press, Great Britain (1988).

Sen, A., "Development: Which Way Now?", in *Economic Journal*, Vol. 93, pp. 745-762 (1983).

Sivard, R.L., *World Military and Social Expenditures*, 1983 (Washington, DC:World Priorities, 1983).

Smith, R., "Military Expenditure and Capitalism," *Cambridge Journal of Economics*, Vol. 1, No. 1 (1977).

Smith, Ronald P., "Military Expenditure and Investment in OECD Countries, 1954-1973," *Journal of Comparative Economics*, Vol. 4, pp. 19-32 (1980a).

Smith, Ron,"The Demand for Military Expenditure," *The Economic Journal*, Vol. 90,pp. 811-820 (1980b).

Summers, Robert and Alan Heston, "A New Set of International Comparisons of Real Product and Price Levels Estimates for 130 Countries, 1950-1985" in *The Review of Income and Wealth*, Series 34, No.1 (March 1988).

Sunkel, Osvaldo, "Transnational Capitalism and National Disintegration in Latin America," *Social and Economic Studies*, Vol. 22, pp. 132-176 (1973).

Tait, Alan A. and and Peter S. Heller. *International Comparisons of Government Expenditure.* IMF Occasional Paper No.10. Washington, D.C.: Internatioanl Monetary Fund, 1982.

Tannahill, Neal R. "Military Intervention in Search of a Development Variable." *Journal of Political and Military Sociology* 3 1976: 219-229.

Taylor, J. "Recent Developments in the Lanchester Theory of COMAST." *Operational Research* 78. K. Haley, ed. Amsterdam: Elsevier North Holland, 1979. 773-806.

Terhal, Peter. "Guns or Grain; Macroeconomic Costs of Indian Defence, 1960-70." *Economic and Political Weekly* 16 (49) 5 December 1981: 1995-2004.

Terrell, L.M.,"Societal Strees, Political Instability, and Levels of Military Effort," *Conflict Resolution* Vol. 15, No.3(1971).

Thurow, Lester, "Who Said Military Dictatorships are Good for the Economy?" *Technology Review* 89 (8) November/December 1986: 22-23.

United Nations Department of Disarmament Affairs. *Reduction of Military Budgets: Refinement of International Reporting and Comparison of Military Expenditures.* New York: United Nations, 1986. Report of the Secretary General   General Assembly Document A/s-12/7.

------. *Reduction of Military Budgets: Construction of Military Price Indexes and Purchasing Power Parities for Comparison of Military Expenditures.* New York: United Nations, 1986. Report of the Secretarly General. General Assembly Document A/40/421.

Vayrynen, Raimo. "The Arab Organization of Industrialization: A Case Study in the Multinational Production of Arms." *Current Research on Peace and Violence* 2(2) 1979: 66-79.

Verner, Joel G. "Budgetary Trade-Offs Between Education and Defense in Latin America: A Research Note." *Journal of Developing Areas* 18 October 1983: 77-92.

Vivekananda, Franklin and Ben E. Algobokhan. "Militarization and Economic Development in Nigeria.: *Scandinavian Journal of Development Alternatives* 6 (2) June 1987:106-121.

Wallace, Michael and Judy Wilson. "Non-Linear Arms Race Models." *Journal of peace Reseach* 15 (2) 1978: 175-192.

Weaver, Jerry L. "Assessing the Impact of Military Rule: Alternative Approaches." *Military Rule in Latin America: Function, Consequences and Perspectives.* Ed. Phillipe C.Schmitter. London: Sage Publications, 1973. 58-116.

Weede, Erich. "Rent Seeking, Military Participation, and Economic Performance in LDC's." *Journal of Conflict Resolution* 30(2) June 1986: 291-314.

West, Robert L. "Improved Measures of the Defence Burden in Developing Countries." *Defence, Security and Development.* Eds. Saadet Deger and Robert L.West. London: Frances Pinter (Publishers), 1987-19-48.

Whynes, David K. *The Economics of Third World Military Expenditure.* London: Macmillan Pr Ltd., 1979. 165.

Wulf, Herbert. "Dependent Militarism in the Periphery and Possible Alternative Concepts." Arms Transfers in the Modern World. Eds.S. Newman and R. Harkavy. New York: Prager, 1979. 246-263.

-------."Developing Countries." *The Structure of the Defense Industry: An International Survey.* Eds. Ball and Leitenberg. New York: St. Martin's Press, 1983, 311-343.

Zuk, Gary and william Thompson. "The Post-Coup Military Spending Question: A Pooled Cross-Sectional Time Analysis." *American Political Science Review* 76 (1) March 1982: 60-74.

Table 24.1 : Share of Third World in Global Security Expenditure, According to Regions, 1950-1984 (in percentages)

Region/Country	1950	1955	1960	1965	1970	1975	1980	1984
Middle East[a]	0.5	0.5	0.9	1.2	2.5	6.9	7.3	7.1
China	0.5	2.9	5.7	8.8	10.3	7.3	7.6	5.6[c]
Far East[b]	2.0	0.8	1.3	1.6	1.8	2.1	3.0	3.1
South Asia	1.2	0.7	0.7	1.2	1.0	1.0	1.2	1.3
Africa	0.1	0.1	0.2	0.7	0.9	2.3	2.5	1.7
South America	1.3	1.0	1.1·	1.1	1.1	1.7	1.8	1.0
Central America	0.5	0.2	0.3	0.3	0.3	0.3	0.4	0.5
All Developing	10.6	6.2	10.2	14.9	17.9	21.6	23.8	21.3

Source : *Stockholm International Peace Research Institute*, World Armaments and Disarmament SIPRI Yearbook, 1972, 1976, 1979, 1982, 1985 (Stockholm: Almqvist & Wiksell, 1972, 1976) and (London:Taylor & Francis, 1979, 1982, 1985).

[a]Includes Egypt.

[b]Excludes China.

[c]Excludes Egypt.

Table 24.2 : Military Spending for Some Developing Countries

$$\frac{\text{Military Exp}}{\text{GDP}} \times 100$$

	1973	1975	1978	1980	1983	1985
Middle East						
Egypt	34.1	35.4	10.2		6.7	5.8
Iran	7.3	13.0	11.2	[5.4]	[2.6]	3.0
Iraq	12.2	11.7		[6.3]	[24.4]	[27.5]
Israel	33.9	26.7	19.8	25.0	23.8	17.7
Saudi Arabia	6.9	(9.7)	[16.3]	(16.6)	(20.3)	(21.8)
Syria	0.7	0.8	0.7	17.3	15.4	15.6
South Asia						
India	3.0	3.3	3.1	3.0	3.1	3.3
Pakistan	6.1	6.1	5.3	(5.7)	(6.9)	(6.8)
Sri Lanka	0.7	0.8	0.7	1.5	1.5	2.6

(Contd.)

Table 24.2 (*Contd.*)

	1973	1975	1978	1980	1983	1985
Far East						
Indonesia	[3.7]	[4.1]	4.3	[3.8]	[3.9]	[3.0]
Korea, South	3.7	4.3	5.7	5.9	5.6	5.2
Africa						
Algeria	(1.7)	1.8		2.1	1.9	1.7
Angola				12.8	16.5	28.4
Ethiopia	1.9	4.5	6.5	8.5	9.0	(8.9)
Ghana	1.4	1.7		0.3	0.5	0.5
Libya	(4.9)	(5.7)	(12.1)	[10.0]	[14.5	
Mozambique				5.6	10.7	11.7
Nigeria		3.7	5.4	2.5	1.9	1.3
South Africa	2.2	3.3	4.7	3.9	3.7	3.7
Tanzania	3.0	3.8	(6.1)	3.7	3.5	3.3
Central America						
Costa Rica	0.5	0.6	(0.6)	0.8	(0.8)	(0.7)
Mexico	0.7	0.7	0.6	0.6	0.5	(0.7)
South America						
Argentina	1.2	(1.7)	2.1	6.4	(4.6)	3.5
Brazil	1.4	1.1	1.1	(0.7)	(0.8)	(0.8)
Chile	3.5	3.9	5.8	6.7	8.0	7.6
Venezuala	1.7	1.9	1.5	2.7	(2.9)	(2.0)

( )   uncertain data

[ ]   very uncertain data

Source : SIPRI (1992)

Table 24.3 : Socio Economic Information about şome Developing Countries

	Population Thousand	Agriculture Index (1980 = 100.0)	Industry Index (1980 = 100)
	1985	*1985*	*1985*
Middle East			
Egypt	48,503	193.6	117.5
Iran	43,000	-	-
Iraq	15,000	-	-
Israel	4,233		
Saudi Arabia	11,508	152.2	93.9
Syria	10,458	171.7	138.2
South Asia			
India (pop. in mil)	765	145.6	149.7
Pakistan	96,180	149.0	143.6
Sri Lanka	15,837	192.0	117.6
Far East			
Indonesia	163	153.1	156.8
Korea, South	41,056	130.1	128.5
Philippines	55,819	237.4	238.6
Thailand	51.683	87.5	133.1
Africa			
Algeria	27,718	165.8	127.8
Angola			
Ethiopia	42,271	126.5	105.7
Ghana	12,737	655.7	1,428.0
Kenya	20,353	164.7	150.7
Libya	3,764		
Mozambique	13,791	241.0	241.0
Nigeria	99,669	209.9	132.1
South Africa	31,593	163.2	172.0
Sudan	21,931	382.0	382.0
Tanzania	22,242	317.1	201.0

(Contd.)

Table 24.3 (*Contd.*)

	Population Thousand	Agriculture Index (1980 = 100.0)	Industry Index (1980 = 100)
	1985	1985	1985
Central America			
Costa Rica	2,490	467.9	513.1
Mexico	78,524	-	-
Nicaragua	3,276	526.8	597.6
South America			
Argentina	30,564	-	-
Brazil	136	-	-
Chile	12,074	-	-
Venezuala	17,317	167.2	152.2

	GNP (K Millions) US $	Value of Imports (mil U.S. $)	Value of Export (mil.US $)
	1985	1985	1985
Middle East			
Egypt	63,840	10,581	5,193
Iran	282,000	11,635	13,435
Iraq	57,640	10,556	
Israel	31,150	9,752	6,256
Saudi Arabia	89,270		21,428
Syria	24,380	3,487	1,759
South Asia			
India	231,700	17,295	9,465
Pakistan	29,150	5,890	2,739
Sri Lanka	2,63	1,988	1,333
Far East			
Indonesia	61,290	14,230	18,711
South Korea	93,660	31,119	30,283
Phillipines	32,130	5,445	4,629
Thailand	39,900	9,239	7,121

Table 24.3 (*Contd.*)

	GNP (K Millions) US $	Value of Imports (mil U.S. $)	Value of Export (mil.US $)
	1985	1985	1985
Africa			
Algeria	63,810	9,974	13,034
Angola	10,330		
Ethopia	4,582	993	333
Ghana	4,493	731	663
Libya	29,950		
Mozambique	1,361	424	
Nigeria	23,760	8,890	12,566
South Africa	73,640	28,299	40,395
Sudan	8,025	1,114	595
Tanzania	2,729	1,028	255
Central America			
Costa Rica	3,930	1,098	976
Mexico	139,200	13,994	22,108
South America			
Argentina	70,650	3,814	8,396
Brazil	256,100	14,300	25,600
Chile	5,659	3,300	3,823
Venezuala	35,170	8,234	14,660

	Gov't Expenditure in billions US $	Arms Import / Total Import x 100		
	1987	1975	1980	1985
Middle East				
Egypt	31,830	10.0	11.3	15.1
Iran	65,700	12.0	3.3	15.5
Iraq		12.0	17.9	39.8
Saudi Arabia	83,810	4.0	6.0	16.1
Syria	12,630	19.0	77.7	40.3

(*Contd.*)

Table 24.3 (*Contd.*)

	Gov't Expenditure in billions US $	Arms Import / Total Import x 100		
	1987	1975	1980	1985
South Asia				
India	52,300	3.0	5.6	14.3
Pakistan	6,995	3.0	7.9	8.0
Sri Lanka	2,156	1.0	0.5	1.6
Far East				
Indonesia	14,410	0	3.1	1.0
Korea, South	19,110	3.0	2.2	1.4
Phillipines	4,306	1.0	0.7	0.5
Thailand	8,862	2.0	3.8	1.5
Africa				
Algeria	25,900	1.0	6.9	4.9
Angola				45.6
Ethiopia	1,808	12.0	100.4	78.0
Ghana	638	1.0	0	0
Libya		13.0	38.4	29.5
Mozambique	265		21.3	
Nigeria	3,348	1.0	0.5	3.6
South Africa	20,720	2.0	0	0.1
Sudan	1,272	0	6.3	5.2
Tanzania	629	2.0	6.4	4.9
Central America				
Costa Rica	930	0	0	1.8
Mexico	36,250	0	0.1	0.2
South America				
Argentina	21,760	1.0	2.0	4.7
Brazil	72,210	1.0	10.5	0.3
Chile	5,659	2.0	4.9	0.7
Venezuala	10.240	1.0	1.1	5.3

Source: SIPRI (1992)

Table 24.4 : World Military Expenditure, In Constant Price Figures. Figures are in US $m, at 1988 prices

Country	1982	1986	1991	ΔY_1 (%)	ΔY_2 (%)
India	6325	9006	9003	42.4	0.3
Pakistan	1767	2459	2862	39.2	16.4
Japan	21291	25924	31083	21.8	19.9
South Asia	8416	12099	12577	43.8	4.0
Far East	42514	48894	58732	15.0	20.1
U.S.	240616	305076	264383	26.8	-13.3
Central America	3698	4098	4019[*]	10.8	-1.9
South America	14724	12224	11376[*]	-17.0	-6.9
Middle East	76183	61274	54432[*]	-19.6	-11.2
Israel	7314	4318	3909	-41.0	-9.5
Syria	3526	2573	3134	-27.0	21.8
Iraq	21952	16531	7414	-24.7	-55.2
Iran	10230	9339	6125	-8.7	-34.4
Egypt	5442	5013	3183	-7.9	-36.5
Yugoslavia	2137	2491	1376	16.6	-44.8
Eastern Europe[**]	324	634	682	95.7	7.6

ΔY_1 = Percent Change in Expenditures Between 1982 and 1986.

ΔY_2 = Percent Change in Expenditures Between 1986 and 1991.

[*] Denotes 1991 figures were not available. In this case, the previous year's figure was used.

[**] Albania, Bulgaria, Czechoslovakia, Hungary, and Poland.

U.S.S.R. figures were not available.

Source : Sipri (1992)

Table 24.5 : World Trade In Major Conventional Weapon System, 1987-91

Seller / Recipient	USSR	USA	France	UK	China	FRG	Czech.	Italy	Brazil	Others	Total
India	13871	-	882	1516	-	254	-	-	-	1039	17562
Pakistan	-	795	33	158	1027	-	-	19	-	267	2299
USA	-	-	4	417	128	201	-	50	-	599	1392
South Asia	-	24	9	-	515	-	-	3	-	286	837
Japan	-	9537	49	164	-	-	-	-	-	0	9570
China	497	113	163	5	-	-	-	-	-	19	797
USSR	-	-	-	-	-	-	2277	-	-	736	3013
Brazil	-	591	492	8	-	107	-	0	-	50	1248
Mexico	-	270	25	16	-	-	-	-	-	26	337
Argentina	-	-	45	-	-	444	-	52	64	80	685
Israel	-	4475	-	-	-	19	-	-	-	72	4566
Syria	3180	-	-	-	-	-	-	-	-	267	3447
Iraq	7049	283	719	-	703	41	125	43	815	542	10320
Iran	715	-	-	-	1390	-	234	-	25	498	2862
Egypt	-	4121	803	3	-	14	-	253	149	117	5460

Figures are value of major conventional weapon systems transferred, in U.S. $m., at constant (1990) prices.

Figures many not add up to totals due to rounding.

Source : SIPRI (1992)

Table 24.6: China's arms exports, 1981-88

Index: 1981 = 100

Year	SIPRI	CRS	ACDA
1981	100	100	100
1982	240	291	305
1983	298	352	362
1984	368	445	436
1985	308	141	143
1986	400	255	247
1987	667	491	200
1988	613	592	..

Sources :   Index constructed from constant price in *SIPRI Yearbook* 1989: *World Armaments and Disarmament* (Oxford University Press: Oxford, 1989), table 6A.2; Grimmelt, R., *CRS Report Supplier Trends in Conventional Arms Transfers to the Third World by Major Supplier, 1980-1988* (Library of Congress: Washington, DC, 1989); US Arms Control and Disarmament Agency, *World Military Expenditures and Arms Transfer 1988* (ACDA: Washington, DC, 1989)

Table 24.7 : Official Chinese defence spending, 1975-90

Year	Milex, current (b. yuan)	Share of CGE[a] (%)	Share of NI[b] (%)	Milex, constant (b. 1988 yuan)
1975	14.25	17.4	6.7	27.88
1976	13.45	16.7	5.5	25.92
1977	14.90	17.7	5.6	28.33
1978	16.78	15.1	5.6	31.65
1979	22.27	17.5	6.7	41.17
1980	19.38	16.0	5.3	33.37
1981	16.80	15.1	4.3	28.20
1982	17.64	15.3	4.1	29.02
1983	17.71	13.7	3.7	28.60
1984	18.08	11.7	3.2	28.41
1985	19.15	10.4	2.7	26.91
1986	20.13	8.6	2.6	26.43
1987	20.98	8.6	2.3	25.32
1988	21.80	8.1	1.9	21.80
1989	24.55[c]	8.4	1.9	20.46
1990	28.90[c]	11.5	—	23.10

[a]CGE: central government expenditure.

[b]NI: national income.

[c]Planned figure.

Source : SIPRI (1992)

Table 24.8 : *Countries*

1 ARG = Argentina
2 BEN = Benin
3 BGD = Bangladesh
4 BOL = Bolivia
5 BRA = Brazil
6 CHL = Chile
7 CHN = China
8 CMR = Cameroon
9 COG = Congo The People's Republic
10 COL = Colombia
11 ECU = Ecuador
12 GAB = Gabon
13 GHA = Ghana
14 GMB = Gambia
15 GTM = Guatemala
16 HND = Honduras
17 HVO = Burkinafaso
18 IDN = Indonesia
19 IND = India
20 LKA = Sri Lanka
21 MEX = Mexico
22 MYS = Malaysia
23 NGA = Nigeria
24 NIC = Nicaragua
25 NPL = Nepal
26 PAK = Pakistan
27 PHL = Philipines
28 PNG = Papu New Guinea
29 SLE = Sierraleone
30 SUR = Suriname
31 TCD = Chad

32   THA = Thailand

33   TUR = Turkey

34   TZA = Tanzania

35   VEN = Venezuela

Source : Sipri (1992)

Table 24.9 : *Variables*

(% Change between 1980 & 1985)

Ragrva.......	=	Constant 1980 price Value Added in Agriculture (local currency)
Rbirth......	=	Crude Birth Rate (per 1000 population)
Rconsp.....	=	Constant 1980 Price Private Consumption at etc. (local currency)
Redt......	=	Total External Debt (US $ outstanding at end of year)
Rexp......	=	Constant 1980 price Exports of Goods and NF Services (local currency)
Rexpgs....	=	Exports of Goods and Services (US $ BoP)
Rexppif.....	=	Fuel Export Price Index (1980 = 100. US dollar-based)
Rexpipp.....	=	Nonfuel Primary Products Export Price Index (1980 = 100 US $-based)
Rexppi......	=	Export Price Index fob (1980 = 100 US dollar-based)
Rfertr.....	=	Total Fertility Rate
Rfoodp.....	=	Food Production Per Capita (1987 = 100, US $)
Rgdy.....	=	Constant 1980 price Gross Domestic Income (local currency)
Rggovce.....	=	%gnp- Government Current Expenditure as % of gnp
Rggovcr.....	=	%gnp- Government Current Revenue as % of gnp
Rgnpagr......	=	%gnp- Constant 1980 price Value Added in Agriculture

Rgngny..... = %gnp- Constant 1980 price Gross National Income

Rgnpcon..... = %gnp- Constant 1980 price Private Consumption etc.

Rgnexp...... = %gnp- Constant 1980 price Exports of Goods and NF Services

Rgngdy..... = %gnp- Constant 1980 price Gross Domestic Income

Rgnimp..... = %gnp- Constant 1980 price Import of Goods and NF Services

Rgnpimc..... = %gnp- Constant 1980 price Capacity to Import

Rgnman.... = %gnp- Constant 1980 price Value Added in Manufacturing

Rgnp.... = Constant 1980 price National Product (local currency)

Rgny..... = Constant 1980 price Gross National Income (local currency)

Rgovcr.... = Government Current Revenue (local currency)

Rgovce..... = Government Current Expenditure (local Currency)

Rimp.... = Constant 1980 price Import of Goods and NF Services (local currency)

Rimpcp..... = Constant 1980 price Capacity to Import (local currency)

Rimppi.... = Import Price Index cif (1980 = 100 US dollar-based)

Rimpgs.... = Imports of Goods & Services (US $ BoP)

Rinvf..... = Net Foreign Direct Investment (US $,BoP)

Rinfmr...... = Infant Mortality Rate (per 1000 infants)

Rlabagr..... = Labor Force Agriculture (%)

Rlabf..... = Labor Force Female (%)

Rlabm...... = Labor Force Male (%)

Rlabfpr.... = Labor Force Participation Rate (proportion of LABF)

Rldebt.....	= Long-Term Debt (US $)
Rlife......	= Life Expectancy at birth (years)
RLIT....	= Percent Literate
Rloans....	= Net Long-Term Loans (US $ as per IBRD DRS)
Rmanva.....	= Constant 1980 price Value Added in Manufacturing (local currency)
Rmanemp....	= Manufacturing Employment (1980 = 100)
Rmanre....	= Manufacturing Real Earnings per Empl. (1980 = 100)
Rmane....	= Manufacturing Earnings as % of Value Added
Rmanro....	= Manufacturing Real Output per Empl (1980 = 100)
Rmanxp....	= Manufacturing Export Price Index (1980 = 100 US dollar-based)
Rmexp.....	= Military Expenditures (US Dollars)
Rnewsp....	= Newspaper Circulation (per 1000 Population)
Roex.....	= Energy Exports
Rof....	= Energy Consumption for Agriculture
Roim....	= Energy Imports
Ror....	= Energy Consumption for Residential Use
Ros....	= Indigenous Production of Energy
Rotl....	= Energy Consumption for Transportation
Rpo.....	= Price of Oil
Rpopac	.... Age-Cohort Population (proportion, ages 15-64)
Rpdens....	= Population Density
Rpopni....	= Population Natural Increase Rate (per 1000)
Rpophys....	= Inhabitants per Physician
Rrad.....	= Number of Radios Sets (per 1000 population)
Rschp.....	= School Enrollment Ratio, primary school (%)
Rtelec....	= Energy for Electricity Generation Consumption as % of TOTFC
Rtlab.....	= Total Labor Force (millions)

Rtoted.....	= Total Energy Requirement
Rtotl...	= Energy Consumption for Transportation as % of TOTFC
Rtotfc.....	= Total Final Consumption of Energy
Rtotod....	= Total other Uses, Energy Consumption
Rtv......	= Number of Television (per 10000 population)

Table 24.10 : *List of Variables, Units, and Sources for India 1950-1985*

POP : Population ; Millions ; Source : International Financial Statistics ; 1982 p. 260 ; 1987 p. 394.

GDP : Gross Domestic Product ; Billions of Rupees ; Base : 1980 prices ; Source : Int. Fin. Stat. ; National Acts 1986 ; 1987, p. 394, Int. F. Stat.

GDPCRRT : Gross Domestic Product ; Billions of Rupees ; Current Prices ; Source : Int. Fin. Stat. ; 1982, p. 260 ; 1987, p. 394.

GROSSFCF : Gross Fixed Capital Formation ; Billions of Rupees ; Current Prices ; Source : Int. Fin. Stat. ; 1951-58 from IBRD Deport 1987, p. 395 ; National Accts, 1986.

MILITEXP : Military Expenditure ; Billions of Dollars ; Current Prices ; Source : U.S. ACDA.

MILIBURD : Military Burden ; Source : 1964-85 from SIPRI, 1950-63 based on Benoit & estimates.

EXPORT : Billions of Rupees ; Current ; Source : Int. Fin. State. ; 1987. p. 394.

IMPORT : Billions of Rupees ; Current ; Source : Int. Fin. Stat. ; 1987, p. 394.

GOVTEXP : Govt. Expenditure ; Billions of Rupees ; Current ; Source : Int. Fin. Stat. ; 1987, p. 394.

MILXCNST : Military Expenditure by Pakistan : Billions of Constant Dollars ; Base : 1987 prices ; Source : U.S. Arms Control and Disarmament Agency USACDA 1977-87.

MILXCRRT : Military Expenditure by Pakistan ; Billions of Current Dollars ; Source : USACDA.

EDXPCRRT : Educational Expenditure ; Thousand Millions ; Rupees ; Current Prices ; Source : National Accts, p. 688.

MANUINDX : Manufacturing Index ; Base : 1980 = 100 ; Source : Int. Fin. Stat. ; 1987.

AGRINDX : Agricultural Index ; Base : 1985 = 100 ; Source : Benoit.

WLDMILXP : World Military Expenditure ; Billions of US$; Base : 1980 prices ; Source : SIPRI

USSRA : Weapon Exports by USSR, *Yearly Figures* ; Base : 1975 prices ; Billions US$ SIPRI.

USSRB : Same ; *Five-year moving averages.*

USAA : Weapon Exports by USA, *Yearly Figures* ; Base ; 1975 prices ; Billions US$$; SIPRI.

USAB : Ditto ; *Five-year moving averages.*

NATINCME : National Income ; Billion of Rupees ; Current Prices ; Int. Fin. Stat. ; 1987, p. 394.

GDPDEF : GDP Deflator ; Int. Fin. Stat. ; 1987, p. 394.

VOLEXP : Volume of Exports ; Base : 1980 = 100 ; Source : Int. Fin. Stat.

VOLIMP : Volume of Imports ; Base : 1980 = 100 ; Source : Int. Fin. Stat.

TRADEBAL : Trade Balance ; Millions US$$; Negative = debit (E-I) ; Current : Int. Fin. Stat. ; p. 394.

AITI : Arms Imports ÷ Total Imports ; *India* ; Source : US Arms Control & Disarm. Agency. (USACSA).

AXTX : Arms Exports ÷ Total Exports ; *India* ; Source : USACDA.

FORCES : No. of People in Armed Forces—*India* ; Thousands ; Source : USACDA.

ATIIPAK : Arms Imports ÷ Total Imports ; *Pakistan* ; Source : SACDA.

AXTXPAK : Arms Exports ÷ Total Exports ; *Pakistan* ; Source : SACDA.

AITICHIN : Arms Imports ÷ Total Imports ; *China* ; Source : USACDA.

AXTXCHIN : Arms Exports ÷ Total Exports ; *China* ; Source : USACDA.

FORCESPK : No. of People in Armed Forces—*Pakistan* ; Source : USACDA.

TRANSCOM : Transportation and Communication : Billions of Rupees ; Constant 1970 prices. World Tables.

FORNAID : Finance from Foreign Aid : Billions of Rupees ; Int. Fin. Stat. ; 1987, p. 394.

WAR : War

EMBARGO : Embargo

Chapter 25

Marketing the Peace Process in the Middle East: The Effectiveness of Thematic and Evaluative Framing in Jordan and Israel

Nehemia Geva, Allison Astorino-Courtois and Alex Mintz

Introduction

Despite dramatic progress towards peace in the Middle East, domestic opposition to a comprehensive agreement remains high. Further progress towards an Arab-Israeli peace depends, at least in part, on the ability of national leaders to obtain public support for their initiatives. The history of the conflict in the region suggests that although moves toward peace are very likely to benefit society as a whole, there have been numerous instances of missed opportunities to achieve peace in that region. In fact, in a classic piece, Mancur Olson (1971) emplains this type of collective action problem where even under expectations of clear gains, collective goods may not be provided in large groups - when those actions involve risk and uncertainty. The essential difficulty is that different groups often see less chance of receiving substantial proportions of the collective gain, and according to the logic of the 'free-rider' are less interested in acting in what is the society's common interest.

Given this difficulty, how might Arab and Israeli political leaders encourage their constituents to continue to support the collective good of a more peaceful regional relations?

The purpose of this chapter is to examine the prospects and possibilities for marketing the Arab-Israeli peace process. Without doubt, the pro-peace marketing effort in the Middle East is compounded by the extreme variance of popular opinion concerning an Arab-Israeli settlement. Leaders in each of the states involved must cope with multiple population segments that differ over both value preferences and the degree to which different 'externalities' of peace are considered salient to their concerns. Expressed in theoretical terms, we are concerned with modes of social preference formation, or more exactly, with how certain preferences might be encouraged over others.

We begin with a brief introduction to the concept of political marketing. This is followed by a discussion of the effects of political framing on attitudes towards peace. Next, is the description of the research design and the presentation of the empirical findings. Finally, several theoretical and policy implications are entertained.

Political Marketing

To the extent that marketing occurs in international conflict resolution[1] policical leaders have tended to discuss peace negotiations most often in terms of the national security pay-offs of an accord. This has been particularly so in the case of Arab-Israeli reations. For years the (official) argument has been that achievement of economic growth, improved social welfare, and other dividends resultant from a peace accord are possible only following satisfaction of security concerns. However, whether or not they are aware of the process, national leaders with multi-decade government tenures(King Hussein in Jordan, Yitzhak Rabin in Israel, Yassir Arafat of the PLO and Hafez al Assad in Syria all fall into this category) have at times exhibited political skill in marketing government decisions and policy in ways acceptable to their constituents. Although this sensitivity to the desires and perceived needs of constituents clearly accords with the general "marketing concept",[2] the notion of political marketing has most commonly been applied in analyses of electoral politics, especially in U.S. presidential campaigns (Newman, 1994:6) The applicability of a marketing oriented approach has been absent however, from the the literature on conflict resolution.

The objective of *political* marketing activities is to affect the political behaviors and attitudes of the targeted audience. If, for example, the product Arab and Israeli leaders are offering is *peace*, then the objective is to encourage support for a peace accord. From a marketing perspective, developing the campaign to encourage support for such an accord involves communication about that product *as well as* an awareness on the part of national leaders of the needs (or opinions) of various constituent groups. It also involves a strategic effort to improve the product and/or communications about it so as best to satisfy those consumer needs (O'Shaughnessy, 1990:.2; Crane, 1972). That is, marketing a peace initiative involves a dual concern with attributes of peace and with communicating these to the groups to which they most appeal. Again, from a marketing perspective it becomes clear that nations can not focus simply on the popular needs satisfaction represented by a negotiated accord, but must also succeed in communicating the benefits of that agreement to domestic groups.

Thus, at the core of political marketing activities are the important processes of popular needs assessment, as well as the transfer of political information to the public. Consider, for example, that in international peace negotiations agreements are sometimes settled for interests other than those that local populations identify among their key concerns. Upon returning home, national leaders are obliged to encourage public support by providing constituents modes of interpreting the international agreements in domestically acceptable terms. However, this mutual communication process itself inevitably involves *framing*, or establishing the boundaries and reference points for the policy debate.

Marketing Communications : *Framing*

Theorists have suggested that the group that can succeed in defining the bounds of the political debate will have a significant advantage in encouraging popular support for its program (e.g., Schattsneider, 1960; Riker, 1982; Riker 1986; Jones, 1993). One way national leaders can attempt to market a political outcome such as a peace initiative is by *framing* the discussion of it.

In brief, by *framing* we mean that leaders introduce organizing themes into the policy debate that affect how the public views a political issue (see Geva and Mintz, 1994). Far from base manipulation, it can be argued that such activities are especially feasible in the area of foreign and security affairs where the public has relatively limited independent knowledge about events (Kegley and Wittkopf, 1991: 300; see also Jones, 1993).

Most commonly frames are introduced by highlighting the salience of certain attributes or dimensions of a particular policy or decision choice (e.g. the peace dividends from a negotiated settlement) while discounting others (e.g. the likehood of territorial concessions associated with pursuing the peace negotiation strategy). In addition, political leaders can portray political. outcomes in loss-gain terms. In this way, national leaders can attempt to define policy debates according to vocabulary and by using concepts favourable to their own positions. For example, Arab and Israeli leaders might frame *peace* as a necessary risk to avoid potential losses, such as the loss of the lives of citizens killed in war, or the loss of further territory. To the extent that there is public acceptance of this frame of reference, it should help mobilize support for peace. Thus, framing can help leaders convince the public that their best option is to make concessions to the enemy if that is what is required to sign a peace treaty (Geva and Mintz,1994).

In fact, extensive research in social cognition devoted to human information processing and cognition has highlighted the effects of an individual's previous knowledge on how he or she interprets, understands and incorporates new information. (For its specific context the decision literature has adopted the term *frame*[3]). The frame that is active during information processing was found to affect which parts of reality are perceived, i.e. which information is perceived relevant for further processing), how information is interpreted and valued, and finally the choice of particular options (c.f. Fiske and Taylor, 1991; Wyer and Srull, 1989). A number of experimental studies demonstrate empirical evidence of framing effects in that the way the decision situation was described significantly affected support for policy choices related to that situations (Frisch, 1993; Kahneman and Tversky, 1984; Quattrone and Tversky, 1988; Mintz and Geva, 1993; Geva and Mintz 1993 Astorino Courtois. 1994). Furthermore, Geva, True and Mintz (1993) find that politicians are aware of these framing effects: they are interested in strategic manipulation of political issues because they expect to change outcomes by doing so. Recently Geva and Mintz (1994) identified two general forms of framing: *thematic* and *evaluative*. Thematic framing involves content based communications - generally directed by national leaders to the public - that concern policy (product) attributes and\or the introduction of organizing themes into the policy debate. In this context the frame serves as a focal lens sensitizing the citizen to specific elements of the political environment.

It also affects the salience of the attributes national leaders have emphasized in encouraging support for preferred policy positions. Hence, the thematic aspect of farming can influence popular attitudes by prioritizing the content considered during the policy debate.

Framing political information can also have an evaluative effect where the frame serves as a reference point to which the external environment is compared. In this sense, the frame acts as an evaluative anchor in the assessment of the environment, and can shift the meaning of the policy debate. Hence, a certain frame might skew perceptions of peace options for example, as either 'rosier', or more grim than the objective situation warrants (Geva and Mintz, 1994). For example, Tversky and Kahneman (1986) have found that simply framing decision choices in terms of losses rather than in terms of gains increases risk-prone behavior and the choice of risky alternatives. Similarly, Geva and Mintz (1994) find that support for an Arab-Israeli peace accord among Israeli respondents was affected when the outcome of an accord was phrased in terms of gains rather than losses.[4] That is, respondents' support of a peace settlement with Syria that would result in Israel retaining part of the Golan territory, was different than their support for an accord (qualitatively the same situation) that was represented as requiring an Israeli withdrawal from the same part of the Golan Heights.

Our analysis addresses the issue of framing the Arab-Israeli peace process at the theoretical and empirical levels:

I. At the *theoretical-conceptual level* the paper evaluates the utility of employing derivations based on framing effects within the context of prospect theory (Tversky and Kahneman, 1986). Recent criticism has questioned the plausibility of utilizing prospect theory to account for political manipulation. For example, Wittman (1991) noted that framing effects have been obtained mainly in experiments that manipulated issues that were rather insignificant and trivial to the subjects. Moreover, critics argue that issues used in tests of prospect theory represent isolated instances that do not offer opportunities for learning that occurs in the 'real world'. The present study, in contrast, tests framing effects on a topic that has toped the national agenda for a relatively long time, and that the populace in the Middle East are very involved in. Furthermore, our research introduces to the study of framing two new wrinkles:(a) the distinction between two types of frames (thematic and evaluative) and the interaction between them; and (b) how evaluative frames interact with the initial reference points people have about an issue. The latter point relates to the argument that evaluative frames affect choices by shifting the reference points used to evaluate the alternatives.

II. At the *applied/policy* level we examine the effectiveness of different frames imposed on the national debates with respect to the Arab-Israeli peace process. Specifically, we explore whether the conventional policy debate that focuses primarily on security advantages of a peace initiative (in marketing terms, on the security attribute/dimension of peace), is in fact the most likely to encourage domestic support. We do so with an evaluation of the extent to which the different dimensions of peace encourage popular support for the peace process among three group: Jewish and Arab students in Israel, and Arab students at the University of Jordan in Amman. The question of whether different groups respond differently to communication in the context of highly salient political issues, or whether the social relevance and centrality of the Arab-Israeli conflict diminish differential reactions of the groups, has important policy implications. We believe that a political marketing-oriented approach to the study of the Arab-Israeli peace process will shed new light on why peace accords might ultimately succeed as well as the reasons they might eventually fail. Further, we argue that this type of analysis has important and timely implications for domestic efforts to promote any settlement. In the next section we describe the general design that was used in our field experiments.

Research Design[5]

The study employed between-groups factorial experimental designs[6], in which 198 Israeli-Jewish, 201 Israeli-Arab and 165 Jordanian students responded to items pertaining to the Arab-Israeli peace process.[7] Specifically, the subjects were asked to indicate the extent of their support for a peace agreement. The items in the questionnaires introduced both foreign and domestic 'externalities' of peace. Each of the items was presented in one of two formats in order to tap the evaluative frame effects suggested by prospect theory (see review by Frisch, 1993). As will be described below, two evaluative frames were used: gain vs. loss, and the ratio-difference framing.

The research material was adapted to the focal concerns of the peace process in Jordan and Israel respectively, so that comparable and symmetric experimetal contexts can be constructed. Thus, for the Israeli participants, the peace process in the Middle-East addressed the negotiations between Israel and Syria. For the Jordanians, the peace was identified as the one involving relations with Israel.

Contextual differences between the two countries also affected the operational definition of the initial reference point of the participants. A variety of pools conducted in Israel indicated that those affiliated with the Labor-led coalition parties are more supportive of the negotiations with Syria than those affiliated with the Likud-led bloc. Hence, the initial reference point regarding the support for the peace was defined in Israel in terms of affiliation with coalition parties ('pro' peace agreement) or with the opposition parties. The different political structure in Jordan required a specifically designed item to categorize respondents as relatively supportive of the peace process, or against it.

Results

a. Framing the Certainty of the Product: Lasting Peace

Mistrust among the nations in the Middle East was cultivated in a rather long sequence of hostilities, wars and other expressions of terror. One of the first issues that confronts the public with regard to any proposed peace agreement in the Middle East is whether it is trustworthy, or in other words, whether the agreement will prevail in the long run. In this regard the average Arab or Israeli citizen is in a situation similar to that of a consumer buying a product that can neither be tested for its long (and even short) term quality, nor returned.

According to Tellis (1987), such uncertainty about quality of the product promotes, "a low price" consumer stratgy. Under this strategy consumers attempt to minimize loss. In the context of interstate peace negotiations, such a "low price" strategy would make the public hesitant to support agreements that imply certain national concessions or costs.

The first experimental study in this project, thus, tested the effects of evaluative framing of the *certainty* associated with a peace agreement on the extent of support for the agreement by subjects who have different positions i.e. different reference points) on this issue. Specifically, the framing manipulation consisted of presenting a question in one of two forms. In one frame (gain or success) respondents were presented with a forecast that there is a 90 percent chance that the peace agreement would be implemented and respected. The second frame (loss or failure) stated that there was a 10 percent chance that the peace agreement would be breached. The degree of certainty was the same in the two presentations, though the reference points differed.

Table 25.1 shows the means of support of the different groups of subjects for the peace initiatives framed in a loss (failure) versus gain (success) perspective.

Table 25.1 : *Support of the Peace Initiative by Jordanian and Israeli Subjects as a Function of Evaluative Framing and the Initial Reference Point*

		Jordanian Subjects	Israeli Jewish Subjects
10% change of failure	Pro-Accord	5.55	5.93
	Anti-accord	3.30	3.50
90% chance of success	Pro-Accord	6.53	7.89
	Anti-accord	2.70	2.20

The findings reveal that the evaluative (gain-loss) frame itself failed to produce a main-effect among the Jordanian and Israeli-Jewish students. It did however, operate in conjunction with the initial reference point. As evident in Figure 25.1(a) the evaluative frame interacted with the initial political reference point of the Israeli-Jewish students, $F_{(1,46)} = 4.39$ p < .05. Framing the certainty of the peace in terms of success increased the discrepancy of support for the peace agreement between students affiliated with the coalition and opposition parties compared to the difference in support of these two groups when the certainty was framed in terms of failure. In other words, those opposed to the peace initiative showed a higher level of support for the agreement when the small prospects for failure were mentioned than when high prospects for success were associated with the agreement. Those who were a-priori more inclined towards peace demonstrated an opposite trend. Figure 25.1(b) shows a similar framing - reference-point interaction among the Jordanian students $F_{(1,161)} = 3.39$ p<0.67.

The framing effect on the actual choice is thus mediated by the relation of the newly introduced frame (e.g. gain vs. loss) to the respondents' initial reference point. This interpretation is compatible with Puto's proposition that reference points that are used in evaluating marketed products are modifiable (Puto, 1987). This further suggests that prospect theory may not be applicable to questions of meaningful political choice without consideration of *a-priori* preferences. Both Jordanian and Israeli-Jewish students less sympathetic to the peace process were more likely to be swayed by a frame minimizing losses, while those in favor were encouraged by heightened prospects of gain.

The results for Israeli-Arab subjects are somewhat different, though they are compatible with our theoretical argument concerning the framing effect. To begin with, the extent of support for the peace agreement among these students is extremely high (M=9.70). Since the establishment of the state of Israel in 1948, this group has faced a dual loyalty problem. On the one hand, as citizens of Israel they are expected to be loyal to the state, while on the other hand they can not necessarily reject familial ties across the borders. This dual loyalty exacerbated the tensions associate with their minority status. For Israeli-Arabs peace in the region implies a win-win situation. Thus, it is not surprising that for this group evaluative framing of the certainty of the peace accord yielded results parallel to the 'pro-peace' groups in Israel and in Jordan, regardless of their initial political orientation. That is, the Israeli-Arab students as a group indicated a higher degree of support for the peace agreement when it was phrased in terms of 90 per cent sucess than when it was framed as 10 per cent failure $F_{(1,44)}=3.46$ p<.05.

It is interesting to note that there is similarity in the pattern of support for a bilateral negotiated peace agreement between the Jordanian and Israeli-Jewish students. Moreover, although the groups responded to conditions involving a different adversary (Israel and Syria, respectively), we did not obtain a significant difference in Israeli-Jewish versus Jordanian cell means (M=4,52 for the Jordanian students and M=4,88 for the Israeli Jewish students.[8] Also, as expected, the extent of support for a peace agreement is higher among both Jordanians and Israeli-Jews who were initially in favor of the peace process than it was for those against it (M=6.91 vs. M=2,85 in the Israeli-Jewish group, and M=6.04 vs. M=3.00 among the Jordanian subjects).

In the next section we focus on the multi-dimensional aspects involved in marketing peace in Israel and Jordan.

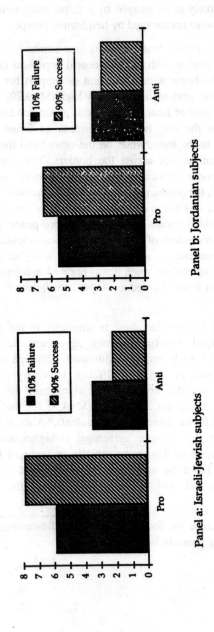

Fig. 25.1 : Support of the Peace Agreement among Jordanian and Israeli Students as a Function of the Political Reference Point and Evaluative Framing

Panel a: Israeli-Jewish subjects

Panel b: Jordanian subjects

b. The Dimensionality of Peace

In this section we evaluate the implications of thematic framing that is, highlighting different attributes/dimensions, for the evaluation of the peace agreement. We have already mentioned that a main concern in the 'marketing' of peace in the Middle-East is to convince the public that risks are worth taking as the 'product' (peace) will not be breached. However, a complementary approach to gaining support for the peace process involves highlighting the positive 'externalities' of peace along various dimensions. Therefore, in the second phase of this project we analyze the effectiveness of two types of thematic frames: one emphasizing domestic externalities of peace, and the other the foreign policy consequences of peace.

A variety of 'product attributes' or *dimensions* can be associated with a peace agreement. For instance, peace may raise expectations of economic prosperity. The abundance of literature on the impact of domestic factors on foreign policy behaviour suggests a number of dimensions (economic, political etc.) that can be used strategically in an attempt to market peace agreements (Mintz and Russett, 1992). The empirical question is which are the dimensions that will encourage support for a Middle East peace? Relatedly, how effective are these thematic frames in conjunction with evaluative framing?

From a theoretical point of view, the focal research question relates the relevance of the dimensions to respondents to the effectiveness of framing. As previously noted, much of the criticism of prospect theory (c.f. Wittman, 1991) has centered on the fact that the issues subjected to experimental analyses are of little or no personal significance to the subject. Consequently, analyses conducted to date, at best, indicate that framing effects may be obtained on relatively irrelevant dimensions. However, research in social cognition (e.g., Fiske and Taylor, 1991) suggests that the relevance of a dimension may increase the sensitivity of the respondent to situational variations, and thereby raise the likelihood of obtaining framing effects on highly significant dimensions rather than on insignificant ones.

One 'foreign policy' and two 'domestic policy' dimensions were introduced as 'externalities' of peace to the Jordanian and Israeli subjects.[9]

The evaluative framing in this part of the study was based mostly on the psychophysical 'ratio-difference principle'.[10] According to Quattrone and Tversky (1988:728) "the ratio-difference says, that the impact of any fixed positive difference between two amounts increases with their ratio." Applied in the context of labor statistics the authors demonstrated a situation described in terms of employment or unemployment with the same difference but with ratios of different magnitude affected respondents' approval of various social policies.

The foreign policy dimension in this study was phrased identically for all three groups of subjects (Israeli-Jews, Israeli-Arabs and Jordanian students). Specifically, subjects were asked to indicate their support for the peace accord given a reduction (or increase) in the *extent of international restrictions (or opportunities) the nation has as a consequences of the negotiated peace agreement.*

The Israeli subjects encountered as the domestic externalities an economic and a social dimension. The economic dimension was operationalized as the rate of employment/unemploymeent in the country. For example, the Israeli subjects indicated their extent of support for the peace process in response to the following one of the two frames; *"The prestigious Institute of Economic Studies in The Hague (an independent research institute in the Netherlands) which is known for its realistic projections, estimates that following the signing of a peace treaty with Syria (which will be accompanied with reduction of the security burden) Israel should expect an economic growth that will be expressed in an increase in the employment rate from 85 percent to 92 percent [a decrease of unemployment from the current 15 percent to 7.6 percent]"*

The social dimension was operationalized in terms of changes in social cohesion/friction in the country. Similarly to the manipulation of the evaluative frame for the economic dimension the 'ratio-difference' was presented as an increase in a social cohesion index from 63 to 83, or as a decrease in social friction from 37 to 13 on a 100 point scale.

For the Jordanian students the domestic externalities of peace were (a) the previously described economic (employment) dimension;[11] and(b) an item that deals with freedom of access across the former 'closed borders'. In terms of the peace 'externalities' this dimension covers both an economic aspect, but also has a religious connotation, as it implies the possibility of visiting the holy sites in Israel.

Table 25.2 shows the means of support for the peace agreement of respondents with different initial positions (reference points) to the peace process as a function of the thematic and evaluative frames.

Table 25.2 : *Support of the Peace Initiative as a Function of Dimension, Evaluative Frame and Initial Reference Point*

	Jordanian Subjects		Israeli Jewish Subjects		Israeli Arab Subjects	
	Pro	Anti	Pro	Anti	Pro	Anti
Foreign Policy						
5% increase in international opportunities	6.46	2.49	6.93	3.33	10.00	10.00
5% decrease in international restrictions	6.58	2.64	5.94	4.86	10.00	10.00
Economic						
5% increase in employment	6.77	2.81	5.44	3.67	10.00	10.00
5% decrease in unemployment	7.21	2.78	7.61	2.13	10.00	10.00
Social						
5% increase in social cohesion			7.09	3.14	9.67	9.63
5% decrease in social friction			6.30	3.40	9.86	10.00
Access						
From Jordan to Israel	6.97	2.45				
From Israel to Jordan	5.77	1.88				

The Foreign Policy Dimension

The foreign policy dimension framed the consequences of an Arab-Israeli peace accord in terms of either an *increase in international opportunities* from 63 to 83 on a bogus index, or as a *decrease* from 37 to 17 on the index representing *international restrictions*. According to the ratio-difference principle, evaluative framing would be considered a better consequence than a smaller change in proportion, would imply that a greater proportional change (despite the fact that the change is of the same difference). As shown in Table 25.2, our findings do not support the prospect prediction for the *foreign policy* dimension. The evaluative frame had no main effect nor did it interact with the initial political stand. Changes in the support of peace resulted only from the original reference point of the subjects (univariate Fs for Israeli Jewish students and Jordanian students are respectively, $F_{(1,46)} = 7.29 p < .01$ and $F_{(1,159)} = 122.49 p < .001$). As in the previous phase of the research, support for a peace agreement among Israeli Jewish and Jordanian students was at the mid-scale (M = 5.32 for the Jews and M = 4.54 for the Jordanians). The Israeli Arab students were extremely supportive of a peace accord.

These results coincide with other studies that suggest that public opinion is not very sensitive to variations in foreign policy when those policies are perceived not to be of hedonic relevance i.e. have direct bearings and consequences for the people's daily life (Geva and Andrade, 1993).

Domestic Dimensions of Peace

For the Israeli Jewish subjects, in the context of the *economic* dimension, the (evaluative) ratio-difference frame interacted with the initial political orientation, $F_{(1,44)}=6.32$ p<.05. However, evaluative framing had no effect in the context of the social dimension.

Among Jordanian students, when the policy choice was framed in terms of the *access* dimension, the evaluative frame interacted with the students' predisposition towards peace. It did not, however, yield any effect when combined with the economic dimension.

Finally, support for the peace agreement among Israeli Arab students yielded a main effect of the ratio-difference effect along the social cohesion dimension.[12] In this group, too, the evaluative frame was non-significant in the context of the economic dimension.

In sum, the analyses that were conducted separately for subjects of each of the two countries show that highlighting a specific externality of peace did not produce significant changes in support of the peace process. Moreover, the evaluative frame by itself had no direct effect on support for the peace process. Yet, on several domestic themes, or externalities of peace, the evaluative frame interacted with the initial political position of the subject in influencing the extent of support for the peace process. A posteriori analysis (Geva and Mintz, 1994) suggests that these effects occurred along the dimension that has the highest hedonic relevance for each of the three groups the economic dimension for the Israeli-Jews, the social cohesion for the Israeli Arab students, and access to Jerusalem for the Jordanian students. In cases of thematic frames with lower hedonic relevance, only the initial reference point of the subjects significantly influenced support for a peace agreement.

Implications

The findings presented in this study have a number of theoretical as well as policy-related implications.

Theoretical Implication

The application of prospect theory to a problem of international conflict resolution has raised several interesting challenges to dominant conceptual propositions. First, we demonstrate the feasibility of framing in the context of a heated, and highly relevant political debate in two very diverse societies. Second, our findings also suggest boundary conditions, and an important theoretical link in the debate over the validity of the notions of prospect theory versus those of the rational choice/ expected utility paradigm (Arrow, 1982). Specifically,

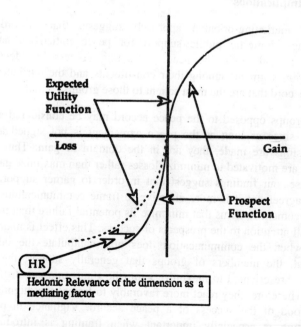

Fig. 25.2 : The Impact of Hedonic Relevance of a Dimension on the Prevalence of Expected Utility and Prospect Functions

the dynamic that determines whether a decision maker's response will correspond to an expected utility versus a prospect model appears to be the function of the *hedonic relevance* of the *thematic frame*. We have already discussed the utilities of distinguishing between thematic and evaluative frames. The thematic frame, or the underlying theme along which evaluative (gain/loss) frames are imposed may vary in their relevance to the decision maker of respondent. The impact of the personal significance or hedonic relevance of a given dimension on consumer choice was demonstrated in a variety of contexts (Jones, 1993; Geva and Andrade, 1993; Rowe and Puto, 1987). Our findings show that the prospect predictions with regard to the effects of gain-loss framing work only when the thematic dimension imposed on the situation has hedonic relevance to the decision maker. However, when the dimension has low hedonic relevance, gain-loss framing may not work and the expected utility curves are more appropriate to describe the effects (see Figure 25.2).[13]

Policy Implications

A marketing-oriented approach suggests that in order to encourage domestic public support for peace initiatives, national leaders should begin the peace marketing process by identifying population segments among their constituents and the attributes of a peace accord that are the most salient to those groups.

Groups opposed to the peace accord may be considered as loss averse, since as long as the peace agreement is not signed and no concessions are made, they are in the domain of gain. Thus, these groups are motivated to minimize losses rather than maximize gains. In this case, our findings suggest that in order to garner support for a peace agreement it is more effective to frame communications about the agreement in terms that minimise its potential failure than in terms that call attention to the prospects of success. This effect is maintained even when the communication does not manipulate the odds. In contrast, the members of groups that generally support the peace process are oriented to maximize gain (the status quo is considered a loss). Therefore, they react more favorably to frames that highlight the likelihood of the success of a peace accord. Vigilance in political marketing is especially important when framing is introduced to hedonically relevant dimensions.

References

Arrow, K. 1982. "Risk Perception in Psychology and Economics." Economics Inquiry 20:1-9.

Astorino-Courtois, A. 1994. "Transforming International Agreements into National Realities: Marketing the Arab-Israeli Peace Process in Jordan." *mimeo.*1994.

Beach, L. 1990. *Image Theory: Decision Making in Personal and Organizational Contexts.* Chichester, England: Wiley.

Crane, E. 1972. *Marketing Communications, Decision Making as a Process of Interaction Between Buyer and Seller,* 2nd Ed. New York: John Wiley & Sons, Inc.

De Bondt, W.F.M., and R. Thaler. 1985. "Does the Stock Market Overreact?" The Journal of Finance 60:793-805.

Fiske, S., and S, and S. Taylor. 1991. *Social Cognition.* New York: McGraw Hill, Inc.

Frisch, D. 1993. "Reasons for Framing Effects." *Organizational Behavior and Human Decision Processes* 54:399-429.

Geva, N., and Andrade. 1993. Foreign and Domestic Policy Decisions: An Experimental Assessment of Public Support. Presented at the Annual Meeting of the Southern Political Science Association, Savannah, Georgia, November 3-6, 1993.

Geva, N., and A. Mintz. 1994. "Framing the Options for Peace." *mimeo.*

Geva, N., J: True, and A. Mintz. 1994. "Farming Democratic Peace Paper presented at the annual meeting of the International Studies Association, Washington D.C., 1994.

Jones, B. 1994. *Serial Shift: Attention and Choice in Politics.* In press at University of Michigan Press.

Kahneman, D., and A Tversky. 1984, " Choices, Values and Frames." American Psychologist 4:341-350.

Kegley, C., and E. Wittkopf. 1991. *American Foreign Policy: Pattern and Process.* New York: St Martin's press

Klein, G. 1989. "Strategies of Decision Making. "*Military Review* May 56-64.

Mintz, A., and N. Geva 1994. *The Poliheuristic Theory of Decision.* Forthcoming

Mintz, A., and B. Russett. 1992. "The Dual Economy and Arab-Israeli Use of Force." *In Defense, Welfare and Growth,* eds. S. Chan, and A. Mintz. London: Routledge.

Mitchel, T., and L. Beach L. Beach. 1990. "Do I Love Thee? Let Me Count... toward an Understanding of Intuitive and Automatic Decision Making." *Organizational Behavior and Human Decision Processes* 47:1-20.

Newman, B. 1994. *The Marketing of the President, Political Marketing as Campaign Strategy.* Thousand Oaks, C.A: Sage Publications.

Olson, M. 1971. *The Logic of Collective Action: Public Goods and the Theory of Groups.* Cambridge, MA: Harvard University Press.

O'Shaughessy, N. 1990. The *Phenomenon of Political Marketing.* New York: St. Martin's Press.

Puto, C. 1987. "The farming of Buying Decisions." *The Journal of Consumer Research* 14:301-315.

Quattrone, G., and A. Tversky. 1988. "Contrasting Rational and Psychological Analyses of Political Choice." *American Political Science Review.* 83:719-736.

Riker, W. 1982. *Populism Against Liberalism: A Confrontation between the Theory of Democracy and the Theory of Social Justice..* San Francisco: W.H. Freeman, Co.

Riker, W. 1986. *The Art of Political Manipulation.* New Haven, CT: Yale University Press.

Rowe, D., and C. Puto. 1987. "Do consumers' reference points affect their buying decision?" *Advances in Consumer 14:188-192*

Schattsneider, E.E. 1960. *The Semisovereign People.* New York: Holt, Rinehart and Winston.

Tellis, G. 1987. "Consumer Purchasing Strategies and the Information in Retail Prices." *Journal of Retailing* 63:279-97.

Tversky, A., and D. Kahneman. 1986 "Rational Choice and the Farming of Decisions" *Journal of Business* 59:251-284.

Wittman, D. 1991. "Contrasting Economic and Psychological Analyses of Political Choice: An Economist's Perspective on Why Cognitive Psychology Does Not Explain Democratic Politics." In *the Economic Approach to politics: A Critical Assessment of the theory of Rational Actor*, ed. K.R. Monroe, Harper Collins.

Wyer, R, and T. Srull. 1989. *Memory and Cognition in its Social Context.* NJ: Hillsdale, Erlbaum.

Notes

* This research was supported by a grant from the Jewish-Arab centre at the University of Haifa, Israel.

1. It is important to note that although marketing theory is applicable to almost any political exchange, there are clearly vital differences between marketing a peace process and marketing consumer goods (see O'Shaughnessy, 1990; Newman, 1994). Not the least among these differences are the costs and value trade-offs involved.

2. Namely, that a seller or service provide must both understand the needs and/or desires of the customer, and position his product as meeting those needs (Newman, 1994: p.6; De Bondt and Thaler, 1985).

3. Beach, (1991); Klien, (1989); Mitchel and Beach, (1991); Tversky and Kahneman, (1981).

4. But this effect was part of the interaction of the frame with the initial political position of the respondents.

5. For a more detailed description of the research design see Astorino-Courtois, 1994; and Geva and Mintz, 1994.

6. The differential accessibility of subjects in Israel and Jordan necessitated the use of a somewhat different design in Jordan. There, a split-plot design was used, as subjects were assigned randomly to one of two questionnaires that included all the questions (dimensions). Similarly to the design used in Israel, subjects responded to only one version (evaluative frame) of the questions.

7. While the use of university students may be considered to lower the potential generalizability of the results, we suggest that if framing 'works' on a relatively

educated and highly involved segment of the population, it should also work on other segments of the population as well.

8. Unweighted means are presented and were used in the ANOVAs

9. Due to contextual differences between the two countries one of the domestic dimensions had to be changed for the Jordanian respondents.

10. With the exception being the item administered in Jordan, pertaining to access to Jerusalem, which was based on the psycho-social differences represented by Jordanian access to Israel vs.Israeli access to Jordan.

11. The wording of the item was adapted to the Jordanian-Israeli circumstances, rather than to Israel and Syria.

12. We gave already explained that among the Israeli Arab group there were no a-priori differences in the disposition towards the negotiations about the peace as all of them were extremely supportive.

13. It has been already stipulated that the main differences between the expected utility models and the prospects theory models is that the latter predicts a steeper curve in the domain of loss than in the domain of gain (represented by the solid line in Figure 2)

Chapter 26

Some Reflections on a Comprehensive System of International Peace and Security

Dietrich Fischer

Mikhail Gorbachev, in an article in *Pravda* of September 17, 1987, on "The Reality and guarantee of a Secure World," called for the creation of a Comprehensive System of International Peace and Security. The article was actually a speech he had prepared to present before the United Nations General Assembly on his first visit there, but he could not attend and deliver it when the United States postponed the planned Washington summit. The proposed system would seek to redress threats not only from war, but also from hunger, poverty, pollution and human rights violations. In his book *Perestroika* (1987, especially p.231) he expanded on that subject. This important idea still has not received the attention it deserves. The following is a modest attempt to reflect on some of the principles that might be applied in designing such a system. Many people, from different parts of the world and with different concerns, must become involved in this discussion to make the idea become reality.

Military aggression represents only one among many threats to human security. A range of dangers to human life and welfare, whether deliberate or not, and their underlying causes will be examined here. One fundamental cause of insecurity can be found in deficient feedback systems. Such systems are essential to maintain or achieve a desirable state and can be observed in all healthy organisms, balanced ecological systems, and secure societies. They consist of the three main components: agreement on a goal, methods to measure deviations from the goal, and corrective mechanisms to bring the system closer to its goal state if it has deviated. This chapter explores various possible ways in which such systems can break down, and how these defects may be corrected. These principles are applied to discuss ways to improve security in a broad sense, not only in terms of preserving peace, but also regarding economic development and protection of a livable environment.

1. SIX SOURCES OF INSECURITY

In the past, discussions about national and international security have focused primarily on the threat of military aggression across borders and and on military means to meet such threats. This threat must not be underestimated, but it represents only one of many dangers we face.

Threats to human life also include hunger, poverty, disease, and natural or human-made disasters, such as destruction of the ozone layer that protects us from carcinogenic ultraviolet radiation. Many more people die from preventable diseases than from war. The world economy operates far below its full potential, and its output is very unequally distributed. Exhaustible resources are being used up rapidly, pollution is generated at a rate that in many cases exceeds nature's capacity to render it harmless, and many species are becoming rapidly extinct, while the human population may soon exceed a sustainable level. Many people are denied elementary rights and freedoms.

Not all dangers to security are caused intentionally. In fact, few people favour war, poverty, pollution or oppression. Why do we observe so much of all of these? Some believe that the basic cause is our lack of knowledge. Others see the problem in human selfishness, or short-sightedness, or deficiencies in our legal systems, or conflicting interests and values. There is no single cause. All of these factors and several others play a role as sources of avoidable human suffering and insecurity.

It is interesting that all of these causes of problems can be regarded as various defects of effective feedback systems. Harold Chestnut (1986) has pointed out that insights from systems control theory, which have long been applied successfully to many engineering tasks, have rarely been used to address social problems. Of course, social problems are generally far more complex and difficult than technical problems, but given the enormous problems we face, we should be open to anything that may shed some light on possible improvemnents. A systems approach allows the integration of insights from many different disciplines into a coherent framework. It looks systematically at dangers to security and surveys potential corrective measures, exploring where a minimum intervention can have a maximum effect. Because of the emphasis on reducing deviations from a desired goal under changing external conditions, the approach proposed here may be called 'adaptation theory' (Fischer 1993). Of course, this is only one among many different angles from which one can look at these problems.

Any viable system, whether an ecological system, the human body or a society, needs numerous automatic feedback mechanisms to remain in a healthy state and to adapt to a changing environment. In nature, prey-predator systems keep the size of animal populations in check. An essential feedback mechanism in our body is the nervous system. The feeling of pain, for example, ensures that we protect injuries until they can heal. Leprosy, a disease of the nervous system, has the consequence that patients no longer feel pain. Therefore, they do not notice minor injuries to their limbs, keep using them and ultimately lose them.

In human society, feedback mechanisms include the legal system, which is designed to deter crime and aggression; the price mechanism, which helps guide production and distribution, where necessary corrected through taxes and subsidies; independent news media and democratic elections, which inhibit abuses of power.

Billions of years of slow evoluation have produced the earth's biosphere where all the various species are in a delicate balance. Modern science and technology have given us the power to upset that balance, and we have promptly begun to do so. We must now quickly supplement nature with some deliberate social regulatory mechanisms to preserve or restore a healthy environment, or human survival is at risk.

A regulatory feedback mechanism consists of a goal, a way to defect deviations from the goal, and methods to bring the system closer to the goal if it has deviated. In a legal system, for example, the laws represent the goals, the role of courts is to determine whether a law has been violated, and the police and corrections system are there to enforce the laws.

A feedback system can be defective in six ways:(1) There may be no agreement on the goal (a question of conflict resolution); (2) even if the goal is clear, deviations from it may not be noticed (a question of observation and communication); (3) even if deviations are detected, those who can correct the, may have no incentive to do so, because others are affected (a question of externalities, and also of ethics, whether we care about each other); (4) the proper incentives may be available, but too late, and if we do not look ahead, we may fail to prevent the problem (a qestion of future planning); (5) even if given timely feedback, people may behave irrationally, out of old habits, prejudice, hatred or other feelings (a question of psychology and culture); (6) someone may be fully aware of a problem and wish to correct it, but not know how or lack the necessary means(a question of science, technology and education). To illustrate these six defects, consider the destruction of the ozone layer.

(1) *Conflict over goals*: Protecting the ozone layer is certainly a desirable goal, but there may be conflicts over how to share the costs.

(2) *Lack of feedback*: For a long time, it was not known that chloro-fluoro-carbons (CFCs) and other industrial waste gases damage the ozone layer.

(3) *Distorted feedback*: The damage we cause to the ozone layer affects five billion other people slightly, but ourselves negligibly. If we all were to carry our personal small ozone shields over our heads, we would probably take better care of them.

(4) *Delayed feedback*: The problem is compounded by the fact that CFCs released today slowly damage the ozone layer for decades to come. If we were to feel the pain from skin cancer the moment we used a spray can containing CFCs, as we feel instant pain when we burn ourselves with a match, CFCs might never have been used.

(5) *Rejected feedback*: Even if people are made aware of the consequences of their behavior and given incentives to do the right thing, old habits sometimes die hard.

(6) *Ignorance or lack of means*: Some people may wish to phase out the use of CFCs, but do not know any alternative, or cannot afford safer but more expensive substitutes.

Similar mechanisms can also be identified as sources of other environmental problems, as well as war, poverty and human rights violations. To illustrate that all six of these defects must be overcome to solve a problem, consider the question of preventing overpopulation. First, agreement is needed on what represents a desirable level of the human population. Current esttimates of a sustainable population size vary widely. Second, an accurate census is needed periodically. Third, parents must have an incentive not to have too many children. If children are the only source of economic security in old age, parents may care little what effect their decision has on the size of the world population. Fourth, incentives must be provided without undue delay. Fifth, it is necessary to overcome "irrational" attiudes, for example that having many children is a source of prestige, or that practicing birth control is a sin. Otherwise, moral exhortations or economic incentives are ineffective. And sixth, if a couple has decided to practice birth control, it must have access to information how to do so, and must be able to afford it. If even one of these six links is missing, success is elusive.

Table 26.1 lists the six basic defects of regulatory feedback systems, with examples and potential remedies. Section 2 applies some of these general principles to environmental protection, Section 3 to economic development, and Section 4 to peace.

Table 26.1 :Some potentiai remedies against the six basic defects in social feedback systems

Source of Problems	Examples	Potential remedies
Conflict over goals (indecision)	Border conflicts; disagreement over fair distribution of resources; clash between developers and ecologists	Prevention and resolution of conflict through fact-finding, mutually beneficial cooperation, removing sources of conflict, negotiation, mediation, arbitration and legal procedures
Lack of feedback (unawareness)	Miscalculations as cause of wars; lack of information about poverty and famines;	Research, education, dissemination of information through scholarly channels and

(Contd.)

Table 26.1 (*Contd.*)

Source of Problems	Examples	Potential remedies
	damages to environment may be unknown	mass media; satellite monitoring to supervise disarmament agreements and to detect threats to the environment
Distorted feedback externalities (selfishness)	Good or bad consequences of decisions affect others; undemocratic forms of government; security dilemma	Ethical norms, m .al education; Law (penal and remunerative); taxes and incentives; greater equality; democratization; nonoffensive defense
Delayed feedback short sightedness	'Fallacy of last move' as cause of arms races and aggression; low saving and investment; over-exploitation of nature	Planning for the future; foresight; increased savings and investment; accelerated feedback; education in critical thinking; intergenerational ethic
Rejected feedback ("irrational" behavior)	Megalomania, anger, hatred, excessive fear, prejudice, racism as causes of civil and international wars; some customs are economically wasteful and ecologically unsound	Constitutional checks and balances; disarmament; expanding international contacts at all levels; global education to improve the understanding of other cultures and to overcome prejudices; taboos

(*Contd.*)

Table 26.1 (*Contd.*)

Source of Problems	Examples	Potential remedies
Lack of remedies (ignorance)	Inadequate defence capacity; lack of resources and production technology; ignorance about ecologically sound production methods	UN Peacekeeping Force; disarmament and economic conversion; more research; development assistance; technology transfer

2. PROTECTING THE ENVIRONMENT

In considering remedies for the six defects, we first refer again to the problem of ozone depletion, as an example that is also typical for other environmental problems.

2.1 Conflict over goals

The best way to solve conflicts of interest is often to find 'win-win' solutions that give each side what it values most. Concerning the dispute about cost sharing for the replacement of CFCs, the developed countries can better afford research to find alternatives to CFCs. For developing countries, food poisoning from lack of refrigeration may currently be a greater health hazard than skin cancer from increased ultraviolet radiation penetrating a thinner ozone layer. If the developed countries make available technology for refrigeration without the use of CFCs to the developing countries, both sides gain what is most important to them: the developed countries obtain a significant reduction in the use of CFCs at comparatively low costs, and the developing countries gain access to a harmless refrigeration technology without a severe financial burden.

2.2 Lack of feedback

Some scientists warned already in the 1960s that CFCs damage the ozone layer, based on computer simulations, but only when satellites revealed an ozone hole over Antarctica in the mid 1980s did the international community react.

Systematic research into potential dangers facing humanity is urgently needed. Carl Sagan (1983) has pointed out that all the major threats to the survival of humanity—the greenhouse effect, the destruction of the ozone layer, and nucler winter—have been discovered by accident. He wondered how many other potential dangers may still be unknown.

Awareness of a problem is a necessary condition for solving it, but not always sufficient. Heinrich Pestalozzi, the founder of public education in Switzerland, believed that eradicating illiteracy would make the world's problems disappear, because people could inform themselves about the causes of their misery and ways to overcome them. We now know that he was over-optimistic.

2.3 Distorted feedback

The principal cause of many environmental problems is that those responsible for them are often different from those who suffer the consequences, due to *externalities*. There are two basic approaches to correct this problem, the moral and the legal one. The moral approach seeks to persuade people to care more about each other. Galtung (1980,p.379) proposed an extension of the Hippocratic oath of the medical profession to similar codes of ethics for the whole range of professions. "Such codes should contain... a pledge to devote oneself above all to the satisfaction of true human needs... and a dedication to the whole world, to serve human needs everywhere, particularly those most in need." Education to social responsibility—the core of most religions—plays an important role in making our world more liveable, but is not always sufficient by itself.

Another strategy to deal with externalities is the legal approach, which assumes that people are basically selfish, but seeks to make it in their self-interest to do what serves the public good, by punishing those who do harm to others. There is an interesting asymmetry: whereas education uses praise and criticism, economic policy uses subsidies and taxes, law now relies only on punishment. One could equally well imagine that those who make an exceptional contribution to the public good receive recognition and support for their endeavors. There is a precedent: Someone who has discovered a new cure for a disease or made a significant contribution to world peace may win a Nobel Prize. But this is about as if once a year we were to select 'the criminal of the year' from around the world for exemplary punishment, while every other criminal got away free. We would hardly consider this an adequate legal system. Today's penal law ought to be complemented with a form of 'remunerative law'. Numerous psychological studies have found that people respond better to encouragement and rewards than to criticism and punishment ('Why Job Criticism Fails: Psychology's New Finidings, 'New York Times, July 26,1988, p.C1). Yet law still relies almost exclusively on punishment. A tranisition from punitive to remunerative law could be as significant as the shift from slavery, where the motivation for work was the fear of punishment, to wage labor, where the main motivation for work is the expectation of a reward.

Applying this principle to the problem of ozone depletion, one could offer a reward to those who first develop a harmless and inexpensive substitute, in addition to fining or closing down companies that continue to damage the environment.

We also need tax reform. Taxes are now levied mainly on hard work and creative ideas that help meet human needs. Instead, we should tax harmful activities, such as pollution and resource depletion (Baumol and Oates, 1975). Paradoxically, this would lower overall taxes, because it would help eliminate problems at the source. To see this, imagine that gasoline was available for free and at year's end the total bill would be paid out of taxes. We would not pay less for gasoline, but a lot more, because many people would waste it. The same thing is now happening to clean air and water. As long as we continue to treat them as if they were free rescources, we pay dearly for the consequences—with higher taxes, higher health care costs and more cancer deaths.

An important remedy to overcome externalities is democracy, so that those who bear the consequences of policy have a voice in shaping it. The least democratic countries, in which the press is censored, have also the worst environmental records. A free press that can expose a government's negligence or wrongdoing, and democratic elections that enable people to replace bad leaders, are essential to a society's health. They play a role analogous to that of white blood cells, which can detect and eliminate disease germs before they multiply and spread through our body. A weakened immune system, for example due to AIDs, can lead to death.

2.4 Delayed feedback

We are responsible to leave our descendants a livable planet. The U.S. Environmental Protection Agency, in evaluating the advisability of anti-pollution legislation, has assigned a value of $1 million to a human life, discounted at 10 percent per year. This implies that after a few generations a human life is worth less than one cent. Yet plutonium has a half-life of 24,000 years. Ramsey (1928) argued that applying discounting to inter-generational comparisons is unethical.

If feedback is delayed, this requires future planning. Governments often react only to crises. At the 1992 Earth Summit in Rio, the United States, the largest emitter of carbon dioxide that causes global warming, was the only major industrial country that refused to agree to a limit. If we wait until we can measure the effects of global warming, such as rising sea levels, it will be too late to prevent it. We cannot suddenly freeze huge quanities of ocean water and deposit them back onto the melting polar ice caps. Waiting until a problem becomes obvious before taking action is comparable to driving a car with closed eyes, waiting until we hit an obstacle or drive over an abyss, instead of anticipating and avoiding dangers.

2.5 Rejected feedback

For most of human history, the best way to survive was to follow traditions handed down over generations. In a stable environment, such behavior served us well, but in the rapidly changing world of today it has become dysfunctional. We have to re-evaluate traditions and adjust them if they are counterproductive. Behavior can be changed most easily at a very early age, before it has become habitual. When Louis Pasteur first saw microbes under a miscroscope and discovered the cycle by which we inflect ourselves with cholera, dysentery or typhus under unyhygienic conditions, education to personal hygiene from early childhood became commonplace. Now that we have discovered how environmental pollution can cause cancer, education to some form of 'global hygiene' has become equally important.

Boulding (1989) has emphasized the important role of taboos, of certain forms of behavior that are considered unacceptable, sometimes at such a deep, even unconscious level that no written laws are necessary to punish them.

2.6 Ignorance or lack of means

The best way to eliminate the use of CFCs and other ozone-depleting gases is through research into harmless substitutes. When the Carter administration proposed a ban on CFCs in spray cans in 1978, the chemical industry was in uproar, claiming that no substitutes were available,but later when the ban was passed, a harmless substitute was soon found. Necessity is the mother of invention.

Voltaire said that freedom is the only good that is used up when it is not used. This also applies to knowledge. Unlike physical resources, which must be taken from someone to be given to someone else, knowledge, once discovered, can be copied without limit at almost no additional costs. For that reason, it is perhaps the most under-utilized resource. If the least resource-, energy-and labour-intensive and the least polluting production technology known anywhere were generally available, everybody could be much better off. The millitary research laboratories from the cold war could now focus on solving pressing global problems of the environment, development and peacekeeping. They could be linked into a global 'Network of Applied Technical Universities and Research Establishments' (NATURE). bringing together former enemies in mutually beneficial projects.

3. ECONOMIC DEVELOPMENT

3.1 Conflict

A typical source of conflict is the pricing of raw materials. Exporters prefer higher and importers lower prices, resembling a zero-sum game. Yet both sides share a common interest in eliminating excessive price fluctuations, for example, by maintaining buffer stocks that are replenished when prices drop below a trigger point, and released when prices rise too high. Since the sales price exceeds the purchase price, such a program can usually pay for itself, or even yield a profit to be shared among importers and exporters. Price stability facilitates future planning for all. Such 'win-win' solutions are most likely to help achieve mutual agreement.

3.2 Lack of feedback

If information about people's suffering does not reach those who can help, help will not be forthcoming. For example, a famine in China in the early 1960s, from which an estimated 10 million people died, was long hidden from the outside world because of strict government censorship. If pictures of starving people had been shown around the world, aid could have been mobilized to save them.

One of the main obstacles to development is corruption. As long as it is easier to get rich by controlling the army and police than by producing goods in demand, it is more tempting to plot coups than to invest. Corruption thrives in a climate of secrecy, where officials can make arbitrary decisions that are not subject to public scrutiny. Openness and competition render corruption impossible. For example, if an official of the central bank can allocate scarce foreign exchange arbitrarily, there is great temptation to give it to friends, or to those who offer the highest bribe. If the available amount is publicly auctioned to the highest bidders, there is no room for corruption. Anyone who offered a bribe could not afford to be simultaneously among the highest bidders. Openness is the enemy of corruption. If the diversion of development aid into private hands is reported in the world press, donor nations will be reluctant to provide more aid. An international people's movement against corruption, *Transparency International,* has recently been founded. It fights corruption by exposing it, in a similar way as Amnesty International has been ιemarkably successful in reducing human rights abuses, or Greenpeace in protecting endangered species and embarrassing polluters.

3.3 Distorted feedback

If companies that run a deficit receive subsidies and those that operate efficiently must hand over their profits to the state, as in some former centrally planned economies, there is little incentive to be efficient. The savings and loan crisis in the United States has similar roots. Savings and loan institutions did not have to hand over their profits to the government, but the government protected them against losses. This is like saying to a child in a gambling casino, 'Here is $ 100, go and play. If you win, keep it. If you lose, come back for more money.' This is an invitation to reckless risk-taking, and the consequences were foreseeable, even if not foreseen.

Unequal income distribution distorts the reflection of true needs in a market, so that it is more profitable to produce food for the pets of the rich than for the children of poor people. Such distortions need to be corrected, through income redistribution, or by subsidizing essential goods so that everybody can afford them, and taxing luxury items to raise the revenue necessary to pay for the subsidies.

3.4 Delayed feedback

Too little is generally invested, whether in plant and equipment or education, because the fruits of the investment may come only many years later. In theory, the prospect of finding a good job as an adult should motivate children to study hard, but it often fails to do so, because playing is more fun. A solution is to make elementary school free and compulsory, and to give grades as an instant feedback for good efforts.

The market often operates too slowly as a regulating mechanism. For example, if there is a shortage of doctors, the price of their services will rise, and this may encourage more students to enter medical school. But by the time they have passed through medical school, completed their practical training and can help relieve the shortage, half a generation may have passed. That adjustment mechanism is too slow. It is necessary for governments to engage in some long range planning to meet society's future needs. This does not mean that particular individuals need to be told to enter medical school, but additional fellowships can attract more students to medical school in time, before a shortage of doctors begins to hurt, while still leaving individuals free to choose any field of study.

3.5 Rejected feedback

Even when economic cooperation would be mutually beneficial, immediately, it may not come about because of old enemy images. Building confidence by starting with small but easily successful joint projects may be the most likely path toward overcoming such prejudice.

Companies may fail to hire the most qualified employees because of racial or class prejudice. Laws prohibiting discrmination may be needed to help overcome such unfair practices.

3.6 Lack of remedies

The greatest obstacle to development is probably a lack of knowledge and resources. Sharing technology can go a long way toward reducing world poverty, but there is also a need for massive transfers of financial resources to meet people's most urgent needs and to build necessary infrastructure projects. The late Jan Tinbergen (in Tinbergen and Fischer 1987,p. 157-58) proposed the creation of a World Treasury. To almost every ministry at the national level, there is some corresponding international organization, except for the treasury. Yet without a treasury, which collects taxes to finance the other ministries, any government would soon collapse.

Funds could initially be raised by auctioning rights to mineral exploration on the deep seabed, outside of any national jurisdiction, to the highest bidder. In this way, the richer countries would automatically tend to pay a higher share of global revenue, without the need for long and difficult negotiations about national assessments. In addition to raising revenue, such an orderly allocation procedure could also help prevent future wars over those resources.

Other potential sources of global revenue might include a carbon tax to stem global warming, and a tax on currency exchanges to help dampen the high short-term volatility of international exchange rates, as proposed by Tobin (1974). Disarmament and economic conversion could make substantial resources available for development (Dumas 1986; Leontief and Duchin, 1983; Sivard 1986). To mention just one example, UNICEF (1990) has estimated that it would cost about $180 million per year to inoculate all of the world's children against the six major childhood diseases, from which nearly three million children under age 5 die each year, less than 10 percent of the price of a single U.S. stealth bomber.

4. PEACE

4.1 Conflict

The basic cause of wars is clearly disagreement over goals, or over strategies to reach them. The ideal approach to war prevention is to reduce conflicts as much as possible through international cooperation. Sherif and Sherif (1969), in a famous experiment in social psychology, could show how inter-group conflicts can be overcome effectively through what they called 'super-ordinate goals,' goals that are in the common interest of two parties, but can only be achieved through mutual cooperation. The same can be observed at the international level. For example, when the United States and the Soviet Union faced the common threat of Hitler Germany, they became allies. But it is not necessary to have common enemy to cooperate. Many global problems can be solved only through global cooperation, in the strong interest of all.

Despite the best efforts toward international cooperation, occasional conflicts are probably inevitable, but they need not lead to violence. There exists a whole range of methods of conflict resolution (Deutsch 1973; Fisher and Ury 1981; Isard and Smith 1982).

The easiest is often if one side make a *unilateral concession.* For example, when a chemical fire near the Rhine river in Basel poisoned the drinking water supplies in neighboring countries, Switzerland did not wait to be taken before the World Court by its neighbours, or even negotiate about the level of compensation, but simply accepted full responsibility for the costs of the cleanup.

If two sides need to take coordinated steps, *negotiations* are the first approach. As Mahatma Gandhi has emphasized, one must see the conflict as a common problem, in whose solution both sides have a role to play. Approaching the other side with the attitude 'I know the solution and all you have to do is to listen to me and do as I tell you' does not work. The solution process must itself resemble the proposed solution. If we seek a situation without violence, we cannot use violent means to achieve it. If we seek an egalitarian solution in which both sides have an equal voice, it cannot be imposed by one side.

Sometimes, a conflict has reached such an acute stage that face-to-face negotiations tend to degenerate into a stream of fruitless mutual accusations. A third party can play a useful role in such cases. Third party involvement includes the offer of good offices, mediation, or arbitration.

The least intrusive role is that of providing good *offices,* for example by hosting a conference between belligerent parties on neutral ground. A skillful *mediator* can often discover win-win solution that bring an improvement to both sides. It is also easier to accept a proposal of a third party without loss of face than to give in to the demands of one's opponent. The most instrusive third party role is that of an arbitrator, whose verdict is binding. To be effective, an arbitrator must seek to build up a reputation of fairness and neutrality.

The International Peace Academy is designed to mediate conflicts before they erupt in war, and to train mediators from around the world. But it has only sixteen staff members, only three of whom can do any field work. Compared with the millions of men under arms, this is totally inadequate. It desperately needs and deserves greater resources. It should not only address the most urgent and intractable conflicts, but should be used on a regular basis, even for minor disputes, so that its staff can gain experience and the world can gain growing confidence in its impartiality and competence.

Mediators play the most useful role if they succeed in making themselves unnecessary, for example by suggesting better procedures and institutions for conflict resolution to the parties involved, so that the parties will be able to resolve future disputes by themselves, without having to rely constantly on interventions by a mediator. Ury et at.(1988) have called such an approach 'dispute systems design'.

If a third party is unsuccessful in helping settle a dispute, the two parties may go before the *World Court.* The World Court needs the power to enforce its decisions (Ferencz and Keyes,1988). Sanctions need not mean the exercise of force. For example, the International Bank of Settlements in Basel, Switzerland, functions effectively in settling disputes between creditors and debtors, not because it has an army to enforce its decisions, but because creditors may never get their loans repaid, and debtors may be unable to secure new loans, unless they adhere to the verdict.

4.2 Lack of feedback

Wars sometimes begin out of misunderstandings or miscalculations. Better communication at many levels can sometimes help clear up potential misunderstandings. Heads of state and other government officials should meet at regular intervals, not only in times of crisis when meetings may become unproductive. Frequent contacts between ordinary citizens are equally important.

An International Satellite Monitoring Agency could play a useful role in detecting preparations for aggression and thus in helping to prevent it. Countries with purely defensive intentions have no reason to hide them. In fact, secret preparations for defense would be practically useless, because they would fail to deter aggression.

Satellites could also monitor the earth's environment and provide early warning about bad harvests, cyclones, changes in flora, fauna and the global climate.

4.3 Distorted feedback

Oskar Morgenstern pointed out that if those who make decisions about war or peace had to fight at the front-line in case of war, there would be fewer wars. National leaders and top military commanders usually try to make sure that they are well protected far behind the front-line or, if necessary, in deep shelters.

Immanuel Kant (1795) argued that governments elected by the people—who would suffer in case of war—would not go to war. This has not been fully born out. Democracies have not only defended themselves, but also intervened militarily in other countries on many occasions. However, so far there has never been any war between two democracies (Doyle 1983; Bremer 1991; Russett 1993). Therefore, if more and more countries have representative governments and decisions about war or peace are no longer taken in secret, war may greatly diminish, if not disappear entirely.

Covert operations to overthrow duly elected foreign governments would hardly take place if there had been an open public debate and vote on them. 'To view these events simply as misjudgments,' writes Michael Shuman (1991,P. 103),'... is to miss the larger point. Scandals and bad judgment are inevitable without vigorous public scrutiny, and the more they are covered up, the worse they become.'

The principle of self-determination—letting those make decisions who bear the consequences—an also help avoid internal conflicts. Whereas the British government has unsuccessfully tried to suppress the conflict between Catholics and Protestants in Northern Ireland with military force, a similar conflict has been ended peacefully in Switzerland by letting the people of the region decide their future by themselves. For many years, the French-speaking Catholic minority in the Jura region of the Canton Bern felt constantly overruled in parliament by the German-speaking Protestant majority. A separatist movement formed and some politically motivated cases of arson began to appear in the 1950s. If nothing had been done, this might have escalated into a violent conflict similar to the one in Northern Ireland. After some delay, the cantonal government of Bern finally agreed to hold a referendum on separation. The first vote, held in the region as a whole, was about evenly split. So the government organized separate votes in the six districts. Three districts had a majority in favor of forming their own Canton and were allowed to do so, while the other three districts preferred to remain with Bern. Everybody felt that they got what they wanted, and the conflict has subsided.

Democracy is no guarantee that people will always make the optimal decision. But if they make a mistake, they have nobody but themselves to blame and can learn from it. However, if a government forces them to do something against their will, and it turns out to be a mistake, they will naturally direct their anger against those authorities. Democracy helps avoid such conflicts.

Another form of externality contributing to wars is the 'security dilemma' (Jervis 1976; Buzan 1983), the fact that if a country acquires more weapons to improve its security, it tends to reduce the security of potential adversaries. One way to break our of this dilemma is non-offensive defense (Afheldt 1976; Fischer 1984a, Galtung 1984, UNIDIR 1990).

It is important to distinguish between two concepts of 'strength' that have long been confused: 'defensive strength' as the ability to resist harm that others may try to inflict on us, and 'offensive strength' as the ability to inflict harm on others. What is most helpful in avoiding war is not to be 'strong' or 'weak' in general, but to be strong in the sense of invulnerability, and weak in the sense of not being a threat to others. The concept of 'balance of power' which is still assigned great importance in theories of international relations, fails to make the essential distinction between offensive and defensive power. A balance of forces is neither necessary not sufficient for stability (Fischer 1982, 1984b).

4.4 Delayed feedback

One cause for the escalation of conflicts and for arms races, which are a major factor increasing the likelihood of war, is the 'fallacy of the last move.' Decision-makers tend to see immediate gains from acquiring larger quantities and more powerful and sophisticated types of arms, or from mobilizing their forces during a crisis, but often overlook the even greater loss to their country's security when their opponents respond in kind. Brams (1985, p. 66-78) shows that if decision-makers look beyond their own next move to an adversary's possible counter-moves, apparently tempting strategies often turn out to be unattractive in the longer run. A more perceptive reading of history can also teach the dangers of vindictiveness. A badly humiliated adversary may seek an opportunity for revenge, leading to future wars, as the experience of Versailles has shown. After World War II, the Marshall plan helped turn the United States' former enemies into allies. Similar massive aid to the former Soviet republics and to Eastern Europe is now needed to help their democratic revolutions succeed.

Rather than waiting for a conflict to erupt into war and then resorting to military force, it is far better to pursue an *active peace policy* that seeks to anticipate and avoid or resolve conflicts long before they lead to war (Fischer 1991). For example, the United States was able to persuade General Pinochet in Chile to hold a referendum on his continued rule and to step from power when he lost it, instead of waiting until a civil war may erupt and then intervening militarily. Such preventive steps are too rare.

4.5 Rejected feedback

Even the best feedback cannot help if a decision-maker chooses to ignore it, out of what others would consider irrational motives. Leaders suffering from megalomania or paranoid fears of conspiracy have often led their countries into war, against the best advice. According to attribution theory (Jones 1973), most people tend to attribute good motives to themselves and bad motives to their adversaries, or to those they may falsely believe to be their adversaries. Jervis (1976,p.170) writes, 'One tends to see what one believes'. Kull (1988) interviewed nuclear strategic planners in East and West and found that they made very different assumptions-unchecked by experience—about how their opponents would behave in a crisis than they themselves would. They were convinced that an opponent would yield to threats, but that they themselves would never do so. Such misperceptions can lead to the escalation of conflicts. Being made aware of such inconsistencies is the first step toward overcoming them.

War has long been justified by the glorification of victory in war in history books. Kenneth Boulding pointed out that the glorification of victory in duelling ended when guns became more accurate and duelling almost inevitably lethal. Modern weapons have made the glorification of war equally obsolete. UNESCO has undertaken a project to write international history textbooks, which are not nationally biased, omitting the vilification of 'enemies'.

4.6 Lack of remedies

Often, those who wish to prevent war most desperately lack the means to do so. A lack of means has two aspects: lack of ideas, and lack of instruments.

To help solve conflicts before they lead to war, Galtung (1980, p.359) proposes that 'during international crises..there is a general scarcity of good ideas.... The suggestion is ...to set up, over a telesatellite communications system... in as many countries as would care to participate... an international panel of experts to tap their brains. He argues that... such discussion programs would constitute a major element of attraction in times of crisis, exactly because people feel... so much victimized by the propaganda machineries they are exposed to (ibid.,p.363).

To deter or reverse aggression, a United Nations Peacekeeping Force can play an important role, at considerably lower costs than if each country maintains its own armed forces. It would be equally wasteful if all the home owners in a small town maintained individual fire engines, instead of combining their resources to form one fire company that can be deployed wherever and whenever it is needed. In addition to dissuading aggression and maintaining cease-fires in civil wars, a standing UN Peacekeeping Force could also help protect lives in case of natural or industrial disasters, in providing temporary food and shelter to refugees, and in restoring a healthy environment. If a UN Peacekeeping Force consists of individually recruited members, rather than national contingents dispatched by governments, their primary loyalty will be to the United Nations, rather than to their own national command structure (Johansen 1990,p.238).

5. CONCLUDING REMARKS

The six basic defects of regulatory feedback mechanisms outlined here can be seen as sources of numerous problems from which the international community suffers. This helps point the way toward general solutions. Similarly, the great variety of natural phenomena in the universe, including life in the earth's biosphere, can ultimately be reduced to complex interactions of the four elementary forces of nature.

The growing interdependence of the world has given rise to a series of global problems that individual states can no longer solve by themselves. Only through worldwide cooperation can we cope with such issues as preventing climate shifts, stemming the international drug trade, or preventing nuclear terrorism. At the same time, improvements in global transportation and communication have made such global cooperation easier.

Many governments are still reluctant to join a global authority to deal with global problems out of fear that they would lose part of their national sovereignty. But that fear is mistaken (Mische and Mische 1977). No country today, for example, has sovereign control over the ozone layer. By joining a global authority that can allocate emission quotas and enforce them, we do not give up control ever our destiny. On the contrary. we gain added control that we do not now posses and could never achieve at the national level.

The first advanced civilizations emerged about 6,000 years ago in the Nile and Euphrates valleys when farmers faced problems that they could not solve alone. To prevent recurrent floods and droughts, it was necessary to build dams to control the flow of those rivers, requiring the organized cooperation of thousands of individuals. This gave rise to the first states, the development of written language, the codification of laws, and a flourishing of science and the arts. Today we face some problems that not even a superpower can solve by itself. Hopefully, this will lead to greater international cooperation before it is too late.

Modern science and technology has given humanity the opportunity to overcome age-old problems of hunger, disease, and poverty. It has also made it possible for us to do far more harm to nature and to ourselves than ever before. The late physicist Richard Feynman said he once met a Buddhist Monk who told him something he never forgot: 'Humanity possesses a key that can open the gates to heaven. But the same key can also open the gates to hell.' The choice is ours.

References

Afheldt, Horst (1976) *Verteidigung und Frieden.* (München : Hanser.)

Baumol, William J., and Wallace E. Oates (1975) *The Theory of Environmental Policy.* (Englewood Cliffs, New Jersey : Prentice Hall.)

Boulding, Kenneth E. (1989) *Three Faces of Power.* (Newbury Park, California : Sage Publications.)

Brams, Steven J. (1985) *Superpower Games : Applying Game Theory to Superpower Conflict.* (New Haven and London : Yale Univ. Press.)

Buzan, Barry (1983) *People, States and Fear : The National Security Problem in International Relations.* (Chapel Hill : The University of North Carolina Press.)

Chestnut, Harold, ed. (1986) *Contributions of Technology to International Conflict Resolution.* Oxford : Pergamon Press.

Deutsch, Morton (1973) *The Resolution of Conflict : Constructive and Destructive Processes.* (New Heaven : Yale University Press.)

Doyle, Michael (1983) 'Kant, Liberal Legacies, and Foreign Affairs,' *Philosophy and Public Affairs,* Vol.12, No.3-4, pp. 205-35, 323-53.

Dumas, Lloyd J. (1986) *The Overburdened Economy.* (Berkeley : University of California Press.)

Ferencz, Benjamin B., with Ken Keyes (1988) *Planethood.* (Coos Bay, Oregon : Vision Books.)

Fischer, Dietrich (1982) 'Invulnerability Without Threat : The Swiss Concept of General Defense,' *Journal of Peace Research,* Vol.19, No. 3, pp. 205-225.

Fischer, Dietrich (1984a) *Preventing War in the Nuclear Age.* (Totowa, New Jersey : Rowman & Allanheld.)

Fischer, Dietrich (1984b) 'Weapons Technology and the Intensity of Arms Races,' *Conflict Management and Peace Science,* Vol. 8, No. 1, pp. 49-69.

Fischer, Dietrich (1991) 'An Active Peace Policy,' in Shuman, Michael, and Julia Sweig, eds., *Conditions of Peace : an Inquiry.* (Washington, DC : The Exploratory Project on the Conditions of Peace.)

Fischer, Dietrich (1993) *Nonmilitary Aspects of Security : A Systems Approach.* (Aldershot, England and Brookfield, Vermont : Dartmouth Publishing Co.)

Fisher, Roger, and William Ury (1981) *Getting to Yes : Negotiating Agreement Without Giving In.* (Boston : Houghton Mifflin.)

Galtung, Johan (1980) *The True Worlds : A Transnational Perspective.* (New York : The Free Press.)

Galtung, John (1984) *There Are Alternatives ! Four Roads to Peace and Security.* (Nottingham, England : Spokesman.)

Gorbachev, Mikhail (1987), *Perestroika : New Thinking for Our Country and the World.* (New York : Harper & Row.)

Isard, Walter, and Catherine Smith (1982) *Conflict Analysis and Practical Conflict Management Procedures.* (Cambridge, Mass.: Ballinger.)

Jervis, Robert (1976) *Perception and Misperception in International Politics.* (Princeton : Princeton University Press.)

Johansen, Robert C. (1990) "Toward Post-Nuclear Global Security : An Overview," in Weston, Burns H., ed., *Alternative Security : Living Without Nuclear Deterrence.* (Boulder, Colorado : Westview Press.)

Jones, Edward (1973) *Ingratiation : An Attributional Approach.* (Morristown, New Jersey : General Learning Press.)

Kant, Immanuel (1795) *Zum ewigen Frieden.* Königsberg. Translated as *Perpetual Peace* (1796). Reprinted 1972.

Kull, Steven (1988) *Minds at War : Nuclear Reality and Inner Conflict of Defense Policy-Makers.* (New York : Basic Bokks.)

Leontief, Wassily, and Faye Duchin (1983) *Millitary Spending : Facts and Figures, Worldwide Implications and Future Outlook.* (Oxford : Oxford University Press.)

Mische, Patricia and Gerald (1977) *Toward a Human World Order.* (New York : Paulist Press.)

Ramsey, Frank P. (1928) "A Mathematical Theory of Saving," *Economic Journal.*

Russett, Bruce (1993) *Grasping the Democratic Peace : Principles of a Post-Cold War World.* (Princeton : Princeton University Press.)

Sagan, Carl (1983) "Nuclear War and Climatic Catastrophe," *Foreign Affairs*, Vol. 62, No. 2, pp. 257-92.

Sherif, Muzafer, and Carolyn Sherif (1969) *Social Psychology.* New York : Harper & Row.

Shuman, Michael (1991) "A Separate Peace Movement : The Role of Participation," in Shumna, Michael, and Julia Sweig, eds., *Conditions of Peace : an Inquiry.* Washington, DC : The Exploratory Project on the Conditions of Peace.

Sivard, Ruth Leger (1986) *World Military and Social Expenditures, 1986.* (Leesburg, Virginia : World Priorities.)

Tinbergen, Jan, and Dietrich Fischer (1987) *Warfare and Welfare : Integrating Security Policy into Socio-Economic Policy.* (Brighton : Wheatsheaf.)

Tobin, James (1974) *The New Economics One Decade Older.* (Princeton : Princeton University Press.)

UNICEF (1990) *The State of the World's Children 1990.* (Oxford : Oxford University Press.)

UNIDIR (1990) *Nonoffensive Defense : A Global Perspective.* (New York : Taylor and Francis.)

Ury, William L., Jeanne M. Brett and Stephen B. Goldberg (1988) *Getting Disputes Resolved : Designing Systems to Cut the Costs of Conflict.* (San Francisco : Jossey-Bass.)

Chapter 27

Peace Dividend, Five Years After the End of the Cold War

Albrecht Horn

THE PEACE DIVIDEND IN THE CONTEXT OF INTERNATIONAL SECURITY

A. Security policies, military expenditures and socio-economic development

The international system comprises sovereign nation-states with legitimate national security interests. The traditional national military security concept is based on the assumption of external threats. The protection of the sovereignty and territorial integrity of the nation-state is the core of legitimate national security concerns.[1] Inter-state conflicts threaten the international security. The maintenance of international security is based on national security strategies and collective security arrangements. The policy objective is to ensure international and national security with a low level of security expenditures. Military expenditures are the essential core of security expenditures.[2]

Military expenditures, as core of the security expenditures, sustain legitimate security policies based on realistic assessments of external (threats assessment, determination of the potential needed to meet the threats).[3] The level and composition of military expenditures reflects the assessment of the external and internal security environment. Positive changes in the security environment like reduced levels of inter-state conflicts, bi- and multilateral disarmament agreements, reduction of domestic conflicts permit the reduction of military expenditures for national security purposes and thus lead to savings.[4]

The determinants of the level of military expenditures comprise a variety of political, strategic, security and economic factors which reflect the concrete security environment. Any change in these factors affects the level of military expenditures and thus determines the scope of possible savings (peace dividend).[5]

Economic factors like tight budgets act as a constraint on the level of military expenditures. The realization of a peace dividend requires positive changes in the determinants of military expenditures, specifically in the functioning of collective international security regimes. National and international security, disarmament and socio-economic development are closely related. The level of military expenditures, indicates the state of the national and international security situation and has economic implications. This linkage is manifold:

(*i*)   Legitimate security policies and concomitant expenditure levels ensure a stable political framework for socio-economic development by removing the external threats to sovereignty and territorial integrity. They can also reduce the level of domestic conflict.

(*ii*) The level of security expenditures, as part of central government expenditures, affects other macroeconomic variables. Excessive military spending has a negative effect on socio-economic development. Legitimate security concerns have to be met at lowest possible levels of expenditures.

(*iii*) Disarmament measures form a specific element of security policies affecting the security environment and the level of military spending. Military spending levels are also affected by bi- and multilateral security arrangements, conflict prevention and conflict settlement strategies, multilateral interventions in cases of threats to peace and international security as well as by the reduction of domestic conflicts.[6]

The 1987 UN Conference on "The relationship between disarmament and development"[7] addressed this specific dimension of the link between security and socio-economic development.

The Final Declaration stressed:

(*i*) the contrast between global military expenditures and socio-economic needs, especially in developing countries;

(*ii*) that over-armament, excessive military spending and under-development constitute threats to national and international security;

(*iii*) that military expenditures and socio-economic development compete for the same finite resources both at the national and international level;

(*iv*) that considering resource constraints in both developed and developing countries, reduced world military spending could contribute significantly to socio-economic development;

(*v*) that increased security provides for reduced military spending and this contributes to socio-economic development. Reduced military spending, undiminished security at low levels of military spending permits to address non-military, socio-economic and ecological threats to security;

(*vi*) that the effective implementation of the collective security provisions of the Charter of the United Nations would enhance international security. This could reduce the need for member states to seek security by exercising their inherent right of individual and collective self-defense. The judgement as to the level of arms and military expenditures essential for security rests with each nation. However, the pursuit of national security regardless of its impact on the security of others can create overall international insecurity. Security should be established at lower levels of armaments;

(*vii*) that non-military, socio-economic and ecological threats endanger increasingly national and international security. The level of global military spending in pursuit of security interests in disproportionally high compared for instance with other government expenditures and levels of official development assistance (ODA);

These basic premises-formulated in 1987-are still relevant.[8]

(B) Changes in the international security environment after the end of the Cold war[9]

The end of the East-West conflict has led to dramatic changes in the international security system. This affects the need for military spending levels and opens opportunities for tangible reductions. The transition from a bipolar to a multipolar security regime has reduced the dangers of a nuclear East-West conflict and has ended proxy wars in developing regions conducted in the context of the East-West confrontation. At the same time constraints on regional conflicts have been removed and the potential for new conflicts has increased (inter-state as well as intra-state conflicts). The need for multilateral security policies (increased role for the UN in peacemaking, peacekeeping and peace enforcement) has grown. Cooperative and collective security systems are increasingly required and gradually take shape.

During the Cold War, the foreign policy and security concerns of the major powers fostered high military budgets and the growth of defense industries and arms transfers. The costs of these policies were extremely high. This was also the case for many developing countries drawn into the East-West conflict. This trend caused serious imbalances between military budgets and other public expenditures vital for economic growth and development.

Despite conflicting trends the changes in the security system after the end of the Cold War and the cessation of the East-West conflict created a basis for reductions of military spending and realisation of savings (peace dividend).

Developed Countries

In industrialised countries, especially the U.S. and Western Europe, the removal of the threat perceived to come from the Warsaw Pact opened possibilities for substantial reductions in military spending. At the same time new emerging, mostly ethnic conflicts, in the former Soviet Union and Yugoslavia, as well as uncertainties about future european security structures prevented a more drastic reduction.

The security environment for the former member states of the Warsaw Pact changed dramatically. The alliance collapsed and the perceived threat from NATO disappeared. At the same time, the breakdown of the system of centrally planned economies and the transition to market based economic systems exerted economic pressure to reduce bloated military budgets. The emergence of conflicts between some of the new states exerted counter-pressure.

Developing Countries

The collapse of the Cold War Security System has not only altered relationships between the US and the former Soviet Union, but global power arrangements as a whole. The post Cold War security doctrines have also shaped security policies in developing countries. Some of these countries were previously drawn into geostrategic spheres of influence and often were engaged in proxy-wars.

The end of the Cold War had new and enormous implications for the majority of small and medium-sized states belonging to the non-aligned movement. New types of intervention threaten their legitimacy and security. The old 'low intensity wars' fought are giving way to new types of conflicts conducted by regional powers. Non-military threats also undermine the economic base of security. Old regional rivalries continue also without the influence of outside powers. New instabilities have emerged and caused increased multilateral efforts to implement global or regional collective security regimes. The serious economic situation has also exerted pressure to reduce military budgets in developing countries even without specific changes in security structures.

The East-West conflict had, at least partially, also a restraining impact. With the removal of this restraint, there is potential for new types of conflicts based on ethnic rivalries and socio-economic tensions. This requires adequate national and international economic policies as well as innovative global and regional security regimes and conflict prevention and solution strategies to avoid these conflicts and new emerging threats to the national security of many developing countries.

These contradicting trends with regard to the security environment of developing countries affects military spending patterns. Positive changes in the security structures and economic pressures determine the basic trend of military budget reductions despite the existence of counteracting factors.

The Peace Dividend, Expectations and Realities

Definition, measurement, magnitude and allocation of the Peace Dividend

Nature and definition

The public discussion on magnitude, allocation and economic consequences of the Peace Dividend suffers from a lack of clear assumptions and definitions. The main deficiencies are:

> (*i*) The insufficient clarification of determinants of military expenditures, their change and the quantification of the impact on the magnitude of savings.

(*ii*)   The imprecise definition of the term peace dividend and its reduction to international transfers (reduction of military spending in developed countries and use of part of released resources for increases in official development assistance).

(*iii*) The identification with a simple reallocation from defence towards other types of government spending (simple trade-off). This simplified approach neglects the necessary adjustment processes leading to short-term costs and long-term benefits. The macroeconomic implications of changes in military spending levels and the implications for international trade and financial flows need further refinement of the models used to determine these relationships.

(*iv*)   There was no sufficient analysis of the different options for allocating the Peace Dividend (at the national and international level) and the resulting economic consequences. In the public discussion, the simplistic trade-off between military and social expenditures dominated without sufficient regard for the complex linkages and the trade-off between short-term adjustment costs and long-term benefits.

The term *Peace Dividend* has to be more precisely defined in order to conduct a realistic analysis of the determinants, magnitude and economic consequences.

The *National Peace Dividend* comprises savings, resulting from reductions in military spending at the national level. These reductions are caused by positive changes in the determinants of military expenditures. The National Peace Dividend has to be realized in developed and developing countries depending on positive changes in the security situation (international and domestic).

The *International Peace Dividend* refers to reductions of military spending in developed countries and use of part of the released resources for increases in official development assistance. The International Peace Dividend thus relates to the international dimension (international transfers). This specific international dimension of the peace Dividend depends on the political willingness to contemplate such additional international transfers and the availability of adequate transfer mechanisms. Beside of additional international transfers (developmental peace dividend) there are also international positive externalities resulting from national peace dividends (allocation of part of the peace dividend towards deficit reduction leads to reductions in interest rates).

The *Disarmament Dividend* (as part of the Peace Dividend) refers to reductions in military spending at the national level made possible through specific uni-, bi- and multilateral disarmament agreements as part of security policies. The size of potential savings depends upon the specific nature of the agreement. The allocation of the Disarmament Dividend can be nationally or internationally. The Disarmament Dividend can thus form part of the National or International Peace Dividend depending on the allocation modalities.

A more sophisticated determination of the Peace Dividend would count the short-term adjustment cost vis-a-vis the long-term benefits. This is at present not feasible due to lack of data and methodological difficulties.

Measurement, magnitude and allocation

According to the definitions, the measurement of the National Peace Dividend is based on the comparison of national military budgets for different years.[10] A base year is chosen (in this case the budget levels before the end of the Cold War). The assumption of changes in the determinants (security situation) leads to reductions of military spending in the following years. The National Peace Dividend is determined by the addition of the reductions in the years following the base year. The savings (reductions) are derived from a scenario where the military spending levels of the base years would have continued.

Taking the base year 1987 and trends until 1994 show the following patterns:[11]

Industrialized countries[12]

-	Actual military spending 1987-1994	5.933 bin US Dollars
-	Military spending level if the 1987 spending patterns would have continued	6.740 bin US Dollars
-	Reductions (savings) National Peace Dividend 1987-1994	807 bin US Dollars

Developing Countries

-	Actual military spending 1987-1994	1.034 bin US Dollars

-   Military spending level if the
    1987 spending patterns would
    have continued   1.160 bin US Dollars
-   Reductions Peace Dividend
    (savings), National 1987-1994   126 bin Dollars

The total Peace Dividend (sum of National Peace Dividends) thus amounted to 933 bin US Dollars from 1987-1994.

If one assumes a further 3 per cent reductions from 1995-2000, under the condition that further improvements in the security system are possible (positive trend in determinants), 459 bin US Dollars (386 bin US Dollars in developed countries and 73 bio US Dollars in developing countries) additional Peace Dividend seems a realistic target. This indicates the potential amounts to be released from military budgets under certain assumptions.

The *allocation* of the National Peace Dividends comprises inter alia the following *options*:

(*i*)   Allocation toward other types of public expenditures within unchanged overall budget levels.

(*ii*) Allocation for reductions in budget deficits.

(*iii*) Allocation for the purpose of reductions of tax levels.

(*iv*) Allocation for increasing official development assistance (international allocation).

In reality, there are specific national allocation patterns, combining the basic options, taking into account to the magnitude of the National Peace Dividend and the national priorities for allocations. These patterns have to be examined in order to determine the macroeconomic implications for different countries. (s. Chapter III)

The allocation for international transfers depends on the political willingness to use part of the released resources for increasing official development assistance. The stagnating levels of officials development assistance indicate that no such additional international transfers have occurred.[13] The International Peace Dividend has, therefore, not materialised, despite potential reductions in military spending levels in developed countries after the end of the Cold War. The reasons for this negative trend are the preponderance of national allocation priorities (reduction of budget deficits, social spending needs in developed countries) and a certain degree of "donor fatigue" due to insufficient positive effects of official development assistance flows in the past.

B. Expectations, realities and prospects with regard to the Peace Dividend

Expectations

The expectations concerning the magnitude of the Peace Dividend after the end of the Cold War were unrealistically high. The expectations of the magnitude of potential savings and their use were based on general assessments and not on an exact analysis of the determining factors, their possible change and the resulting reductions in military spending levels.

Unrealistic expectations referring to the use (allocation) of the Peace Dividend neglected such allocation options like use for reductions in budget deficits and adjustment of tax levels. The prevailing view was that there would be a direct reallocation toward social spending within national budgets. This approach also underestimated the short-term adjustment costs and the complex macro- and microeconomic linkages initiated by reductions in military spending.

The identification of the Peace Dividend with the International Peace Dividend (narrow interpretation) caused also misunderstanding and misinterpretations. This false identification forms the basis of many discussions in multilateral fora.

This erroneous interpretation led to conclusions of a 'lack of the peace Dividend'.

Realities

The unrealistic expectations concerning magnitude, domestic allocation and international transfers have, in consequence, led to expressions either of a "non-existent Peace Dividends" or "insufficient tangible economic gains". A more *realistic assessment* indicates that:

(i) There was a National Peace Dividend of remarkable proportions already realized as consequence of positive changes in the security environment after the end of the Cold War (the larger part was realized in developed countries).

(ii) The National Peace Dividend was used domestically for a variety of purposes, especially for fiscal consolidation and support for social services, mainly in developed countries.

(iii) The international dimension of the Peace Dividend did not materialize due to domestic pressures in developed countries and some degree of "donor fatigue". The political will to use part of the released resources for international transfers was lacking. This has led to legitimate expression of concern and frustration on the part of the developing countries.

Prospects

The *Prospect* for a further realisation of National Peace Dividends, in developed as well as in developing countries, depends primarily on favourable changes in the international security system and the resolution of inter-state conflicts. Conflicts resolution and conflict prevention strategies, regional and global cooperative security regimes, bi- and multilateral disarmament agreements and refinement of multilateral tools of statecraft (peacemaking, peacekeeping and peace enforcement) can all contribute to positive changes in the determinants of military spending levels and in consequence to tangible reductions in military spending levels. This offers the prospect of further National Peace Dividends of remarkable magnitudes.

The process of reallocation, with its short-term adjustment costs and long-term benefits, needs further analysis in order to specify the costs and benefits of different national allocation options, including possible international consequences (positive externalities, additional transfers).

The international allocation (International Peace Dividend) option needs to be rigorously pursued in order to create the political will and the adequate mechanisms for such additional transfers, including the assessment of their efficient use. This could be a real contribution toward addressing urgent socio-economic problems in developing countries and link security policies and socio-economic development more closely.

THE INTERNATIONAL PEACE DIVIDEND

(A) Military expenditure reductions in developed countries and the flow of official development assistance

The International Peace Dividend was defined as the use of part of savings resulting from military spending cuts in developed countries and their redirection toward development finance (increase of official development assistance).

This problem arose in the context of changes in the security environment after the end of the Cold War and related expectations about potential savings (National Peace Dividend in dividend countries and the international allocation option). This was supposed to link the end of the East-West Conflict with prospects for improving development cooperation.

The link between reductions in the level of military spending and development cooperation, especially development finance, is a long-standing issue in international discussions and negotiations. The issue was discussed in the context of UN activities on the relationship between disarmament, as an specific part of changes in the security policies, and socio-economic development. There have been numerous proposals that link reductions in military spending to development finance, specifically levels of official development assistance. The proposals date back before the end of the Cold War, but this historic event and its implications gave new impetus to these proposals.

France proposed, in 1978, the establishment of a 'disarmament fund for development'. Such fund could be alternatively financed from savings resulting from specific disarmament agreements (disarmament dividend approach), an arms levy (contributions as an agreed rate in proportion to military expenditures) or by voluntary contributions (contributions at the discretion of states, not predetermined as a ratio of disarmament savings or arms expenditures). The 1987 UN Conference on 'Disarmament and Development' agreed to further pursue the analysis of possible reductions of military spending levels and the use for development finance purposes.

The end of the Cold War revived this idea. The concept of the Disarmament Dividend was broadened to that of a Peace Dividend, providing for other specific changes in security regimes beyond. disarmament agreements and resulting savings (reductions of military spending levels in accordance with changes in the determinants of the security regime). The Disarmament Dividend forms a specific part of a broader defined Peace Dividend.

The concept of an International Peace Dividend as the international dimension of the 'Peace Dividend (use of part of the savings in developed countries for international transfers) got new and increased attention. The favourable developments in the security environment, as consequence of the end of the East-West Conflict, were cause for optimism concerning additional sources for development finance. Due to domestic pressures for resource allocation in developed countries and some degree of "general donor fatigue" no additional international transfers materialized despite sizable reductions of military budgets in developed countries.

The focus should be on further positive changes in the security regimes leading to further tangible reductions in military spending levels in developed as well as in developing countries. On this basis,it could be further examined how to create the conditions, modalities and mechanisms also for increased international transfers of released resources. This would be a positive contribution to new approaches in international development cooperation based on mutual interests.

(B) Military expenditure levels in developing countries and commitments for the provisions of official development assistance

Recent developments stress another link namely that between military expenditure levels in developing countries and ODA-commitments on the part of donor countries. This approach is caused by perceptions of excessive military spending and striking disproportions between military and other types of government expenditures, especially social expenditures in some developing countries (distortions in budget priorities).

The new approach proceeds from the fact that prior to the end of the Cold War, official development assistance was often granted without regard to political, economic and social consequences of military build-ups in developing countries. In the Cold War context foreign economic and military assistance supported levels of military expenditures in some developing countries which could not have been sustained by domestic resources.

After the end of the Cold War perceptions and interests have changed. The review of budget priorities and the proportion between military expenditures and other types of government expenditures has become part of policy reviews by international agencies like IMF, IBRD and OECD.

The OECD in the 'Orientations on Participating Development and Good Governance'[14] raises the issue of the military burden for the economies of developing countries. This requires an assessment of security policies which determine levels of military spending and the socio-economic consequences. The sensitive nature of the subject requires a measured approach. The level of military expenditures has to be evaluated in the context of the specific security environment and available options for positive changes. The terms excessive military spending can only be defined in relation to the real security threats facing states as well as in relations to their economic potential (military burden). This requires careful assessments and policy dialogues based on mutual trust. Any approach to impose additional conditionalities (defense conditionality linking ODA commit-ments to specific levels of military expenditures or specific reduction targets) would be counter-productive. A positive linkage, providing rewards and incentives for reductions in military expenditures seems more adequate. The policy dialogue should include the assessment of the security environment, adequate security policies to meet security threats, optimal levels of military expenditures and options for specific reductions, their allocation and economic consequences.

The United Nations should, in accordance with the mandate formulated at the 1987 UN Conference on "Disarmament and Development" and in light of new requirements as outlined in the proposal for an "Agenda for Development" continue:

(*i*) to assess security requirements and levels of military spending taking into account the need to keep military expenditures at the lowest possible levels;

(*ii*) carry out comparative analysis of different types of government spending and analyse ways to correct distortions in budget priorities;

(*iii*) carry out assessments of socio-economic consequences of changes in military spending levels (nationally and internationally);

(*iv*) to assess the implications of changes in security regimes for maganitudes allocation of the released resources (Peace Dividend, determinants, magnitudes, allocation options and economic implications);

(*v*) to analyse options, modalities and mechanisms for the use of parts of the Peace Dividend for additional development finance;

(*vi*) to elaborate specific reward and incentive schemes for the reduction of military expenditures.

This analysis can contribute to a productive policy dialogue on the link between security policies, security expenditures and socio-economic development patterns.

Notes

1. In addition to external threats, internal instability can also undermine national security (domestic conflicts).

2. For conceptual (definition), methodological (accounting, conversion, price indices etc.) issues refer to:
    -   Horn, "Security and Development", UN/DESIPA Working Paper 1994.
    -   Happe, "Military Expenditures", Peace Economics/Peace Science, Vol. 2 No. 1, 1994
    -   Herrrea, "Statistics on Military Expenditures in Developing Countries", OECD 1994.

3. The extended concept of national security, going beyond national military security, includes socio-economic and ecological threats to the socio-economic system of nation-states (national economic security).

4. Their perception of positive changes in the international security regimes leads to expectations of savings which are often termed as "Peace Dividend" (potential or real) or "Disarmament Dividend".

5. The use of the "stock-flow concept" could be envisaged in analyzing the existing stock of military assets and relating it to the "flow of military expenditures". This would presuppose to disaggregate the different elements of global military expenditure figures. (Procurement, R/D P. Wages)

6. The multilateral forms of statecraft like preventive diplomacy and peacemaking, peacekeeping, multilateral economic sanctions and peace enforcement are important tools to ensure effective international security and to contain security expenditures.

7.   . "International Conference on the Relationship between Disarmament and Development", Final Document, New York 1987.

      . "Disarmament and Development" Declaration by the Panel of Eminent Personalities", UN-New York 1986.

      . "Establishment of an International Disarmament Fund for Development", UNIDIR Geneva 1984.

8. Proposals for releasing resources by reductions of military spending at the national level, especially in developed countries, and use of part of the released resources for additional ODA flows (Disarmament for Development Fund based on a Disarmament Divided, arms levy or voluntary approach) have not been implemented due to political differences.

9. This is a short indication of change in the security environment based on the assessment by intenationally recognized experts. This is not the place to go into detailed analyses of specific regional problems. S. Singh/Bernauer "Security of Third World Countries", UNIDIR, Geneva 1993; Institute for strategic Studies "The Military Balance", London 1993/94.

10. An a Priori determination based on single determinants of military expenditures, their change over time and the quantification of the impact on reductions in military spending proves not possible due to methodological difficulties. It remains only to determine the National Peace Dividend by ex-post comparison of military budgets.

11. This is only for illustrative purposes. A more detailed calculation is undertaken in the following chapters (magnitude, allocation, economic consequences).

12. Based on figures on military expenditures as indicated in the 1994 UNDP Human Development Report and the SIPRI Yearbook 1994 (World Armaments and Disarmament).

13. It has also to be taken into account that private capital flows (commercial credits, foreign direct investment, portfolio investment) have gained a relative greater importance vis-a-vis ODA flows, at least for some developing countries. It is also a fact that the lack of an "International Peace Dividend" is due to end of Cold War clientelism.

14. OECD, Paris 1993.

Postscript : Management of Peace, Security and Development

Manas Chatterji

Introduction

Globalization of business, high technology and drastic changes in manufacturing processes are bringing people together. On the other hand, the demise of bipolarity and the growth of nationalism are tearing the fabric of nation states and opening up the old woud. The unfortunate example in Yugoslavia, Somalia, etc. have shown us how ugly violent is the world we live in. The international community faces the challenge of transferring the system of nations as we knew it to a new framework without violence. This process of transformation is interlinked with such factors such as democratic principles, environmental security, free trade, demographic balance, technology transfer, new financial and monetary system, human rights, settlements of refugees and host of other factors.

This does not only require a world wide body like the U.N. to keep peace and make peace, using force if necessary but also a new type of development policy fully funded by the developed countries. Controlling arms production in the developing countries and arms shipment from the development countries are urgently needed. According to some, sanctions should be used if necessary. The subject matter of sanctions is quite controversial. Without taking a stand I shall discuss it general nature, objective and problems

The major powers are reluctant to get involved in any war. To protect their economic and political interest (like anti-terrorism) they use sanctions. In many instances, United Nations (starting with South Africa) took recourse to economic sanctions. Individual countries also applied their own sanctions. Sometimes it is not clear about the motive of the sanctions. The goals of the imposer country may be public or hidden objectives or even some which the imposer itself is not aware of. Besides inflicting a punishment there may be other motives like

(1) setting the rules of acceptable behaviour of a state (2) impress other states about the justification of the step (3) responding to domestic politics or influences public opinion. (4) as an alternative to costly military intervention (5) and use sanctions as a bargaining chip and (6) reacting to the strategic advantage of an adversary.

Tariffs is also a kind of sanction. Although many times sanctions are intended against alleged terrorist countries, it may be mixed with economic motives. So a sanction as well as motive for sanctions are multidimensional. It is a legal question whether an individual country has a basis to declare an unilateral sanction against a country without going through the U.N. This matter needs discussion. If it is against the U.N. charter. international law or international agreements (like GATT) what should be done? There are many other related questions. But the basic question is whether the objectives of a sanction is ever achieved.

Before we discuss the efficacy of the sanction process, one point should be made clear. In most developing countries, the interest of the leaders and decision-makers are not the same as those of the people. So a sanction intended to punish a country may lead to the punishment of the people rather than the leaders and decision-makers. The hypothesis that if people suffer, they will protest and affect the decision process is also not valid in most. cases.

Hufbauer, Schott and Elliott (1990) give a list of .the economic sanctions adopted for the foreign policy goals during 1914-1990. There were instances of successful sanctions when the situation was favorable say (1) UK against former U.S.S.R. (1931) (2) U.S.S.R against Finland (1958) (3) Soviet Union against Lithuania (1990), etc. There were sanctions which did not succeed despite the situation was favorable. Some examples are (1) League of Nations against Italy (1935) (2) U.S.A., U.K. etc. against Japan (1939-41) (3) U.S.S.R. and Eastern block countries against former Yugoslavia (1948-53) (4) U.S. against Cuba (1960) (5) UN against Rhodesia (1966). It is not clear whether sanctions against Iraq is a failure or success. Makio Miyagawa (1992) lists the following conditions for a successful economic sanction both from the point of view of the target and imposing countries.

(1) dependence on trade

(2) size of the economy

(3) availability of substitutes

(4) trade partners

(5) foreign exchange resources

(6) monitoring

(7) economic systems

Factors limiting the efficacy listed by him are (1) no leakage, other countries are not eager to help the target (2) direct and indirect cost to the imposer (3) internal pressure groups (4) legal limitation (5) fear of war (6) shifting the power blocks (7) may strengthen the target through political/social integration rather than disintegration. There are also many methodological issues of sanctions (Hufbauer, Schott and Elliott (1990).

Although sanctions for arms embargo can succeed in the short run, in the long-run it can be counter productive. It may not lead to stability at a lower level of weapons possessed by the developing countries. It can also be used as an excuse by the developed countries. What is needed is a development aid policy to eradicate world wide poverty. This will serve as the carrot and sanctions linking development assistance to military spending can be used as the stick. for that, we need a strong, financially viable United Nations with military power.

The first and foremost task is to reduce conflict. For this purpose I suggest the establishment of an International Training of Conflict Prevention and Management. In this institute civil servants, politicians, leaders and other decision-makers in the national and international arena can be trained how to communicate, mediate, compromise, make peace and keep peace. Very often, this lack of training prevents them from starting or stopping the conflict even if they want to avoid conflict. I hope such an institute will be established in the future.

References

Hufbauer, G.S. Schott, & K.A. Elliott. (1990). Economic Sanctions Reconsidered. Institute of International Economics, World Bank, Washington. Macmillan: London.

Miyagawa Makio. (1992).   Does Economic Sanction Work? Macmillan: London.

Index